D1719553

DNA AMPLIFICATION

CURRENT TECHNOLOGIES AND APPLICATIONS

Edited by:

Vadim V. Demidov
Boston University, USA

and

Natalia E. Broude
Boston University, USA

Inspired by and dedicated to the golden jubilee of the double helix and complementarity principle, which underlie all methods of DNA amplification presented in this book.

Cover illustration: Electron micrograph of DNA amplification product obtained from a DNA plasmid by rolling-circle replication (see Figure 4 page 251).

Copyright © 2004
Horizon Bioscience
32 Hewitts Lane
Wymondham
Norfolk NR18 0JA
U.K.

www.horizonbioscience.com

British Library Cataloguing-in-Publication Data

A catalogue record for this book is available from the British Library

ISBN: 0-9545232-9-6

Description or mention of instrumentation, software, or other products in this book does not imply endorsement by the author or publisher. The author and publisher do not assume responsibility for the validity of any products or procedures mentioned or described in this book or for the consequences of their use.

All rights reserved. No part of this publication may be reproduced, stored in a retrieval system, or transmitted, in any form or by any means, electronic, mechanical, photocopying, recording or otherwise, without the prior permission of the publisher. No claim to original U.S. Government works.

Printed and bound in Great Britain by The Cromwell Press

Contents

Section 3. Isothermal Methods of DNA Amplification

Section 4. DNA Amplification in Detection of Non-DNA Analytes

Contributors

Michael Adler*
Chimera biotec GmbH
Emil-Figge Str. 76a
44227 Dortmund
Germany

Luis Blanco
Instituto de Biología Molecular
"Eladio Viñuela" (CSIC)
Centro de Biología Molecular
"Severo Ochoa" (CSIC-UAM)
Universidad Autónoma
Cantoblanco, 28049 Madrid
Spain

Natalia E. Broude*
Center for Advanced Biotechnology
Boston University
36 Cummington Street
Boston, MA 02215
U.S.A.

Weiguo Cao*
South Carolina Experimental Station
Clemson University
51 New Cherry Street
Clemson, SC 29634
U.S.A.

Connie Chang
Cornell University Medical College
New York, NY 10021
U.S.A.

Cecile Colpaert
Department of Pathology
University Hospital Antwerp
Wilrijkstraat 10
B-2650 Edegem
Belgium

Hirock D. Datta
Molecular Staging, Inc.
300 George Street
New Haven, CT 06511
U.S.A.

Vadim V. Demidov*
Center for Advanced Biotechnology
Boston University
36 Cummington Street
Boston, MA 02215
U.S.A.

John C. Detter*
US DOE Joint Genome Institute
2800 Mitchell Drive
Walnut Creek, CA 94598
U.S.A.

Richard S. Dondero
Molecular Staging, Inc.
300 George Street
New Haven, CT 06511
U.S.A.

Konstantin Ebralidse
Beth Israel Deaconess Medical Center
and Harvard Medical School
Boston, MA 02115
U.S.A.

Michael Egholm
Molecular Staging, Inc.
300 George Street
New Haven, CT 06511
U.S.A.

Abdellatif Errami
Department of Clinical Genetics
Free University of Amsterdam
v.d. Boechorststraat 7
1081BT Amsterdam
Netherlands

Richard Finkers
Plant Research International
WUR, Wageningen
Netherlands
Current address:
Laboratory of Plant Breeding
WUR, Wageningen
Netherlands

Manohar R. Furtado*
Applied Biosystems
850 Lincoln Centre Drive
Foster City, CA 94404
U.S.A.

Christian Harwanegg
VBC-GENOMICS
Bioscience Research GmbH
1030 Vienna
Austria

Tobin J. Hellyer*
BD Diagnostic Systems
54 Loveton Circle
Sparks, MD 21152
U.S.A.

Sjaak van Heusden
Plant Research International
WUR, Wageningen
Netherlands

Seiyu Hosono
Molecular Staging, Inc.
300 George Street
New Haven, CT 06511
U.S.A.

Martin Huber
VBC-GENOMICS
Bioscience Research GmbH
1030 Vienna
Austria

Hidetoshi Kanda
Eiken Chemical Co., Ltd.
1381-3 Shimoishigami
Ohtawara, Tochigi 324-0036
Japan

Konstantin Khrapko*
Beth Israel Deaconess Medical Center
and Harvard Medical School
Boston, MA 02115
U.S.A.

Stephen F. Kingsmore
Molecular Staging, Inc.
300 George Street
New Haven, CT 06511
U.S.A.

Sergei A. Kozyavkin
Fidelity Systems, Inc.
7961 Cessna Avenue
Gaithersburg, MD 20879
U.S.A.

Yevgenya Kraytsberg
Beth Israel Deaconess Medical Center
and Harvard Medical School
Boston, MA 02115
U.S.A.

Andrew M. Kuhn
BD Diagnostic Systems
54 Loveton Circle
Sparks, MD 21152
U.S.A.

Heiko Kuhn
Center for Advanced Biotechnology
Boston University
36 Cummington Street
Boston, MA 02215
U.S.A.

Roger S. Lasken*
Molecular Staging, Inc.
300 George Street
New Haven, CT 06511
U.S.A.

José M. Lázaro
Instituto de Biología Molecular
"Eladio Viñuela" (CSIC)
Centro de Biología Molecular
"Severo Ochoa" (CSIC-UAM)
Universidad Autónoma
Cantoblanco, 28049 Madrid
Spain

Kenneth J. Livak
Applied Biosystems
850 Lincoln Centre Drive
Foster City, CA 94404
U.S.A.

Yasuyoshi Mori
Eiken Chemical Co., Ltd.
1381-3 Shimoishigami
Ohtawara, Tochigi 324-0036
Japan

Aki Morikawa
Winship Cancer Institute
Emory University
1365 Clifton Road NE
Atlanta, GA 30322
U.S.A.

Manfred W. Mueller
VBC-GENOMICS
Bioscience Research GmbH
1030 Vienna
Austria

Michael C. Mullenix*
Molecular Staging, Inc.
300 George Street
New Haven, CT 06511
U.S.A.

James G. Nadeau
BD Diagnostic Systems
54 Loveton Circle
Sparks, MD 21152
U.S.A.

Kentaro Nagamine
Eiken Chemical Co., Ltd.
1381-3 Shimoishigami
Ohtawara, Tochigi 324-0036
Japan

Ekaterina Nekhaeva
Beth Israel Deaconess Medical
Center
and Harvard Medical School
Boston, MA 02115
U.S.A.
PerkinElmer Life and Analytical Sciences
Boston, MA 02118
U.S.A.

John R. Nelson
Amersham Biosciences
800 Centennial Avenue
PO Box 1327
Piscataway, NJ 08855
U.S.A.

Christof M. Niemeyer
Universität Dortmund
Fachbereich Chemie
Biologisch-Chemische
Mikrostrukturtechnik
Otto-Hahn-Str. 6
D-44227 Dortmund
Germany

Tsugunori Notomi*
Eiken Chemical Co., Ltd.
1381-3 Shimoishigami
Ohtawara, Tochigi 324-0036
Japan

Anders O.H. Nygren
Department of Clinical Genetics
Free University of Amsterdam
v.d. Boechorststraat 7
1081BT Amsterdam
Netherlands

Andrey R. Pavlov
Fidelity Systems, Inc.
7961 Cessna Avenue
Gaithersburg, MD 20879
U.S.A.

Nadejda V. Pavlova
Fidelity Systems, Inc.
7961 Cessna Avenue
Gaithersburg, MD 20879
U.S.A.

Lorah T. Perlee
Molecular Staging, Inc.
300 George Street
New Haven, CT 06511
U.S.A.

Olga V. Petrauskene
Applied Biosystems
850 Lincoln Centre Drive
Foster City, CA 94404
U.S.A.

Xiaoquan Qi*
Sainsbury Laboratory
Norwich Research Park
Colney, Norwich NR4 7UH
U.K.

Paul M. Richardson
US DOE Joint Genome Institute
2800 Mitchell Drive
Walnut Creek, CA 94598
U.S.A.

Margarita Salas*
Instituto de Biología Molecular
"Eladio Viñuela" (CSIC)
Centro de Biología Molecular
"Severo Ochoa" (CSIC-UAM)
Universidad Autónoma
Cantoblanco, 28049 Madrid
Spain

Wolfgang M. Schmidt*
VBC-GENOMICS
Bioscience Research GmbH
1030 Vienna
Austria

Jan P. Schouten*
MRC-Holland
Hudsonstraat 68
1057SN Amsterdam
Netherlands

Alexei I. Slesarev*
Fidelity Systems, Inc.
7961 Cessna Avenue
Gaithersburg, MD 20879
U.S.A.

Keith Thornton
BD Diagnostic Systems
54 Loveton Circle
Sparks, MD 21152
U.S.A.

Miguel de Vega
Instituto de Biología Molecular
"Eladio Viñuela" (CSIC)
Centro de Biología Molecular
"Severo Ochoa" (CSIC-UAM)
Universidad Autónoma
Cantoblanco, 28049 Madrid
Spain

Sha-Sha Wang
BD Diagnostic Systems
54 Loveton Circle
Sparks, MD 21152
U.S.A.

Tanisha Williams
Winship Cancer Institute
Emory University
1365 Clifton Road NE
Atlanta, GA 30322
U.S.A.

David M. Wolfe
BD Diagnostic Systems
54 Loveton Circle
Sparks, MD 21152
U.S.A.

Diansheng Zhong
Winship Cancer Institute
Emory University
1365 Clifton Road NE
Atlanta, GA 30322
U.S.A.

Wei Zhou*
Winship Cancer Institute
Emory University
1365 Clifton Road NE
Atlanta, GA 30322
U.S.A.

* Corresponding author

Preface

DNA amplification is the cornerstone of modern biotechnology and it is also a key procedure in numerous basic studies involving DNA molecules. All methods for DNA amplification have rested on the concept of DNA strand complementarity discovered by James Watson and Francis Crick fifty years ago. To an equal extent, these methods became possible with the discovery of DNA polymerases first identified by Arthur Kornberg soon after the Watson-Crick discovery and DNA ligases discovered in 1967 by Martin Gellert, Charles Richardson, Jerard Hurwitz, Robert Lehman and others. Using these enzymes (and later their thermostable variants), a variety of isothermal and temperature-cycling amplification techniques have been developed starting in late 1980s. Among these techniques, Kari Mullis' polymerase chain reaction (PCR) was the first one and it is still the most popular amplification method. Yet, some alternatives to PCR have also successfully invaded the area. The emergence of such methodologies significantly widened the range of approaches for DNA amplification and dramatically changed the abilities of basic and applied researchers in various fields of life sciences. It will not be an exaggeration to say that now no research related to DNA can be performed without the employment of DNA amplification procedures.

Despite the importance of this topic we found to our surprise that only a few books were published that deal with the subject. Moreover, these books cover mostly PCR-based techniques and/or describe the use of PCR and other DNA amplification approaches for specific goals, such as clinical analysis, environmental microbiology, forensics, etc. This information shortage was a major motivation for us to compile a book on a wider range of methods for DNA amplification with emphasis on their diverse applications. Besides, almost twenty years after PCR was invented, we now are witnessing a new stage in the craft of DNA amplification thanks to the introduction of real-time PCR, several powerful non-PCR DNA amplification techniques and microarray technologies. In an attempt to represent the current state-of-the-art our book covers both well-established and newly-developed protocols with promising potential.

Although the book goes far beyond PCR by presenting a number of isothermal assays along with the ligation-based thermocycling approaches, PCR remains, despite some limitations, the dominant diagnostic technique for target DNA amplification and analysis, and recently this primary method has been systematically improved in many ways. That is why a significant part of our book is devoted to new PCR developments. A separate section is devoted to a group of enzymes, both natural and engineered, which are employed for DNA amplification and related purposes. We also present here the use of DNA amplification in the detection of non-DNA analytes. Note that we do not consider *per se* the methods for the detection of amplicons obtained by one way or another except for those few that establish a new potent amplification approach, as in the case of real-time PCR or real-time strand displacement amplification (SDA).

We hope that our book will serve as a practical tool and reference source for a broad audience of academic researchers and industry biotechnologists who rely in their work on DNA amplification techniques. We are very grateful to all the contributors and to the publisher who made this book possible.

Vadim V. Demidov and Natalia E. Broude
May 2004
Boston, USA

Books of Related Interest

Malaria Parasites: Genomes and Molecular Biology	2004
Pathogenic Fungi: Structural Biology and Taxonomy	2004
Pathogenic Fungi: Host Interactions and Emerging Strategies for Control	2004
Bacterial Spore Formers: Probiotics and Emerging Applications	2004
Strict and Facultative Anaerobes: Medical and Environmental Aspects	2004
Foot and Mouth Disease: Current Perspectives	2004
Sumoylation: Molecular Biology and Biochemistry	2004
DNA Amplification: Current Technologies and Applications	2004
Prions and Prion Diseases: Current Perspectives	2004
Real-Time PCR: An Essential Guide	2004
Protein Expression Technologies: Current Status and Future Trends	2004
Computational Genomics: Theory and Application	2004
The Internet for Cell and Molecular Biologists (2nd Edition)	2004
Tuberculosis: The Microbe Host Interface	2004
Metabolic Engineering in the Post Genomic Era	2004
Peptide Nucleic Acids: Protocols and Applications (2nd Edition)	2004
Ebola and Marburg Viruses: Molecular and Cellular Biology	2004
MRSA: Current Perspectives	2003
Genome Mapping and Sequencing	2003
Bioremediation: A Critical Review	2003
Frontiers in Computational Genomics	2003
Transgenic Plants: Current Innovations and Future Trends	2003
Bioinformatics and Genomes: Current Perspectives	2003
Vaccine Delivery Strategies	2003
Multiple Drug Resistant Bacteria: Emerging Strategies	2003
Regulatory Networks in Prokaryotes	2003
Genomics of GC-Rich Gram-Positive Bacteria	2002
Genomic Technologies: Present and Future	2002
Probiotics and Prebiotics: Where are We Going?	2002

Full details of all these books at: www.horizonbioscience.com

Section 1

Enzymes Used in DNA Amplification

1.1

Thermostable Chimeric DNA Polymerases with High Resistance to Inhibitors

Andrey R. Pavlov, Nadejda V. Pavlova, Sergei A. Kozyavkin and Alexei I. Slesarev

Abstract

We have developed and put to use a new technology for production of chimeric DNA polymerases with outstanding thermostability, processivity and resistance to PCR inhibitors. The protein chimeras contain polymerase domains fused with helix-hairpin-helix (HhH) domains derived from topoisomerase V of *M. kandleri*. The advantages of new polymerases allow for cycle sequencing and PCR in high salt concentrations and at temperatures inaccessible for other DNA polymerases. Our approach resulted in TOPOTAQ series of DNA polymerases, which represent an excellent choice for DNA amplification in samples with intercalating dyes (SYBR green, SYBR gold, ethidium bromide, indigo), organic solvents (phenol), and physiological fluids (such as blood and urine).

Background

Introduction of DNA polymerases from thermophilic organisms to laboratory practice made revolutionary changes in biotechnology. The use of thermostable DNA polymerases greatly improved efficiency, simplicity, and specificity of DNA amplification (as high temperature facilitated DNA synthesis on templates with complex secondary structures) and generated numerous powerful methods for assaying low abundance nucleic acids. Yet, despite the very successful application of these proteins, the search for new enzymes with better properties and attempts to modify characteristics of existing enzymes continue. Various techniques pose different and sometimes contradictory requirements to DNA synthesis, which could not be easily achieved with available polymerases. For example, adaptation of RNA templates for the synthesis often requires decreasing stringency of DNA polymerases, which, in turn, impairs fidelity of the synthesis. The ability of some DNA polymerases to structure-specifically bind and pause on the DNA template (Tombline, *et al.*, 1996) using 5'→3' exonuclease domain can decrease fidelity of synthesis (Barnes, 1992); however, removal of the domain decreases processivity and resistance of the enzymes to inhibitors (Lawyer, *et al.*, 1993; Pavlov, *et al.*, 2002).

A number of new natural thermostable DNA polymerases from various sources have been identified, sequenced, cloned and expressed to optimize the DNA amplification. Also, blends of different thermophilic DNA polymerases and blends of DNA polymerases with other proteins are used to attain specific goals in DNA synthesis. Nevertheless, these new polymerases and combination approaches do not satisfy all needs in DNA amplification.

Another way to modify properties of DNA polymerases suggests engineering new enzymes. Traditionally, point-directed mutations of catalytic domains were performed to eliminate 3'→5' exonuclease activity, improve ddNTP utilization needed for DNA sequencing, or alter fidelity of synthesis (Tabor and Richardson, 1995; Bohlke, *et al.*, 2000; Evans, *et al.*, 2000; Patel and Loeb, 2000; Arezi, *et al.*, 2002; Yang, *et al.*, 2002). Also, removal of the entire 5'→3' nuclease domain is easily accomplished in many cases. Success of such alterations in protein structures is partially provided by the exceptional stability of the catalytic domains of the enzymes required for their function at high temperatures.

Recently, a new approach has been successfully applied to construct new proteins by fusing catalytic domains of thermostable DNA polymerases and DNA binding domains of other thermostable enzymes (Pavlov, *et al.*, 2002). The potency of the protein-engineering method was confirmed with *Taq* and *Pfu* DNA polymerases, which belong to different structural protein families. The method allowed to produce polymerases that retain high processivity at high levels of salts (Pavlov, *et al.*, 2002) and other inhibitors of DNA synthesis. In addition, it was demonstrated that attachment of additional domains can greatly increase the thermostability of chimeric DNA polymerases. Here we present several examples demonstrating significant advantages of the engineered enzymes' employment for DNA amplification.

Protocols

Materials

The DNA polymerase chimera TopoTaq contains the HhH-type repeats H-L of *M. kandleri* topoisomerase V (TopoV) fused with the N-terminus of the *Taq* polymerase Stoffel fragment; TaqTopoC1, TaqTopoC2, and TaqTopoC3 enzymes consist of the Stoffel fragment's C-terminus fused with the TopoV's repeats B-L, E-L and H-L, respectively; PfuC2 contains the E-L repeats at the C-terminus of Pfu polymerase (Belova, *et al.*, 2001; Pavlov, *et al.*, 2002). Chimeric polymerases TopoTaq and TaqTopoC1 were expressed and purified as previously described (Pavlov, *et al.*, 2002). FS mutant of TopoTaq construct (TopoTaqFS) was prepared by substituting Phe683 (Tabor and Richardson, 1995) and Arg686 in the TopoTaq sequence for Tyr and Ser, respectively, using conventional PCR and cloning techniques. One unit (1 U) of DNA polymerase activity was determined as the amount of polymerase that produces the same initial rate of DNA synthesis with 1 μM fluorescent primer-template junction (PTJ) duplex substrate at 70°C, in a standard primer extension assay (Pavlov, *et al.*, 2002), as

1 U of commercial AmpliTaq or *Pfu* DNA polymerase (as determined by the manufacturers). AmpliTaq DNA polymerase, 10X PCR Buffer II, and BigDye terminator v2 (BDT v2) kit were obtained from Applied BioSystems (Foster City, CA, USA) and *Pfu* DNA polymerase and 10X Cloned *Pfu* Buffer were purchased from Stratagene Cloning Systems. M13mp18(+) single-stranded DNA and ALF M13 Universal fluorescent primer were from Amersham Pharmacia Biotech (Piscataway, NJ, USA). DNA molecular weight marker X (0.07–12.2 kbp) was from Roche Diagnostics Corporation (Indianapolis, IN, USA). Chimeric polymerases TaqTopoC2 (TOPOTAQ100™) and PfuC2 (PYROTOPO™), TopoTaq amplification buffer, all plasmids and primers were obtained from Fidelity Systems, Inc. (Gaithersburg, MD, USA). Fluorescent dyes SYBR Gold and SYBR Green were purchased from Molecular Probes, Inc. (Eugene, OR, USA), dNTPs were obtained from MBI Fermentas (Lithuania); other chemicals were of highest reagent grade.

Measuring the Rate of Primer-Extension Reactions

The primer-extension assay used for comparative studies of different polymerases was performed with a fluorescent duplex substrate containing a primer-template junction (PTJ). The duplex was prepared by annealing a 5'-end labeled with fluorescein 20-nt long primer with a 30-nt long template:

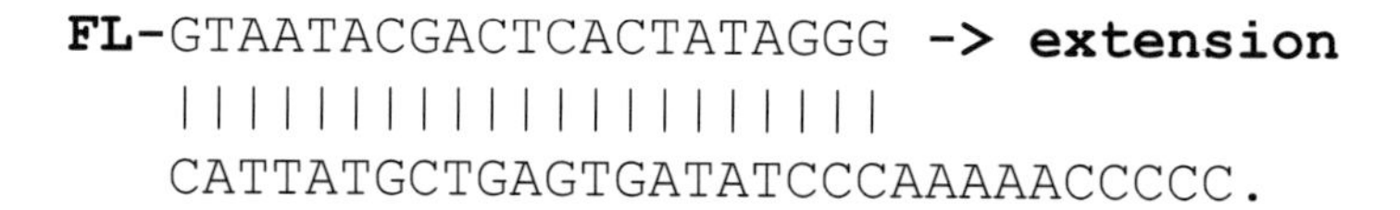

DNA polymerase reaction mixtures (15-20 μl) contained dATP, dTTP, dCTP, and dGTP (1mM each), 4.5 mM $MgCl_2$, detergents Tween 20 and Nonidet P-40 (0.2% each), fixed concentrations of PTJ – duplex, other additions, and appropriate amounts of DNA polymerases in the buffer, as indicated. The corresponding control samples contained all components except the polymerases. Primer extensions were carried out for a preset time at 75°C in PTC-150 Minicycler (MJ Research). For each reaction, 5 μl samples were removed after 3, 6, and 9 min.

of incubation and chilled to 4°C, followed by immediate addition of 20 μl of 20 mM EDTA. The samples were desalted by centrifugation through Sephadex G-50 spun columns, diluted, and analyzed on an ABI Prism 377 DNA sequencer (Applied BioSystems). For each sample, raw data were extracted from the sequencer trace files with the program Chromas v.1.5 (Technelysium Pty Ltd., Australia), and the integral fluorescent intensities for each product peak were calculated. The total amount of fluorescent extended products for each time of incubation was determined, and the initial rates of extension were calculated from the progressive accumulation of the products as described (Pavlov, *et al.*, 2002).

High-Salt Cycle Sequencing of Single-Stranded DNA

Sequencing reaction mixtures (total volume 5μl) contained 2 μl BDT v2 kit, 1 μl 5X TopoTaq amplification buffer with 10 mM $MgCl_2$, 0.5 μl TopoTaq FS (0.6 U/μl), 0.5 μl M13mp18 (+) ssDNA (0.1 μg/μl) with large amounts of NaCl, and 1 μl M13 1 μM forward primer (TGTAAAACGACGGCCAGT). The tubes with the reaction mixtures were placed in a thermal cycler and cycle sequencing was performed as follow:

1. Heating the tubes at 95°C for 2 minutes.
2. Repeating the following for 50 cycles:
   - Rapid thermal ramp to 95°C
   - 95°C for 5 seconds
   - Rapid thermal ramp to 55°C
   - 55°C for 30 seconds
   - Rapid thermal ramp to 60°C
   - 60°C for 4 minutes.
3. Rapid thermal ramp to 4°C and holding until ready to purify.

The reaction mixtures were diluted with 20 μl of 20 mM EDTA. The samples were desalted by centrifugation through Sephadex G-50 spun columns, diluted, and analyzed on an ABI Prism 377 DNA sequencer.

PCR Amplification with Chimeric DNA Polymerases

PCR with TopoTaq, TaqTopoC1, and TOPOTAQ100 DNA polymerases was carried out in microcentrifuge tubes containing 10 μl 2x TopoTaq Amplification buffer with 6 mM $MgCl_2$, 1 μl premixed dNTPs (10 mM each dNTP), 1 μl primer mixture (10 μM each primer), 1 U enzyme, and 20 ng plasmid DNA in total volume 20 μl. The reaction mixtures with *Pfu* and PYROTOPO DNA polymerases contained the same amounts of DNA, primers and dNTPs added to 2 μl 10X Cloned *Pfu* Buffer and supplied with 1 μl of 20 mM $MgCl_2$ in total volume 20 μl; PCR with AmpliTaq was performed with 2 μl 10X PCR Buffer II (without Mg^{2+}) supplied with 1.5 μl 20 mM $MgCl_2$ in the same total volume.

Amplification of DNA targets cloned into plasmid pet21d was carried out with primers GTAATACGACTCACTATAGGG and GCTAGTTATTGCTCAGCGG; PCR from M13mp18(+) DNA was performed with ALF M13 Universal fluorescent primer (**FL**-CGACGTTGTAAAACGACGGCCAGT) and primer CAGGAAACAGCTATGACC (M13 reverse). The tubes with the reaction mixtures were placed in a thermal cycler, and three-step cycling was performed as following:

1. Heating the tubes at 100°C (TopoTaq, TaqTopoC1, TOPOTAQ100, and PYROTOPO) or 94°C (AmpliTaq and *Pfu*) for 2 minutes.
2. Repeating for 30 cycles:
   - Rapid thermal ramp to 100°C (TopoTaq, TaqTopoC1, TOPOTAQ100, and PYROTOPO) or 94°C (AmpliTaq and *Pfu*)
   - Denature: 100°C (TopoTaq, TaqTopoC1, TOPOTAQ100, and PYROTOPO) or 94°C (AmpliTaq and *Pfu*), 40 s
   - Rapid thermal ramp to 50°C
   - Annealing primers 50°C, 30 s
   - Rapid thermal ramp to 72°C
   - Primer extension: 72°C, 2 min (for targets shorter 1 kb) or 2 min/kb (for longer sequences).
3. Rapid thermal ramp to 4°C and holding until ready to purify.

Although the components of the reactions can be combined at ambient temperature, we prefer to mix them on ice to diminish extension of primer dimers and increase specificity of amplification. Heating at 100°C is absolutely essential for fast dissociation from DNA templates and, therefore, for successful cycling of TaqTopoC1, TOPOTAQ100, and PYROTOPO DNA polymerases. For all chimeric DNA polymerases, the concentration of Mg^{2+} should exceed the total concentration of dNTPs and other chelators (such as EDTA) by 1-1.5 mM. Targets for the chimeric polymerases except PYROTOPO should not exceed 2 kb. For most DNA templates one unit of enzyme is sufficient. When amplifying sequences above 3 kb with PYROTOPO DNA polymerase, extra amount of enzyme may be required.

Example 1: Effect of Ionic Inhibitors on the Rate of Primer Extension

Using the PTJ construct (see **Protocols**), we have compared the inhibition of DNA-synthetic activities of AmpliTaq and *Pfu* DNA polymerases with four chimeric proteins consisting of *Taq* or *Pfu* catalytic domain and TopoV DNA binding domains (Pavlov, *et al.*, 2002). The primer-extension reactions were carried out at low concentrations of substrate, and the initial rates were proportional both to total protein and PTJ concentrations. Figure 1 shows the sigmoid curves indicating the cooperative inhibition of the enzymes by salts (Pavlov, *et al.*, 2002). Experimental values of initial polymerization rates were analyzed by nonlinear regression using the following function:

$$\mathrm{v} = \frac{\mathrm{v}_0}{1 + \left(\frac{[I]}{K_i}\right)^{\alpha}} \tag{1}$$

where v and v_0 are initial primer-extension rates with and without inhibitor, respectively, $[I]$ is the inhibitor concentration, K_i is an apparent inhibition constant, and α is a parameter of cooperativity (Pavlov, *et al.*, 2002). Values for K_i and α obtained by the regression analysis are listed in Table 1. As with chlorides and potassium glutamate (Pavlov,

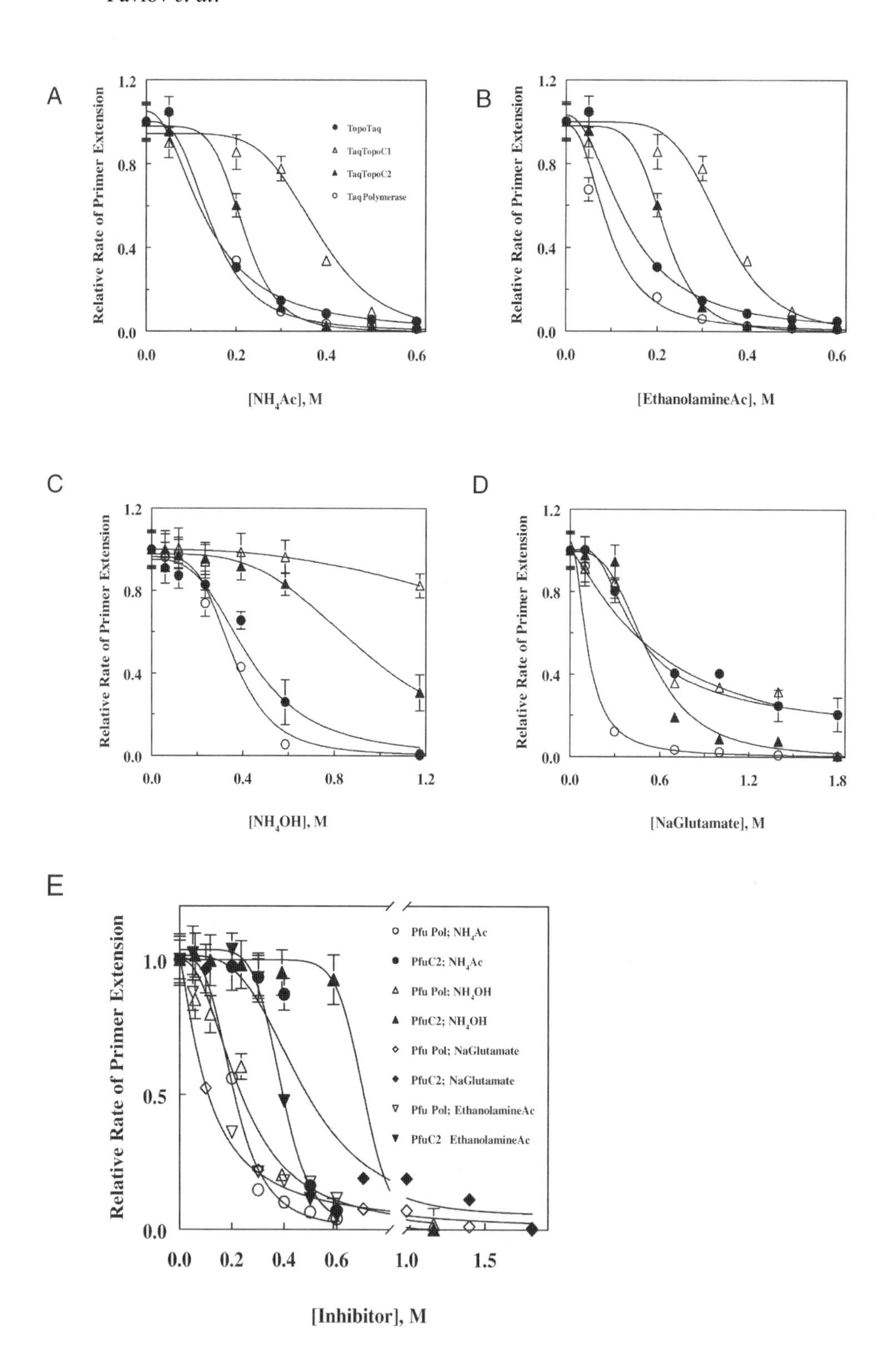
A
B
C
D
E
Relative Rate of Primer Extension
TopoTaq
TaqTopoC1
TaqTopoC2
Taq Polymerase
[NH4Ac], M
[EthanolamineAc], M
[NH4OH], M
[NaGlutamate], M
Pfu Pol; NH4Ac
PfuC2; NH4Ac
Pfu Pol; NH4OH
PfuC2; NH4OH
Pfu Pol; NaGlutamate
PfuC2; NaGlutamate
Pfu Pol; EthanolamineAc
PfuC2 EthanolamineAc
[Inhibitor], M

et al., 2002), chimeric DNA polymerases show increased resistance to acetates. Also, the predominant role of anions in the inhibition is clearly seen (compare ammonium acetate with ammonium hydroxide). Still, the enzymes' inhibition by sodium glutamate quantitatively somewhat differs from that of potassium glutamate studied earlier (Pavlov, *et al.*, 2002), thus demonstrating that cations can modify the inhibiting action of anions to a certain extent as well.

Example 2: Dye-Terminator Cycle Sequencing at High Salt Concentrations

Mutants of *Taq* DNA polymerase with the capacity to incorporate fluorogenic substrates and comprising additional DNA-binding domains could render a favorable opportunity to sequence DNA directly from cell cultures or biological fluids, therefore eliminating costly DNA purification. Also, an ability of DNA sequencing at high concentrations of salt would provide DNA with the enhanced resistance to a high-temperature degradation, thus allowing to significantly increase the number of DNA amplification cycles. Accordingly, we tested the salt resistance of TopoTaqFS (that contains two mutations in the active site to accommodate fluorescently-labeled dideoxynucleotides) in the dye-terminator cycle sequencing reactions carried out with BDT v2 kit in the presence of increasing concentrations of NaCl.

The incorporation of fluorescent nucleotides from this kit could be observed up to 0.4 - 0.5 M NaCl (Figure 2). In contrast, no signal could be generated by the polymerase from the sequencing kit (AmpliTaqFS) even at the lowest salt concentration used (0.2 M NaCl). In agreement with our previous findings (Pavlov, *et al.*, 2002), these results

Figure 1. Effect of ionic inhibitors on initial rates of primer extension reactions catalyzed by *Taq* DNA polymerase, *Pfu* polB, and the hybrid polymerases. The dependencies of the rates for enzymes with *Taq* polymerase catalytic domain are plotted for ammonium acetate (Panel A), ethanolamine acetate (Panel B), ammonium hydroxide (Panel C), and sodium glutamate (Panel D). The dependencies of the rates for enzymes with *Pfu* polymerase catalytic domain are collected in Panel E. Solid lines are theoretical curves of inhibition calculated by using Eq. 1.

Table 1. Inhibition of the DNA elongation activity of chimeric DNA polymerases by ionic inhibitors.[a]

Protein	NH_4Ac		EthAc		NH_4OH		NaGlu	
	K_i (mM)[b]	α[b]	K_i (mM)	α	K_i (mM)	α	K_i (mM)	α
AmpliTaq	144.3 ± 21	3.09 ± 0.3	94.0 ± 2	2.40 ± 0.1	378.1 ± 8	5.72 ± 0.5	120.2 ± 24	1.99 ± 0.2
TopoTaq	298.0 ± 74	2.82 ± 1.1	134.8 ± 13	2.12 ± 0.4	477.9 ± 13	5.08 ± 0.4	318.1 ± 86	0.97 ± 0.2
TaqTopoC1	380.9 ± 19	5.65 ± 1.3	363.1 ± 6	6.91 ± 0.7	2202.7 ± 267	2.46 ± 0.4	583.9 ± 64	1.25 ± 0.4
TaqTopC2	236.8 ± 5	7.55 ± 0.7	216.1 ± 53	5.70 ± 1.1	944.7 ± 23	3.54 ± 0.2	525.8 ± 89	3.21 ± 0.7
Pfu	208.7 ± 20	3.46 ± 0.5	110.9 ± 27	1.25 ± 0.2	235.5 ± 37	2.27 ± 0.6	108.9 ± 39	1.31 ± 0.1
*Pfu*C2	415.1 ± 3	7.48 ± 0.2	390.9 ± 2	6.93 ± 0.1	705.2 ± 62	14.00 ± 7.2	471.0 ± 79	3.63 ± 0.8

[a] Some of these inhibitors (chlorides, acetates and glutamates) are intrinsic in biological samples, whereas the others (ethanolamine and ammonium) can persist as contaminations in oligonucleotide primers.
[b] K_i and α are apparent inhibition constant and cooperativity parameter, respectively.

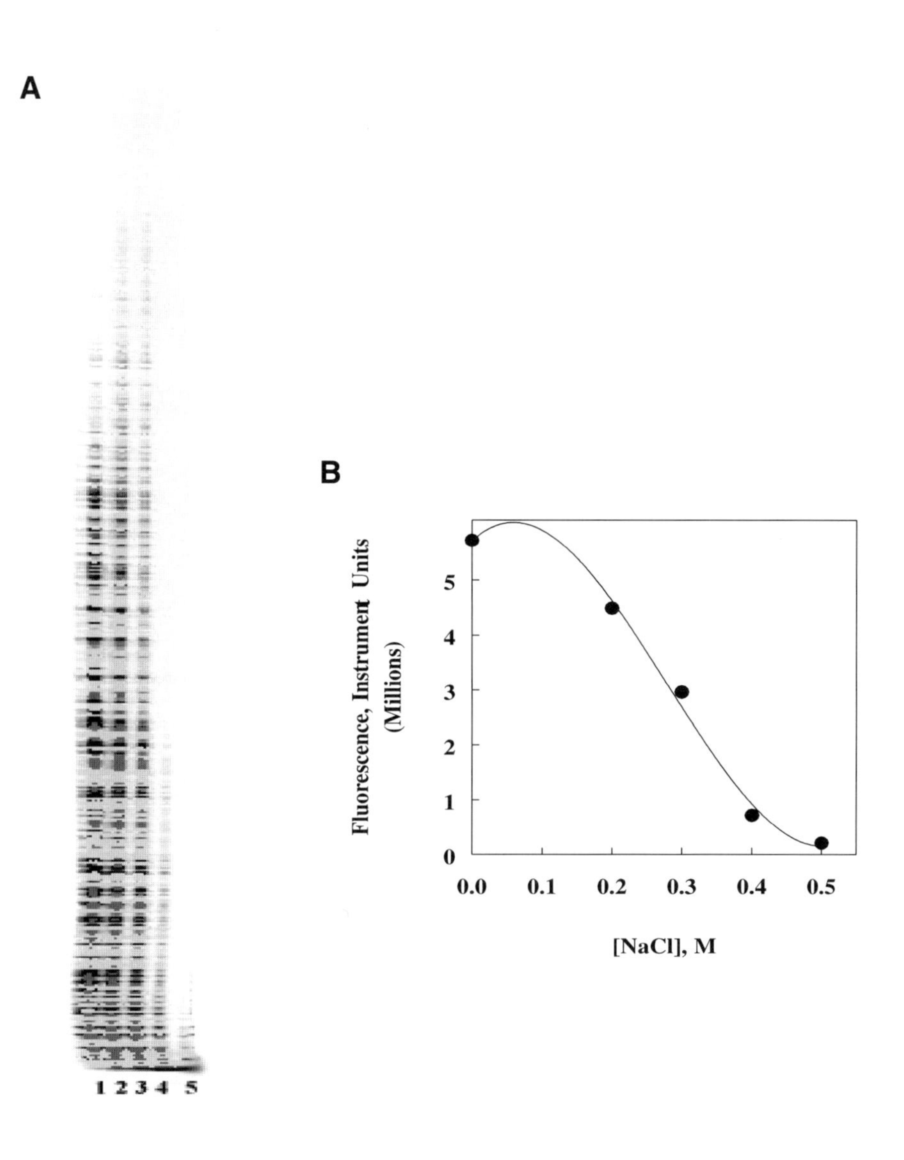

Figure 2. Dye terminator cycle sequencing using TopoTaqFS mutant. *Panel A* displays a urea gel analysis picture for M13mp18(+) sequencing reactions (50 ng ssDNA) with BDT v2 Kit, M13 Forward primer after 50 cycles. Tracks 1-5 show reactions carried out in the presence of 1 M betaine and 0; 0.2; 0.3; 0.4; and 0.5 M NaCl in the sequencing media, subsequently. *Panel B* shows how the integrated fluorescence intensity in the gel tracks decreases with increase of salt concentration.

demonstrate that salts do not inhibit the catalysis of chain elongation by active sites of DNA polymerases, but interfere with the ability of the proteins to remain bound to DNA substrate during synthesis. Thus, tethering the catalytic DNA polymerase domain to the TopoV DNA-binding domains allows for the successful synthesis of DNA at very high level of salt. Moreover, we show that introduction of FS mutations into the active site does not decrease the resistance of the chimeric polymerase to NaCl.

Figure 3. Amplification of 110 nt region of ssDNA M13mp18(+) with ALF M13 Universal fluorescent primer and M13 reverse primer in the presence of 0.25 M NaCl. DNA polymerases used: 1 - AmpliTaq; 2 - TopoTaq; 3 - TaqTopoC1; 4 - TaqTopoC2; 5 - PfuC2; 6 – cloned *Pfu*. The products were resolved on a 10% sequencing gel with ABI PRISM 377 DNA sequencer.

Example 3: PCR Amplification of DNA in the Presence of Common Inhibitors

Attachment of extra DNA-binding domains allows the DNA polymerases to perform DNA amplification at very high concentration of salts, intercalating dyes and other known PCR inhibitors. Figure 3 shows that all hybrid polymerases can amplify short single-stranded DNA fragment (~100 nt) at salt concentration as high as 0.25 M NaCl. High-salt amplification of 0.5 kb and 1.8 kb double-stranded DNA targets by the protein chimeras is presented in Figure 4. The workability of chimeric polymerases in the presence of fluorescent dyes, indigo, phenol, and blood is demonstrated in Figure 5.

Discussion

Connecting together the DNA binding domains and catalytic domains of DNA processing enzymes may greatly improve performance of the hybrid proteins in operational environments that differ from the natural ones. One potential problem with such hybrids is that the attached polypeptide segments may obstruct functioning of the catalytic domains. Although the use of structurally defined domains with especially high internal stability (*e.g.* from proteins of thermophiles), diminishes the probability of misfolding or other adverse effects on conformation of the functional entities (Pavlov, *et al.*, 2002), the effects of interdomain interactions on the catalysis cannot be rationally predicted *a priori*.

Despite the possible complications, our approach aimed at raising the salt tolerance of polymerases and their stability by attaching helix-hairpin-helix (HhH) domains (Pavlov, *et al.*, 2002) derived from topoisomerase V (TopoV) of *M. kandleri* (Belova, *et al.*, 2002) successfully produced the chimeras with expected properties. It resulted in TOPOTAQ series of hybrid DNA polymerases that can be used for cycle sequencing and PCR reactions directly with crude samples in the presence of common DNA polymerase inhibitors, and at temperatures inaccessible for other DNA polymerases. Here we demonstrate that

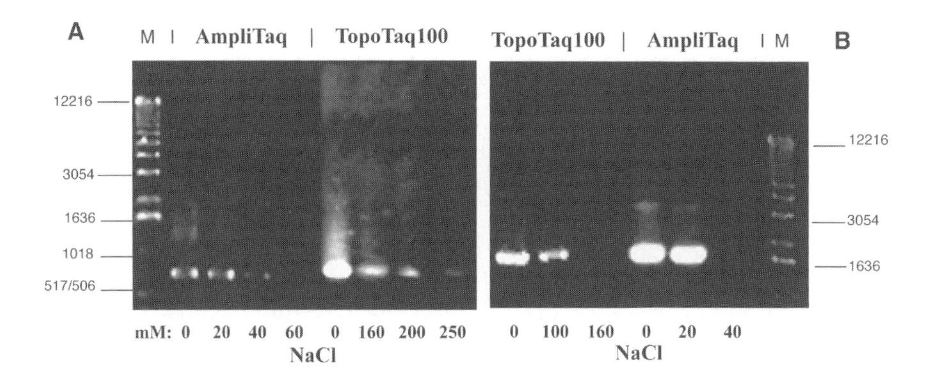

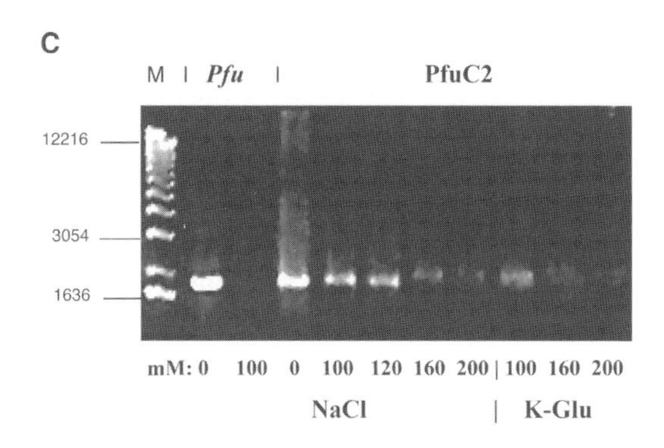

Figure 4. Amplification of DNA in the presence of salts. *Panel A* displays products of PCR carried out by TopoTaq100 polymerase with 0.5 kb target DNA in the presence of NaCl. Products of DNA amplification by AmpliTaq polymerase are shown for comparison. *Panel B* compares efficiencies of PCR amplification of a 1.8 kb target DNA by TopoTaq100 and AmpliTaq in NaCl. *Panel C* shows performance of PYROTOPO polymerase *vs* cloned *Pfu* DNA polymerase in PCR with the 1.8 kb target carried out in salts. The PCR products were resolved on a 1% agarose gels and stained with ethidium bromide. Lanes M contain DNA Markers.

attaching HhH domains to polymerases also increases the resistance of chimeric enzymes to other ionic inhibitors (see Table 1).

Note that the resistance of chimeras to inhibitors depends on the type of the enzymatic construct. These data are consistent with the mechanism of anionic inhibition (Pavlov, *et al.*, 2002), as can be seen from the similarity of the inhibition parameters for the hybrids by acetates and the difference in inhibition by ammonium acetate and ammonium

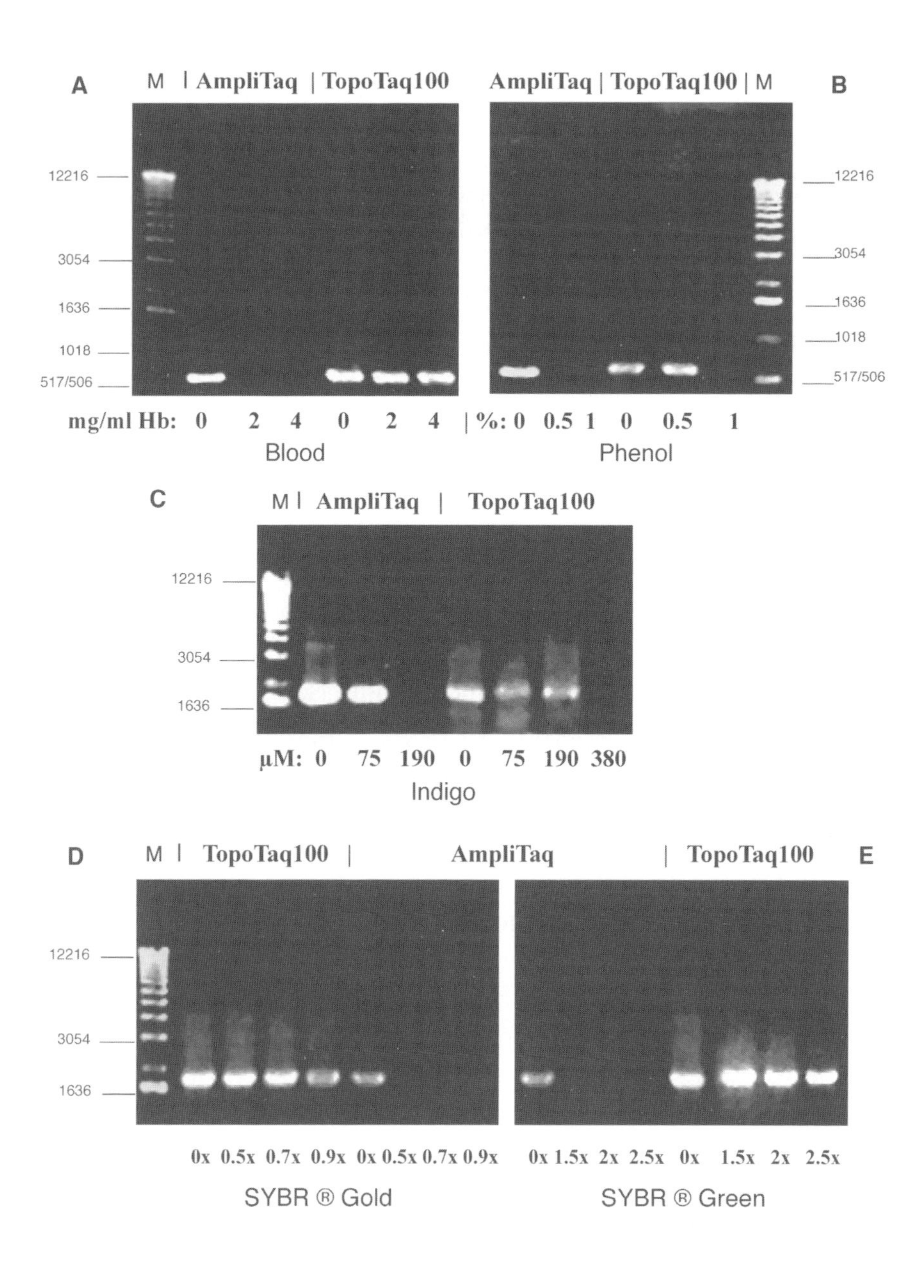

Figure 5. Amplification of DNA in the presence of organic inhibitors. *Panels A* and *B* show the efficiencies of amplification of the 0.5 kb DNA target by TopoTaq100 and AmpliTaq in the presence of blood and phenol, respectively. *Panel C* compares inhibition of PCR amplification of the 1.8 kb DNA target carried out by AmpliTaq and TopoTaq100 with synthetic indigo. *Panels D* and *E* display performance of TopoTaq100 and AmpliTaq polymerases in PCR carried out in the presence of the fluorescent dyes SYBR Gold and SYBR Green.

hydroxide. Also, it seems that pH of the reaction is less important for the polymerase activity than the concentration of anions. However, it appears that cations may modify inhibition by anions, as with glutamate, sodium decreases cooperativity of inhibition, and, in some cases decreases the resistance to the inhibitor. TopoTaq, TaqTopoC2, as well as AmpliTaq have similar sensitivity to the acetate inhibitiors as was reported for chlorides (Pavlov, *et al.*, 2002), but TopoTaq binds less inhibiting acetate ions. *Pfu* and PfuC2 DNA polymerases appear to be less sensitive to acetates than to chlorides, and have similar tolerance to glutamates and acetates.

Additional studies have shown that not every engineered protein hybrid with high thermostability and DNA polymerase activity can be a valuable asset for current biotechnological applications. For example, the TaqTopoC3 chimera, due to its relatively weak interaction with DNA, does not seem to have any properties that make it better than the already known commercial polymerases. On the other hand, because of very tight interaction with DNA, TaqTopoC1 is not efficient in cycling on long templates; however, it could be useful for some other applications. Constructs with fewer HhH motifs attached to the catalytic polymerase domain (TaqTopoC2 and PfuC2) successfully amplify DNA in PCR that can be performed in the presence of salts, phenol, urea, and other inhibitors. Nonetheless, in some cases, they may require more careful design of primers and/or more stringent choice of cycling conditions, as they can cause the non-specific interactions of primers with template DNA. In addition, they require very high temperature of the denaturing step of cycling (~100°C) for fast dissociation from DNA templates. Also, it should be taken into account that PCR amplification at higher salt concentrations presents additional challenges, as salt increases the non-specific annealing of primers and makes more difficult the melting of templates.

Currently, we utilized only a few structural units derived from DNA polymerases and *M. kandleri* TopoV. Other HhH motifs of TopoV (24 in total) are potentially a useful resource for creation of DNA processing enzymes that may yield the adjustable strength of interaction with the substrates. Optimization of the DNA-binding domains and linker segments in the composite proteins could provide fast movement and catalysis within the restricted areas on nucleic acid molecules: synthetic part of protein chimeras that tightly bind DNA via domains attached by long and flexible linkers would restrainedly slide along DNA templates, thus synthesizing oligonucleotide products strictly limited by the length of the linkers. This robust potential may be used in the DNA array technologies where it is so crucial to efficiently synthesize short DNA sequences at harsh conditions without any longer by-products. We therefore hope that such hybrid enzymes could become indispensable tools for molecular-biotechnological applications.

Acknowledgements

We thank Anna Pavlova for help with manuscript preparation. This work was supported in part by DOE and NIH grants to S.A.K and A.I.S.

References

Arezi, B., Hansen, C.J., and Hogrefe, H.H. (2002). Efficient and high fidelity incorporation of dye-terminators by a novel archaeal DNA polymerase mutant. J. Mol. Biol. *322*, 719-729.

Barnes, W.M. (1992). The fidelity of Taq polymerase catalyzing PCR is improved by an N-terminal deletion. Gene *112*, 29-35.

Belova, G.I., Prasad, R., Nazimov, I.V., Wilson, S.H., and Slesarev, A.I. (2002). The domain organization and properties of individual domains of DNA topoisomerase V, a type 1B topoisomerase with DNA repair activities. J. Biol. Chem. *277*, 4959-4965.

Bohlke, K., Pisani, F.M., Vorgias , C.E., Frey, B., Sobek, H., Rossi, M., and Antranikian, G. (2000). PCR performance of the B-type DNA polymerase from the thermophilic euryarchaeon *Thermococcus aggregans* improved by mutations in the Y-GG/A motif. Nucleic Acids Res. *28*, 3910-3917.

Evans, S.J., Fogg, M.J., Mamone, A., Davis, M., Pearl, L.H., and Connolly, B.A. (2000). Improving dideoxynucleotide-triphosphate utilisation by the hyper-thermophilic DNA polymerase from the archaeon *Pyrococcus furiosus*. Nucleic Acids Res. *28*, 1059-1066.

Lawyer, F.C., Stoffel, S., Saiki, R.K., Chang, S.Y., Landre, P.A., Abramson, R.D., and Gelfand, D.H. (1993). High-level expression, purification, and enzymatic characterization of full-length *Thermus aquaticus* DNA polymerase and a truncated form deficient in 5' to 3' exonuclease activity. PCR Methods Appl. *2*, 275-287.

Patel, P.H. and Loeb, L.A. (2000). Multiple amino acid substitutions allow DNA polymerases to synthesize RNA. J. Biol. Chem. *275*, 40266-40272.

Pavlov, A.R., Belova, G.I., Kozyavkin, S.A., and Slesarev, A.I. (2002). Helix-hairpin-helix motifs confer salt resistance and processivity on chimeric DNA polymerases. Proc. Natl. Acad. Sci. USA. *99*, 13510-13515.

Tabor, S. and Richardson, C.C. (1995). A single residue in DNA polymerases of the *Escherichia coli* DNA polymerase I family is critical for distinguishing between deoxy- and dideoxyribonucleotides. Proc. Natl. Acad. Sci. USA. *92*, 6339-6343.

Tombline, G., Bellizzi, D., and Sgaramella, V. (1996). Heterogeneity of primer extension products in asymmetric PCR is due both to cleavage by a structure-specific exo/endonuclease activity of DNA polymerases and to premature stops. Proc. Natl. Acad. Sci. USA. *93*, 2724-2728.

Yang, S.-W., Astatke, M., Potter, J., and Chatterjee, D.K. (2002). Mutant *Thermotoga neapolitana* DNA polymerase I: altered catalytic properties for non-templated nucleotide addition and incorporation of correct nucleotides. Nucleic Acids Res. *30*, 4314-4320.

1.2

Ø29 DNA Polymerase, a Potent Amplification Enzyme

Margarita Salas, Miguel de Vega,
José M. Lázaro and Luis Blanco

Abstract

ø29 DNA polymerase is a 66 kDa monomeric DNA-dependent DNA polymerase responsible for all the mesophilic DNA synthesis reactions required to replicate the 19-kb-long linear dsDNA genome of bacteriophage ø29. To initiate replication at each DNA end, ø29 DNA polymerase catalyzes the incorporation of dAMP to a specific protein, the terminal protein (TP), which acts as a primer. Then, switching from TP-priming to DNA-priming occurs progressively, defining a transition stage in which TP still interacts with ø29 DNA polymerase. Once dissociated from TP, a single ø29 DNA polymerase molecule replicates each DNA strand without dissociation from the template, whereas it produces the displacement of the non-template strand. Due to these two specific properties of ø29 DNA polymerase: high processivity and strand-displacement ability, neither accessory proteins nor helicases are required for the elongating stage of ø29 DNA replication. From a more applied point of view, these two important properties of ø29 DNA polymerase, together with its high fidelity of DNA synthesis, form the basis for the application of this enzyme in an increasing number of *in vitro* procedures for isothermal DNA amplification.

Background

ø29 DNA Replication

The *Bacillus subtilis* phage ø29 has a linear, double-stranded DNA 19,285 base pairs long, with a reiteration of three thymine residues at the 3' ends, and a terminal protein (TP) covalently linked to the two 5' ends. The TP is the primer for the initiation of ø29 DNA replication, that starts from both DNA ends and is catalyzed by the viral DNA polymerase (see Figure 1 from Blanco *et al.,* 1994). As a first step, the ø29 DNA polymerase forms a heterodimer with the TP that recognizes the DNA ends by interaction with the parental TP covalently bound to the DNA. Positioning at the origins is stimulated by the viral histone-like protein p6 (DBP) that forms a nucleoprotein complex at the duplex DNA termini to facilitate their opening (Serrano *et al.,* 1993). The viral DNA polymerase catalyzes the covalent linkage of the initiating nucleotide, dAMP, into the hydroxyl group of a specific serine residue (Ser232) in the TP (Salas, 1991, Salas *et al*., 1996) directed by the second 3' terminal nucleotide of the TP-DNA template.

After a sliding-back step to recover the information corresponding to the 3' terminal nucleotide (Méndez *et al.,* 1992), the phage DNA polymerase itself incorporates up to 9 nucleotides into the TP while complexed with this protein primer, and finally dissociates from TP when nucleotide 10 is inserted onto the nascent DNA chain (Méndez *et al.,* 1997). In addition to the conventional usage of the OH group of a nucleotide to elongate the DNA chain, ø29 DNA polymerase is able to use the OH group of a specific serine residue (Ser^{232}) in the TP primer as the attacking nucleophile to incorporate the initiating nucleotide. Once ø29 DNA polymerase has incorporated the tenth nucleotide to the nascent TP-DNA chain, elongation proceeds by a mechanism of

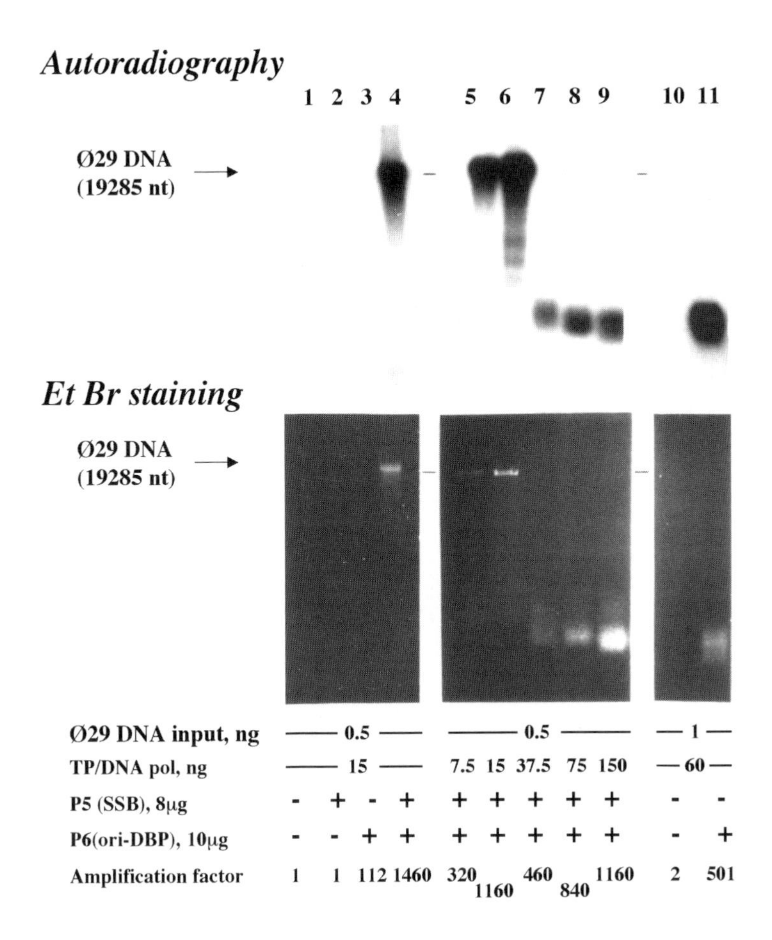

Figure 1. Efficiency of the *in vitro* ø29 DNA amplification with the TP-primed ø29 DNA polymerase depending on experimental conditions (from Blanco *et al.*, 1994; copyright of the Natl. Acad. Sci. USA). All assays were performed essentially as described in Materials and Methods using the indicated amounts of input ø29 DNA. Amplification factor (efficiency of amplification) corresponding to each particular condition was calculated as the ratio between the amount of newly synthesized DNA and the amount of input ø29 DNA.

strand-displacement with a very high processivity (Blanco *et al.,* 1989). In addition to the synthetic reactions, TP-deoxynucleotidylation and DNA polymerisation, ø29 DNA polymerase is able to catalyse two degradative reactions, *i.e.* pyrophosphorolysis and 3'-5' exonucleolysis (see Table 1).

Table 1. Enzymatic activities of ø29 DNA polymerase.

Enzymatic activity	Relevant feature	Biological role
TP-deoxynucleotidylation	OH provided by Ser^{232} in the TP Directed by the second template nucleotide	Initiation of ø29 DNA replication
DNA polymerisation	High processivity Strand displacement	Elongation of ø29 DNA replication
Pyrophosphorolysis	Able to remove a mismatch	Improvement of insertion fidelity (?)
3'-5' Exonuclease	Processive	Proof-reading of DNA insertion errors

Besides ø29 DNA polymerase plus TP and DBP proteins, that are required for initiation of ø29 DNA replication, the viral SSB protein, product of gene 5 (p5), is essential for elongation steps of ø29 DNA synthesis *in vivo* (Mellado *et al.,* 1980); this protein also stimulates the *in vitro* DNA replication by ø29 polymerase enzyme (Martín *et al.,* 1989). An *in vitro* ø29 TP-DNA amplification system has been developed which requires the four ø29 proteins mentioned above. The amplification factor obtained *in vitro* can be over 10^3-fold after 1 hr at 30°C, and the *in vitro* amplified TP-tagged ø29 DNA is fully infective as compared with ø29 DNA isolated *in vivo* (Blanco *et al.,* 1994).

ø29 DNA Polymerase

The ø29 DNA polymerase, product of the viral early gene 2, is a small monomeric enzyme of only about 66 kDa that is able to catalyze both the initiation and the elongation steps of ø29 DNA replication (reviewed by Blanco and Salas, 1996). For initiation, ø29 DNA polymerase binds to TP, thus acquiring an ability to recognize the ø29 DNA termini and to catalyze the formation of a TP-dAMP complex (initiation reaction). After sliding-back and incorporation of 10 nucleotides, ø29 DNA polymerase dissociates from TP, and processive elongation takes place. The number of nucleotides incorporated per single enzyme/DNA binding event is defined as processivity, and this property distinguishes repair and replicative DNA polymerases (for a review, see Kornberg and Baker, 1992). Whereas the high processivity displayed by most replicative DNA polymerases is conferred by interaction with accessory proteins that clamp the enzyme to DNA (for a review, see Kuriyan and O'Donnell, 1993), ø29 DNA polymerase is highly processive (>70 kb) in the absence of any accessory protein (Blanco *et al.,* 1989), and also in DNA replication reactions for which only DNA primers are provided.

Another key property of ø29 DNA polymerase that is essential during ø29 DNA replication, but can also be displayed during a more general DNA synthesis process, is its strong strand-displacement capacity. Using singly-primed ~7 x 10^3-nt-long M13 single-stranded (ss) DNA as template, we could demonstrate that ø29 DNA polymerase is able to synthesize DNA chains greater than 70 kb by a "rolling circle" mechanism, in the absence of any other protein (Blanco *et al.,* 1989). Furthermore, conditions that increase the stability of secondary structure in the template do not affect the processivity and strand-displacement ability of ø29 DNA polymerase.

Materials and Methods

Nucleotides, Proteins and DNA Templates

Unlabeled dNTPs and [α-^{32}P]dATP (3000 Ci/mmol) were obtained from Amersham Biosciences. ø29 DNA polymerase and TP were overproduced in *Escherichia coli* cells and purified as described (Lázaro *et al.,* 1995; Zaballos and Salas, 1989); ø29 double-stranded DNA binding protein (DBP) and ø29 single-stranded DNA binding protein (SSB or p5 protein) were purified as described (Pastrana *et al.,* 1985; Martín *et al.,* 1989). TP-linked ø29 DNA from ø29 *sus14* (1242) virions, isolated as described (Peñalva and Salas, 1982), was used as input template for *in vitro* ø29 DNA amplification experiments. M13mp18 ssDNA was purchased from Amersham Biosciences.

Protocols

TP-Primed DNA Amplification Assay

The first step of ø29 DNA replication consists in the formation of a heterodimer between a DNA polymerase molecule and a free TP molecule, with the further recognition of the replication origins. To optimise the pre-initiation step of ø29 DNA replication, an equimolar (1:1) complex between highly purified ø29 DNA polymerase and TP is obtained by pre-incubation of both proteins in the presence of 20 mM ammonium sulfate. The incubation mixture contains, in 10 μl, 50 mM Tris-HCl (pH 7.5), 20 mM ammonium sulfate, 1 mM DTT, 4% glycerol, bovine serum albumine (0.1 mg/ml), 80 μM each dCTP, dGTP, dTTP and [α-^{32}P]dATP (2 μCi), 10 μg of ø29 DBP, 8 μg of ø29 SSB, 15 ng of the preformed TP/DNA polymerase complex and 20 ng of free TP. As template, 0.5 ng of ø29 TP-DNA is used. Replication starts by adding 10 mM $MgCl_2$ as enzyme activator. After incubation for 1 h at 30 °C, reaction is stopped by addition of 10 mM EDTA. Non-incorporated [α-^{32}P]dATP is removed by filtration through Sephadex G-50 spin columns in the presence of 0.1% SDS. Quantitation of the DNA synthesised *in vitro*, measured as the total amount (in ng) of dNTP incorporated, is based on the Cherenkov-type radioactivity of the excluded fraction. Samples are analysed by 0.7% denaturing agarose gel electrophoresis (McDonell *et al.*, 1977) followed by ethidium bromide staining and autoradiography.

M13 ssDNA Rolling-Circle Amplification (RCA)

To provide a singly primed DNA template for RCA, 18 nM M13mp18 ssDNA was hybridised to 180 nM universal 17-mer oligonucleotide primer (obtained from Isogen) in the presence of 0.2 M NaCl and 60 mM Tris-HCl, pH 7.5 at 90°C during 5 min, followed by cooling to room temperature. The 10 μl RCA-reaction mixture additionally contains 50 mM Tris-HCl (pH 7.5), 1 mM DTT, 4% glycerol, bovine serum albumine (0.1 mg/ml), 80 μM or 0.4 mM each dCTP, dGTP, dTTP and [α-^{32}P]dATP (1 μCi), as indicated, and 160 ng of ø29 DNA polymerase. RCA starts by adding 10 mM $MgCl_2$ as enzyme activator. After incubation at 30°C, at time intervals ranging from 5 to 80 min, the reaction is stopped by adding 10 mM EDTA, 0.1% SDS, and the samples are filtered through Sephadex G-50 spin columns in the presence of 0.1% SDS. The Cherenkov radiation of the excluded fraction is counted and samples are subjected to 0.7% denaturing agarose gel electrophoresis. M13mp18 ssDNA is detected by ethidium bromide staining and then gels are dried and autoradiographed. To determine the stimulatory effect of ø29 SSB protein on ø29 DNA polymerase replication rate by RCA, oligo-primed M13mp18 ssDNA (see above) is incubated with ø29 SSB ranging from 0 to 28 μM before adding DNA polymerase.

Examples

Example 1: TP-Primed ø29 DNA Amplification

Figure 1 shows the results of the *in vitro* ø29 DNA amplification based on the TP-priming. Using the conditions described above, one can see that the use of the four phage proteins (ø29 DNA polymerase, TP, DBP and SSB) yields an amplification factor over the input DNA (0.5 ng) of three orders of magnitude (730 ng of amplified DNA). Importantly, the length of the amplified DNA (~19 kb) corresponds to full-length ø29 DNA. As it can be observed in Figure 1, a delicate interplay among the four proteins and the input DNA is necessary to get a productive amplification, suggesting that an optimal balance between the number of initiations and the ratio of the elongation factors to growing DNA chains is required. We have shown that the presence of SSB protein, with its ability to bind the ssDNA generated during strand displacement, is needed to avoid the formation and accumulation of snapped-back strands containing an inverted duplication of DNA ends (Esteban *et al.*, 1997). ø29 DBP facilitates the opening of ø29 DNA replication origins, increasing the affinity of the TP/DNA polymerase complex for the DNA ends. As shown in Figure 1 (lanes 7-9, 11), under certain conditions, such as the imbalance of initiating/elongating factors or absence of SSB protein, ø29 DNA polymerase is able to switch the DNA strand used as template, producing palindromic sequences (snapped-back strands) containing an inverted duplication of one of the two ø29 DNA termini, that can be exponentially amplified. The biological intactness of full-length amplified ø29 DNA molecules was demonstrated by their ability to produce phage ø29 particles when transfected into *B. subtilis* cells (Blanco *et al.,* 1994).

Example 2: RCA with M13 ssDNA

Figure 2 shows the ability of ø29 DNA polymerase to produce rolling-circle DNA amplification (RCA) using oligonucleotide-primed M13mp18 ssDNA as substrate. Under the conditions described above (see section Protocols), ø29 DNA polymerase is able to replicate the 7,400-nt-long M13mp18 ssDNA in only 5 min. At longer incubation times, replication occurs coupled to strand displacement, generating very long single-stranded concatemeric DNA molecules with a size greater than 70 kb, which is about 10-fold the length of the circular template. Such an "asymmetric" DNA amplification by RCA relies on the high processivity and strand-displacement capacity that are intrinsic to ø29 DNA polymerase. As a control, Figure 2 also shows the limited amount of amplified products obtained with *E. coli* DNA polymerase I Klenow fragment, a polymerase enzyme incapable of strand displacement. Note here that in contrast to Klenow enzyme, elevated salt concentrations, which increase the stability of newly synthesized DNA duplex, do not affect the RCA activity of ø29 DNA polymerase (compare lanes in Figure 2 for these two polymerases that correspond to addition of NaCl).

Addition of ø29 SSB prevents the non-productive binding of ø29 DNA polymerase molecules to the enormous amounts of ssDNA generated in this reaction, particularly at a low enzyme/DNA ratio, thus increasing the efficiency of primer usage by ø29 DNA polymerase (Gascón *et al.,* 2000). This effect results in a stimulation of the overall incorporation of dNTPs up to 20-fold. Moreover, as shown in Figure 3, ø29 SSB has a second effect, increasing the rate of elongation of ø29 DNA polymerase by 2-fold. Such an effect is likely due to the formation of "patches" of SSB on the ssDNA that is being displaced that could contribute to a more efficient unwinding by ø29 DNA polymerase during processive translocation through the duplex DNA.

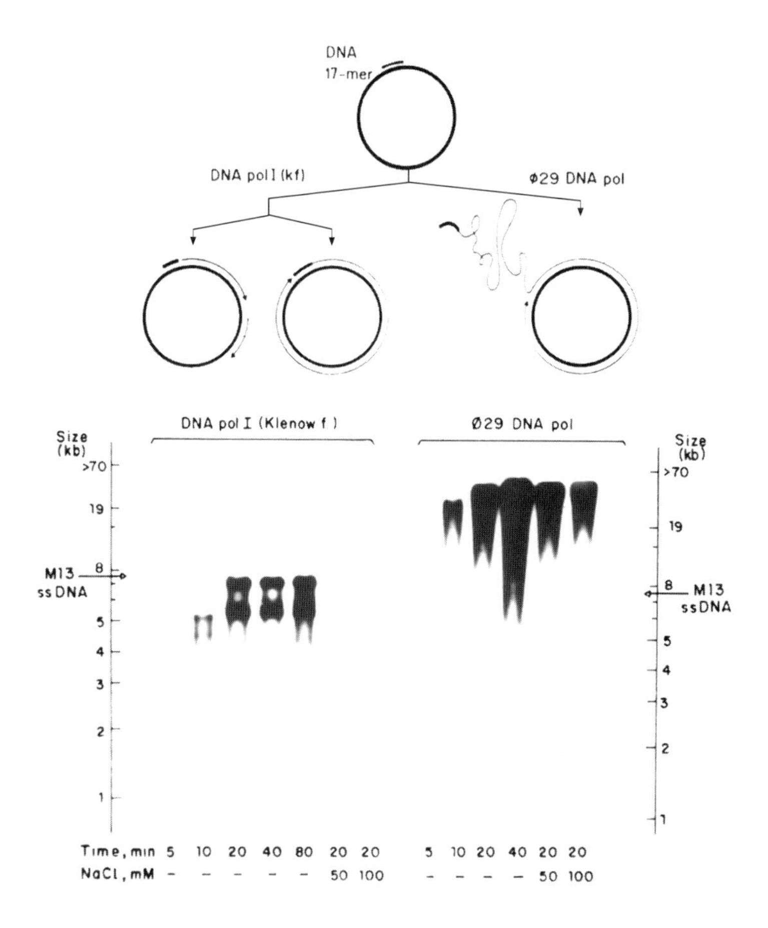

Figure 2. Strand-displacement RCA-type replication of M13 DNA by the oligo-primed ø29 DNA polymerase. The incubation mixture was described in Materials and Methods and contained 0.4 mM dNTPs. This figure presents a scheme of the replicating DNA molecules and the autoradiogram of the replication products (reproduced with permission from Blanco *et al.*, 1989; copyright of the Am. Soc. Biochem. Mol. Biol.).

Discussion

The results presented here along with our numerous studies of ø29 DNA polymerase prove that it is an ideal enzyme for isothermal DNA amplification, particularly to amplify large pieces of DNA, a caveat for commercial thermostable DNA amplification enzymes. This robust potential relies on the remarkable intrinsic processivity of ø29 DNA polymerase that allows ~10^4 polymerization cycles to occur without dissociating from the template (Blanco *et al.,* 1989). To achieve such a high processivity, most DNA polymerases require the assistance of accessory proteins, *e.g.* the ß-subunit of the *E. coli* PolIII holoenzyme or PCNA protein (the eukaryotic analog for this function), that clamp the catalytic complex to DNA (Kornberg and Baker, 1992). Besides, ø29 DNA polymerase can polymerize through duplex DNA, coupling processive polymerization to strand displacement, with an efficiency comparable to that obtained *in vitro* with much more complex sets of proteins, including DNA helicases. Moreover, the fidelity parameters of ø29 DNA polymerase are remarkable, with a base insertion fidelity in the range of 1 error in 10^6-10^7 bases incorporated (Esteban *et al.,* 1993), and boosted by its potent and selective 3'-5' exonuclease activity, whose proofreading contributes to further reduce the misincorporation frequency by about two orders of magnitude (Garmendia *et al.,* 1992).

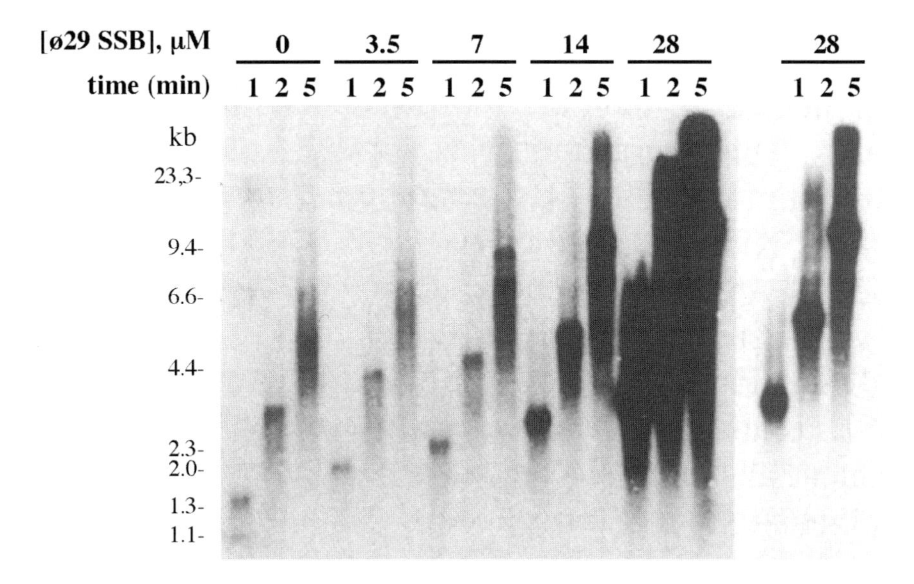

Figure 3. Dose-dependent effect of the ø29 SSB protein (p5) on ø29 DNA polymerase in oligo-primed M13 ssDNA replication. The assay was carried out as described in Materials and Methods and contained 80 μM dNTPs. Two different exposures are shown for the highest SSB protein concentration. The amount of newly synthesized DNA was estimated by densitometric scanning of the autoradiograms, whereas its length was determined by comparison with DNA markers of known size.

Both high processivity and strand-displacement ability of ø29 DNA polymerase are probably related to the high affinity of this enzyme for single-stranded DNA. No clear clues can be derived from amino acid sequence comparisons between ø29 DNA polymerase and other members of DNA polymerase family B. Therefore, a 3D-structural analysis of ø29 DNA polymerase complexed with either TP or DNA (currently under investigation in collaboration with J. Wang's group) may help to elucidate the structural basis responsible for the extraordinary processivity of ø29 DNA polymerase. On the other hand, site-directed mutagenesis studies pointed to the N-terminal domain (containing the 3'-5' exonuclease) as the main enzyme's part responsible for the strand-displacement capacity of ø29 DNA polymerase (Soengas *et al.,* 1992). Thus, it is tempting to speculate that a strand-displacement capacity could be an accesory function of ancestral DNA polymerases running through duplex nucleic acids, and that a SSB domain (as that present nowadays in 3'-5' exonucleases) could confer such a strand-displacement capacity when associated or fused to the catalytic polypeptide. The PCR-like way of the TP-primed isothermal replication of ø29 DNA, in which the two DNA strands are duplicated continuously, could explain why such a strand-displacement capacity was evolutionarily maintained in ø29 DNA polymerase. Conversely, a strand-displacement capacity must be absent in any DNA polymerase responsible for the lagging-strand DNA synthesis.

The results presented in this chapter establish some of the requisites for the development of isothermal DNA amplification strategy based on the bacteriophage ø29 DNA replication machinery that is able to reliably amplify very large (>70 kb) linear segments of DNA. A DNA amplification method based on the protein-primed DNA replication mechanism would require the positioning of a minimal ø29 DNA replication origin at both DNA ends of the dsDNA fragment of interest. The simplest version of an active ø29 DNA origin is constituted by the left 68 terminal bases, that contain the necessary sequence elements to be recognized by the origin-binding protein DBP (Serrano *et al.*, 1989). Thus, after two denaturation/polymerization cycles to position ø29 ori-containing plus template termini-specific "hybrid" primers, and to generate ø29 ori-terminated dsDNA fragments, further cycles of isothermal amplification would be triggered by addition of TP as primer and the remaining ø29 DNA amplification proteins. Then, continuous cycles of TP-primed initiation/elongation/strand-displacement would produce an exponential amplification of the desired DNA duplex at a single temperature. Alternatively, DNA molecules suitable for the TP-primed amplification could be obtained directly (without prior denaturation) by ligation of dsDNA adaptors with ø29 ori sequences to both ends of the DNA molecule to be amplified.

In addition, ø29 DNA polymerase makes possible the generation of concatemeric ssDNA amplicons of a unique and given polarity, as obtained by RCA on circular DNA templates containing the sequence of interest. As shown here, the efficiency of this process could be largely improved by providing stoichiometric amounts of the viral SSB. The resulting amplification products would be advantageous probes/ templates for hybridization/detection and sequencing studies.

A number of related protocols based on the strand-displacement amplification of one or both DNA strands, either on circular or linear templates of variuos sizes, have been recently developed (Lizardi *et al.*, 1998; Dean *et al.*, 2002; Nelson *et al.*, 2002; Detter *et al.*, 2002; Lage *et al.*, 2003). For all these techniques to be applied in a simple way, a DNA polymerase with intrinsic strand-displacement capacity would be a critical requisite. For the moment, ø29 DNA polymerase is the ideal enzyme to this end as it fulfills all the requirements for isothermal amplification of DNA, being the basis for the amplification systems *Templiphi* (www.templiphi.com), and *Genomiphi* (www.genomiphi.com), developed by Amersham Biosciences/ Molecular Staging (see chapters 3.5 and 3.6 in this book).

Acknowledgements

This work has been aided by research grant 5R01 GM27242-24 from the National Institutes of Health, by grant BMC 2002-03818 from the Ministery of Science and Technology and by an institutional grant from Fundación Ramón Areces to the Centro de Biología Molecular "Severo Ochoa".

References

Blanco, L., Bernad, A., Lázaro, J.M., Martín, G., Garmendia, C., and Salas, M. (1989). Highly efficient DNA synthesis by the phage ø29 DNA polymerase. J. Biol. Chem. *264*, 8935-8940.

Blanco, L., Lázaro, J.M., de Vega, M., Bonnin, A., and Salas, M. (1994). Terminal protein-primed DNA amplification. Proc. Natl. Acad. Sci. USA *91*, 12198-12202.

Blanco, L. and Salas, M. (1996). Relating structure to function in ø29 DNA polymerase. J. Biol. Chem. *271*, 8509-8512.

Dean, F.B., Hosono, S., Fang, L., Wu, X., Faruqi, A.F., Bray-Ward, P., Sun, Z., Zong, Q., Du, Y., Du, J., Driscoll, M., Song, W., Kingsmore, S.F., Egholm, M., and Lasken, R.S. (2002). Comprehensive human genome amplification using multiple displacement amplification. Proc. Natl. Acad. Sci. USA *99*, 5261-5266.

Detter, J.C., Jett, J.M., Lucas, S.M., Dalin, E., Arellano, A.R., Wang, M., Nelson, J.R., Chapman, J., Lou, Y., Rokhsar, D., Hawkins, T.L., and Richardson, P.M. (2002). Isothermal strand-displacement amplification applications for high-throughput genomics. Genomics *80*, 691-698.

Esteban, J. A., Salas, M., and Blanco, L. (1993). Fidelity of ø29 DNA polymerase. Comparison between protein-primed initiation and DNA polymerisation. J. Biol. Chem. *268*, 2719-2726.

Esteban, J.A., Blanco, L.,Villar, L., and Salas, M. (1997). *In vitro* evolution of terminal protein-containing genomes. Proc. Natl. Acad. Sci. USA *94*, 2921-2926.

Garmendia, C., Bernad, A., Esteban, J.A., Blanco, L., and Salas, M. (1992). The bacteriophage ø29 DNA polymerase, a proofreading enzyme. J. Biol. Chem. *267*, 2594-2599.

Gascón, I., Lázaro, J.M., and Salas, M. (2000). Differential functional behavior of viral ø29, Nf and GA-1 SSB proteins. Nucleic Acids Res. *28*, 2034-2042.

Kornberg, A. and Baker, T. (1992). DNA replication. 2nd edition. Freeman (San Francisco).

Kuriyan, J. and O'Donnell, M. (1993). Sliding clamps of DNA polymerases. J. Mol. Biol. *234*, 915-925.

Lage, J.M., Leamon, J.H., Pejovic, T., Hamann, S., Lacey, M., Dillon, D., Segraves, R., Vossbrinck, B., Gonzalez, A., Pinkel, D., Albertson, D.G., Costa, J., and Lizardi, P.M. (2003). Whole genome analysis of genetic alterations in small DNA samples using hyperbranched strand displacement amplification and array-CGH. Genome Res. *13*, 294-307.

Lázaro, J. M., Blanco, L., and Salas, M. (1995). Purification of bacteriophage ø29 DNA polymerase. Methods Enzymol. *262*, 42-49.

Lizardi, P.M., Huang, X., Zhu, Z., Bray-Ward, P., Thomas, D.C., and Ward, D.C. (1998). Mutation detection and single-molecule counting using isothermal rolling-circle amplification. Nat. Genet. *19*, 225-32.

Martín, G., Lázaro, J. M., Méndez, E., and Salas, M. (1989). Characterization of the phage ø29 protein p5 as a single-stranded DNA binding protein. Function in ø29

DNA-protein p3 replication. Nucleic Acids Res. *17*, 3663-3672.
McDonnell, M. W., Simon, M. N., and Studier, F. W. (1977). Analysis of restriction fragments of T7 DNA and determination of molecular weights by eletrophoresis in neutral and alkaline gels. J. Mol. Biol. *110*, 119-146.
Mellado, R.P., Peñalva, M.A., Inciarte, M.R., and Salas, M. (1980). The protein covalently linked to the 5' termini of the DNA of *Bacillus subtilis* phage ø29 is involved in the initiation of DNA replication. Virology *104*, 84-96.
Méndez, J., Blanco, L., Esteban, J.A., Bernad, A., and Salas, M. (1992). Initiation of ø29 DNA replication occurs at the second 3' nucleotide of the linear template: a sliding-back mechanism for protein-primed DNA replication. Proc. Natl. Acad. Sci. USA *89*, 9579-9583.
Méndez, J., Blanco, L., and Salas, M. (1997). Protein-primed DNA replication: a transition between two modes of priming by a unique DNA polymerase. EMBO J. *16*, 2519-2527.
Nelson, J.R., Cai, Y.C., Giesler, T.L., Farchaus, J.W., Sundaram, S.T., Ortiz-Rivera, M., Hosta, L.P., Hewitt, P.L., Mamone, J.A., Palaniappan, C., and Fuller, C.W. (2002). TempliPhi, phi29 DNA polymerase based rolling circle amplification of templates for DNA sequencing. Biotechniques *Suppl*, 44-47.
Pastrana, R., Lázaro, J.M., Blanco, L., García, J.A., Méndez, E., and Salas, M. (1985). Overproduction and purification of protein p6 of *Bacillus subtilis* phage ø29: role in the initiation of DNA replication. Nucleic Acids Res. *13*, 3083-3100.
Peñalva, M.A. and Salas, M. (1982). Initiation of phage ø29 DNA replication *in vitro*: formation of a covalent complex between the terminal protein p3 and 5'-dAMP. Proc. Natl. Acad. Sci. USA *79*, 5522-5526.
Salas, M. (1991). Protein-priming of DNA replication. Annu. Rev. Biochem. *60*, 39-71.
Salas, M., Miller, J.T., Leis, J., and DePamphilis, M.L. (1996). Mechanisms for priming DNA synthesis. In: *DNA Replication in Eukaryotic Cells*. M. DePamphilis, ed. Cold Spring Harbor Press, pp. 131-176.
Serrano, M., Gutiérrez, C., Prieto, I., Hermoso, J.M., and Salas, M. (1989). Signals at the bacteriophage ø29 DNA replication origins required for protein p6 binding and activity. EMBO J. *8*, 1879-1885.
Serrano, M., Gutiérrez, C., Freire, R., Bravo, A., Salas, M., and Hermoso, J.M. (1994) Phage ø29 protein p6: a viral histone-like protein. Biochimie *76*, 981-991.
Soengas, M.S., Esteban, J.A., Lázaro, J.M., Bernad, A., Blasco, M.A., Salas, M., and Blanco, L. (1992). Site-directed mutagenesis at the Exo III motif of ø29 DNA polymerase. Overlapping structural domains for the 3'-5' exonuclease and strand-displacement activities. EMBO J. *11*, 4227-4237.
Zaballos, A. and Salas, M. (1989). Functional domains in the bacteriophage ϕ29 terminal protein for the interaction with the ø29 DNA polymerase and with DNA. Nucleic Acids Res. *17*, 10353-10366.

1.3

High-Fidelity Thermostable DNA Ligases as a Tool for DNA Amplification

Weiguo Cao

Abstract

DNA ligases catalyze covalent joining of two DNA strands on DNA template at a nick junction. The strict requirement of base pair complementarity at the nick junction has been exploited for development of ligase-based technologies aimed for detection of sequence variations. After discovery of thermostable ligases, methods employing amplification of diagnostic signal through repeated cycles of denaturation, annealing and ligation have been developed analogous to polymerase chain reaction (PCR). This chapter focuses on principles and practical aspects of ligase chain reaction (LCR) and ligase detection reaction (LDR).

Background

Reaction Mechanism and Fidelity of DNA Ligases

DNA ligases catalyze the formation of a phosphodiester bond at a nick junction on double-stranded DNA (Lehman, 1974). Strand joining is an essential physiological process that occurs during DNA replication, recombination and repair. All of the sequenced genomes from different species contain at least one gene coding for DNA ligase. The chemistry of the strand-joining reaction has been elucidated by early studies on *Escherichia coli* DNA ligase and T4 DNA ligase (Lehman, 1974). A complete ligation reaction involves three steps: enzyme adenylation, substrate adenylation, and nick-closure (Figure 1A). In the enzyme

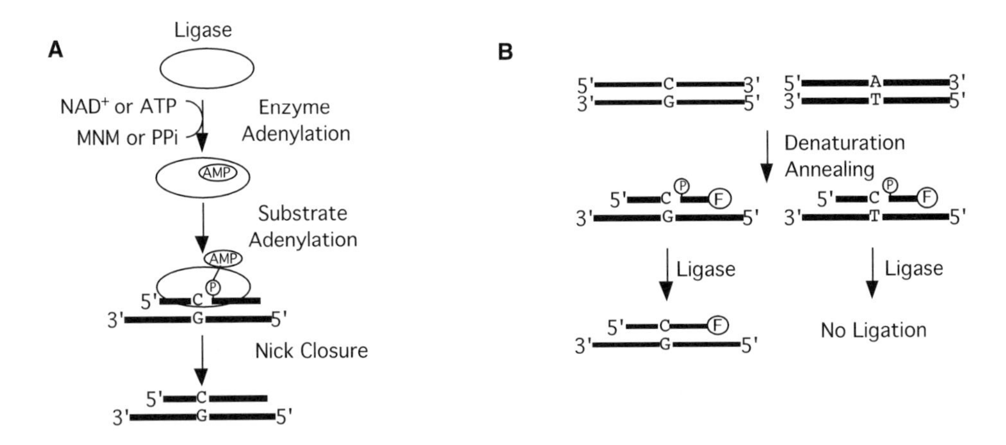

Figure 1. Steps of ligation reaction and principle of ligase-based allele discrimination.
A: Three steps of ligation reaction. In the enzyme adenylation step, an AMP group is transferred from the cofactor NAD or ATP to a lysine residue in the adenylation motif KXDG through a phosphoamide linkage. In the substrate adenylation step, this AMP group is transferred to the 5' phosphate at the nick through a pyrophosphate linkage to form a DNA-adenylate intermediate (AppDNA). In the nick-closure step, a phosphodiester bond is formed to seal the nick and release AMP.
B: Principle of ligase-based allele discrimination. Genomic DNA or PCR products are denatured and annealed with ligation probes. Because of the high fidelity of the DNA ligase at the nick junction, a perfectly matched base pair will be sealed while a mismatch base pair will be left unsealed. The fluorophore (F) attached to one of the ligation probes allows for visualization of the ligation product after electrophoresis on an ABI 373 or 377 sequencer. P, phosphate.

adenylation step, an adenosine monophosphate (AMP) group is transferred from the cofactor ATP or NAD^+ to the ε-amino group of a lysyl residue in the adenylation motif KXDG through formation of a phosphoamide bond. In the substrate adenylation step, the AMP group is transferred to the 5' phosphate at the nick through a pyrophosphate linkage to form a DNA-adenylate intermediate (AppDNA). In the nick-closure step, a phosphodiester bond is formed with sealing the nick and releasing AMP. All three steps in ligation reaction are reversible. Detailed accounts on ligase discovery and chemistry of ligation are covered by earlier reviews (Engler and Richardson, 1982; Lehman, 1974). Several recent reviews dealing with NAD^+- and ATP-dependent ligases are also available (Cao, 2001; Cao, 2002; Cherepanov and de Vries, 2002; Doherty and Suh, 2000; Lieber, 1999; Martin and MacNeill, 2002; Timson *et al.*, 2000; Tomkinson *et al.*, 2001; Wilkinson *et al.*, 2001; Wilson *et al.*, 2003)

Despite differences in the structure of cofactors needed for ligation, sequence alignment, mutagenesis analysis, and three-dimensional structure studies have revealed that both groups of ligases (ATP- and NAD^+-dependent) belong to the same nucleotidyl transferase superfamily, which also includes RNA ligases and RNA capping enzymes (Aravind and Koonin, 1999; Shuman and Schwer, 1995; Sriskanda *et al.*, 1999). It is now known that the members of this superfamily contain conserved protein motifs defining the nucleotide binding sites and also essential amino acid residues for multiple nucleotidyl transfer steps that occur in both ligation and mRNA capping reactions. Ligase-based assays use both types of DNA ligases, *e.g.* ATP-dependent ligases, such as T4 DNA ligase, and NAD^+-dependent ligases, such as *Thermus thermophilus* HB8 (*Tth*) ligase. However, only thermostable ligases from thermophilic microorganisms such as *Thermus* species enable ligase-based amplification.

All ligase-based detection methods rely on the strict requirement of perfect complementarity at the ligation junction allowing distinguish a single base pair difference (Barany, 1991a; Barany, 1991b; Jarvius *et al.*, 2003; Landegren *et al.*, 1988; Wu and Wallace, 1989a). High-fidelity ligases seal a nick only when the bases at the ligation junction form a perfectly matched base pair, while the mismatched base pair remains unsealed (Figure 1B). Thus, a mutation signal is discriminated by a ligation event.

Ligase fidelity differs substantially among different ligases. The fidelity of T4 DNA ligase is one to two orders of magnitude lower than that of ligases from thermophilic bacteria such as *Thermus* species (Tong *et al.*, 1999). The *Tth* ligase appears to require a greater protein-DNA interface for ligation, as it fails to ligate hexamers while T4 and T7 DNA ligases effectively ligate hexamers (Pritchard and Southern, 1997). DNA ligases in general maintain higher fidelity when a mismatch base pair is located at the 3' side of the nick rather than at the 5' side (Luo *et al.*, 1996; Tomkinson *et al.*, 1992; Wu and Wallace, 1989b). However, different mismatch base pairs are ligated by different ligases with different efficiencies. For example, a purine-purine mismatch such as G/G at the 5' side is a good substrate for the *Aquifex aeolicus* ligase and the *Vaccinia* virus ATP ligase, but not for the *Tth* ligase (Luo *et al.*, 1996; Shuman, 1995; Tong *et al.*, 2000). At the 3' side of the nick junction, a T/G mismatch is more prone to misligation by the *Tth* ligase than other mismatches (Luo *et al.*, 1996). Mismatches several base pairs away from the nick junction also show reduced ligation by the *Tth* ligase, suggesting that this ligase is sensitive to base pair mismatches within the distance of the protein-DNA interface (Pritchard and Southern, 1997; Tong *et al.*, 1999).

Many ligases including thermostable ligases can tolerate substitution of Mg^{2+} for Mn^{2+} as the metal cofactor; however, generally the ligation fidelity is lower with Mn^{2+} (Tong *et al.*, 2000; Tong *et al.*, 1999). Improved fidelity has been achieved through protein engineering of *Thermus thermophilus* ligase and finding of natural variants of enzyme by screening of *Thermus* isolates (Luo *et al.*, 1996; Tong *et al.*, 1999). Biochemical studies on a NAD ligase from the hyperthermophilic bacterium *Aquifex aeolicus* demonstrate that the ligation fidelity is controlled at both the second step (substrate adenylation) and the third step of ligation (nick-closure) (Tong *et al.*, 2000).

The extreme sensitivity of ligases to mismatches at the nick junction has been used for development of ligase-based technologies aimed for detection of sequence variations. After discovery of thermostable ligases, methods employing amplification of diagnostic signal through repeated cycles of denaturation, annealing and ligation have been developed.

Principles of LCR and LDR

The exponential amplification power of PCR relies on the fact that each primer extension product serves as a template in the next run of DNA synthesis. Likewise, the basic premise of LCR is to generate a ligation product that serves as a template in the next run of DNA ligation (Figure 2A). LCR requires two pairs of ligation primers complementary to two strands of duplex DNA. Each pair of primers is annealed to the DNA template with formation of a nick, which is ligated by DNA ligase if there is a perfect complementarity at the nick junction. If there is a single nucleotide mismatch, the ligation reaction fails. Thus, allele-specific discrimination is achieved. In the second ligation cycle, the two pairs of ligation primers are annealed to both the original templates and the ligation products from the first cycle and then ligated.

The continuous thermocycling using thermostable DNA ligases enables exponential signal amplification. It was soon recognized that self-annealing of the two pairs of LCR primers generated DNA duplex with blunt ends, which may be ligated by DNA ligase. To prevent blunt end ligation and increase LCR sensitivity, high salt concentrations have been used, which indeed decreased background (Wu and Wallace, 1989a). Gap-LCR differs from conventional LCR by designing two pairs of LCR primers complementary to the target but with a gap between the juxtaposed primer pair. Strand joining and subsequent ligation requires a gap-filling reaction catalyzed by DNA polymerase, thus, preventing blunt end ligation (Abravaya *et al.*, 1995). Abbott Laboratories (Abbott Park, IL) has developed a Gap-LCR-based semi-automated assay which has been used for the detection of *Mycobacterium tuberculosis, Chlamydia trachomatis*, HIV, and other pathogens, as well as for genotyping in the beta-2 adrenergic receptor gene (Ausina *et al.*, 1997; Kehl *et al.*, 1998; Ruiz-Serrano *et al.*, 1998; Young *et al.*, 2000; Yu *et al.*, 2001; Zanchetta *et al.*, 2000).

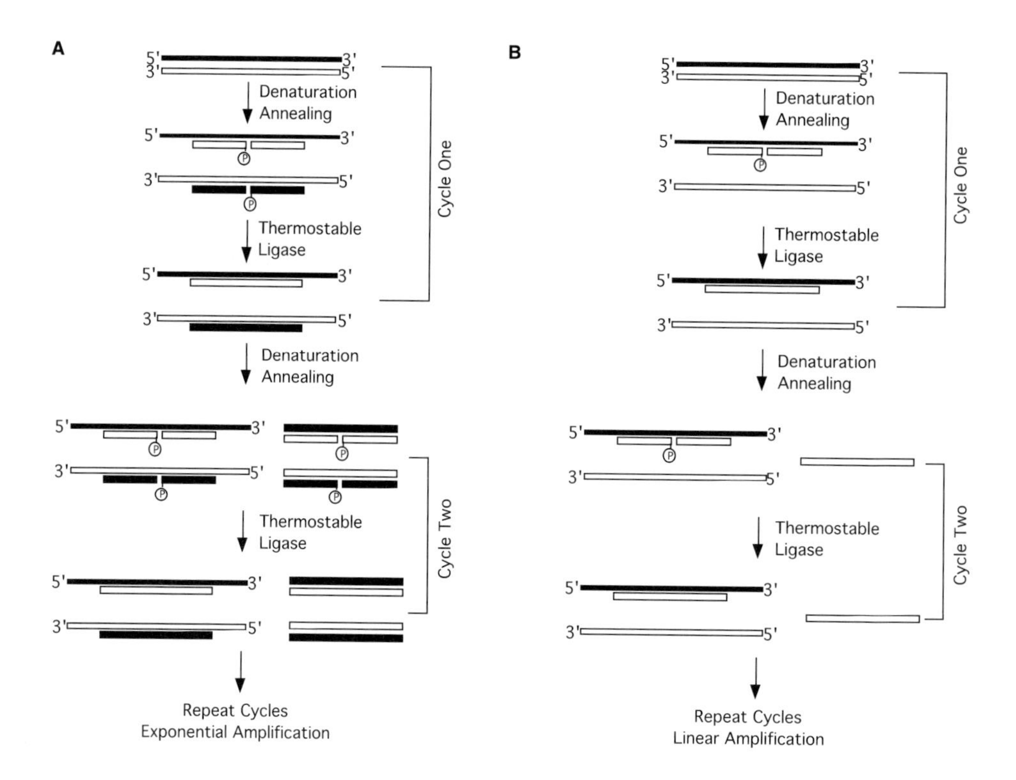

Figure 2. Principles of ligase-based amplification methods.
A: Ligase Chain Reaction (LCR). In the first cycle, two pairs of primers are annealed to the two strands of denatured DNA. Thermostable DNA ligase seals the nicks to yield two ligation products that serve as templates in subsequent cycle. Thus, exponential amplification is achieved.
B: Ligase Detection Reaction (LDR). In the first cycle, one pair of primers is annealed to the complementary strand of the denatured DNA. Thermostable DNA ligase seals the nick and yields one ligation product. Continuous cycles of annealing and ligation result in linear amplification of the signal.

Ligase Detection Reaction (LDR) uses only one pair of ligation primers targeted at one DNA strand (Figure 2B). As a result, LDR amplifies target sequence with linear kinetics but avoids the template-independent ligation associated with LCR. The high sensitivity offered by ligase-based techniques is ideally suited for identifying mutations on a large background of normal sequences. This situation is happening in studies of cancer, when cancer cells containing somatic mutations may be immersed in a vast excess of normal cells. It is also important in studies of HIV or other rapidly mutating viruses, in which a drug-resistant mutation may coexist with other variations of the viral genome. LDR primers may be tagged with unique sequences (zip code) to facilitate high throughput detection by universal arrays or microspheres (Favis *et al.*, 2000; Gerry *et al.*, 1999; Iannone *et al.*, 2000). Recently, a method combining direct LDR on non-amplified genomic DNA with single pair of labeled primers capable of Fluorescence Resonance Energy Transfer (spFRET) has been reported in a study aimed to detect low abundant cancer mutations (Wabuyele *et al.*, 2003).

In addition to detection of known mutations in DNA sequences, ligase-based techniques have been developed for scanning of unknown mutations (Huang *et al.*, 2002; Zhang *et al.*, 1998; Zhang *et al.*, 2002). Ligase-based techniques are also applied to detection of RNA or proteins molecules (Fredriksson *et al.*, 2002; Nilsson *et al.*, 2000; Yeakley *et al.*, 2002; Zhang *et al.*, 1998). The following sections describe two ligase-based amplification protocols, LCR and LDR, adapted from the original reports (Barany, 1991a; Khanna *et al.*, 1999).

Protocol 1: Ligase chain reaction (LCR)

Materials and Methods

To perform LCR, genomic DNA should be isolated using conventional methods like phenol/chloroform extraction and ethanol precipitation or by using QIAmp Tissue kit (Qiagen). A detailed protocol for DNA isolation from cell lines and paraffin sections is described in step 1. LCR can be performed using commercially available thermostable DNA ligases and thermocyclers. LCR primers should be designed with Tm's about 66-70°C, which is just at or slightly above the temperature of ligation. LCR primers may be labeled using radioisotopes (^{32}P) or fluorophores and the LCR products can be visualized using a phosphorimager or fluoroimager, respectively, after denaturing gel electrophoresis. If oligonucleotides are labeled with fluorophores such as FAM or TET, the reaction products can be also analyzed on automated DNA sequencers (*e.g.* ABI 377 DNA sequencer, Applied Biosystems). The details of analysis using automated DNA sequencers are described in protocol 2. In a multiplex reaction, the ligation products may be distinguished from each other by designing ligation products with different length or by using different fluorescent signals.

Step 1: Genomic DNA Extraction from Cell Lines and Paraffin Sections

DNA was isolated from the cell lines using standard protocols employing proteinase K/SDS treatment. To prepare DNA from paraffin-embedded archival tumors, which represent a consecutive series of primary colon cancers removed by surgical resection at a single institution, 10 μm tissue sections were cut from the paraffin blocks, and morphologically distinct regions were microdissected to minimize stromal contamination. Samples were purified from paraffin via sequential extraction with xylene, 100% ethanol and acetone and dried under vacuum. The pellet was incubated overnight with proteinase K (200 μg/ml in 50 mM Tris pH 8.5, 1 mM EDTA and 0.5% Tween 20) at 55°C. After heating at 100°C for 10 min, debris was removed by centrifugation and the supernatant was stored at 4°C (Korn *et al.*, 1993). Greater purity of DNA was achieved by phenol/chloroform extraction and ethanol precipitation (Sato *et al.*, 1990), or by using the QIAamp Tissue Kit (Qiagen).

Step 2: ^{32}P Labeling of Oligonucleotides

LDR primers should be purified by polyacrylamide gel electrophoresis or high performance liquid chromatography (HPLC) before labeling. Many commercial providers such as Integrated DNA Technologies (IDT) offer purification services. One of the LCR primers (15 pmole) is 5' end labeled in 20 μl of 30 mM Tris-HCl (pH 8.0), 10 mM $MgCl_2$, 0.5 mM EDTA, 5 mM dithiothreitol and 400 μCi of [γ-^{32}P] ATP (6,000 Ci/mM, New England Nuclear), by addition of 15 units of T4 polynucleotide kinase (New England Biolabs, Beverly, MA). After incubation at 37°C for 45 min, unlabeled ATP was added to a final concentration 1 mM, and incubation was continued for additional 2 min at 37°C. The reaction was terminated by addition of 0.5 μl of 0.5 M EDTA, and inactivation of polynucleotide kinase at 65°C for 10 min.

Unincorporated ^{32}P was removed by gel filtration through Sephadex G-25 column pre-equilibrated with TE buffer. Specific activity ranged from $7x10^8$ to 10^9 cpm/μg oligonucleotide.

Step 3: LCR

For LCR, equimolar amounts of labeled oligonucleotide (200,000 cpm, 40 fmoles) and unlabeled allele-specific oligonucleotides were incubated in the presence of target DNA (ranging from 10^7 molecules to less than one molecule per tube) in 10 μl of 20 mM Tris-HCl (pH 7.6) containing 100 mM KCl, 10 mM $MgCl_2$, 1 mM EDTA, 1 mM NAD, 10 mM dithiothreitol, 4 μg salmon sperm DNA, and 15 units of *Tth* DNA ligase (New England Biolabs). Note that the *Thermus aquaticus* DNA ligase available from NEB is the same enzyme as Tth HB8 ligase (Barany and Gelfand, 1991). Reactions were incubated at 94°C for 1 min followed by 65°C for 4 min, and this cycle was repeated 20 - 30 times.

Step 4: Electrophoresis

Ligation mixtures (4 μl, 40,000-80,000 cpm/lane) were denatured by boiling for 3 min in 45% formamide prior to loading onto the gel. Electrophoresis was in a 10% polyacrylamide gel containing 7 M urea in a 1 x TBE buffer. After fixing the gel in 10% methanol with 10% acetic acid solution, gels were dried and exposed overnight at –70°C with Kodak XAR-5 film with a Cronex intensifying screen (Du Pont). Alternatively, if fluorescent labels are used in LCR, the products may be separated and analyzed using automated gene sequencing instruments (see protocol 2 for the details).

Example: Detection of Sickle Cell β^s-Globin Genotype

The normal β^A and sickle cell β^S genes differ by a single A/T transversion that leads to a change of a glutamic acid for a valine in the hemoglobin β chain. Diagnostic oligonucleotides containing the

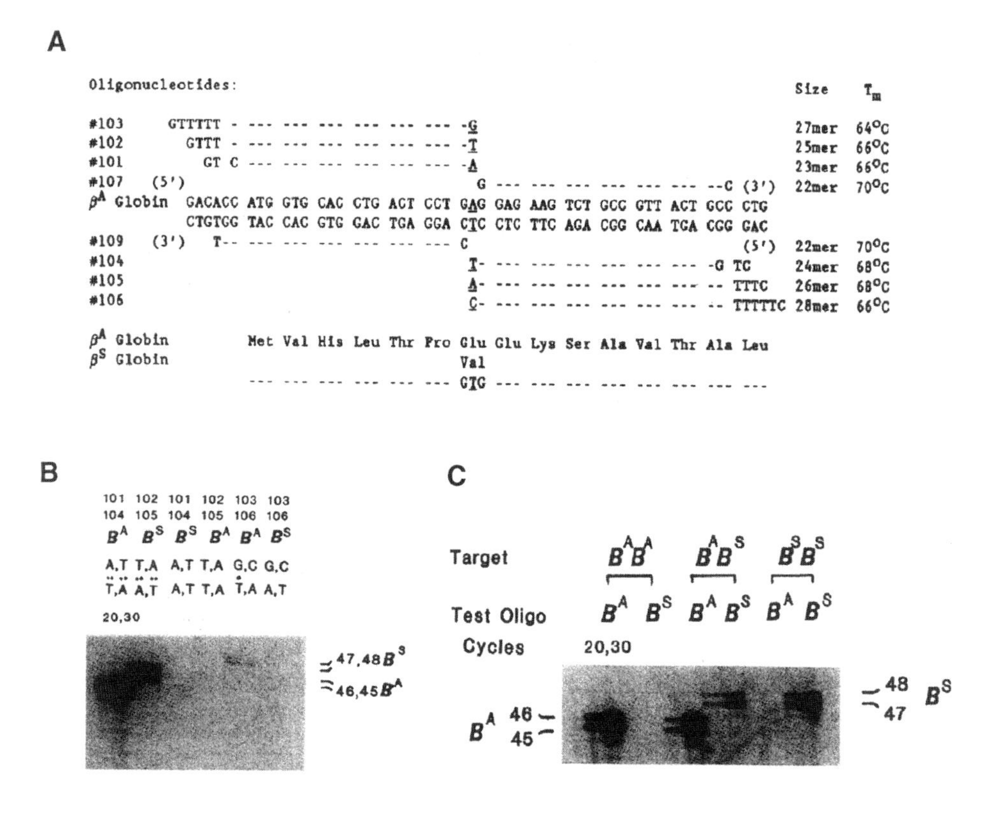

Figure 3. Detection of sickle cell allele by LCR.
A: Nucleotide sequence of a fragment of a β-globin gene and of the oligonucleotide probes used to discriminate $ß^A$ and $ß^S$ globin genes. Oligonucleotides #101 and #104 detect the $ß^A$ target, while #102 and #105 detect the $ß^S$ target when ligated to labeled oligonucleotides #107 and #109, respectively. Oligonucleotides #103 and #106 were designed to assay the efficiency of ligation of G:T or G:A and C:A or C:T mismatches using $ß^A$ or $ß^S$ globin gene targets, respectively.
B: Specificity of LCR using cloned β-globin fragments. Electropherograms obtained after 20 and 30 cycles of LCR, separation of the products on denaturing 10% PAGE and overnight exposure with Kodak XAR-5 film. Numbers of allele-specific oligonucleotides and globin genotype are indicated above each lane.
C: Detection of ß-globin alleles in human genomic DNA using LCR. DNAs from normal ($ß^Aß^A$), carrier ($ß^Aß^S$), and sickle cell ($ß^Sß^S$) individuals were subjected to LCR in two separate tubes containing ^{32}P-labeled oligonucleotides (#107 and #109), and either $ß^A$ test oligonucleotides (#101 and #104) or $ß^S$ test oligonucleotides (#102 and #105). Ligation products of 45 and 46 bp and 47 and 48 bp indicate presence of the $ß^A$ or $ß^S$ globin gene, respectively. LCR was performed for 20 and 30 cycles in all reactions. The length of the LCR products is indicated in base pairs. Reproduced from Barany, 1991a with permission.

3' nucleotide unique for each allele were synthesized with different length 5' tails (Figure 3A). Upon ligation to the invariant ^{32}P-labeled adjacent oligonucleotide, the individual products could be distinguished when separated on a polyacrylamide denaturing gel and detected by autoradiography (Figure 3B).

To detect ß-globin alleles in human genomic DNA, blood DNA from the normal ($ß^Aß^A$), carrier ($ß^Aß^S$), and sickle cell ($ß^Sß^S$) individuals was tested using allele-specific LCR. With target DNA corresponding to 10 μl of blood, $ß^A$ and $ß^S$ alleles could be readily detected using allele-specific LCR (Figure 3C). As observed with plasmid-cloned target DNA (Figure 3B), the efficiency of ligation (and hence detection) is somewhat less for the $ß^S$- than that for the $ß^A$-specific oligonucleotides. This difference may be a function of the nucleotide sequence at the ligation junction, or of the particular oligonucleotide (with different 5' tails) used in these LCR experiments. Nevertheless, the results demonstrate feasibility of direct LCR for allelic discrimination of blood samples without primary PCR.

Protocol 2: PCR/LDR

Ligation detection reaction (LDR) is very similar to LCR in terms of methods, reagents, and equipment necessary for the experiment. Primers design and labeling and DNA isolation is performed exactly how it is described in Protocol 1. There is, however, substantial difference in sensitivity of LCR and LDR because of different kinetics of product formation (exponential in LCR versus linear in LDR). To achieve comparable level of sensitivity, DNA template in LDR is pre-amplified using conventional PCR. LDR products can be visualized using the same methods as described for LCA. The PCR/LDR protocol below entails use of fluorescently labeled primers and detection of LDR products by GeneScan analysis on an ABI DNA sequencer. LDR products may also be detected by universal arrays or microspheres. Interested readers should consult the original papers for details (Favis *et al.*, 2000; Gerry *et al.*, 1999; Iannone *et al.*, 2000).

Step 1: PCR

PCR pre-amplifications were carried out in 50 μl of 10 mM Tris/HCl buffer pH 8.3 containing 10 mM KCl, 4 mM MgCl2, 250 μM dNTPs, 1 μM forward and reverse primers, and 1-50 ng of genomic DNA. The set of primers to amplify exons 1 and 2 of the K-*ras* gene contained: Ex.1.3 forward, 5' AAC CTT ATG TGT GAC ATG TTC TAA TAT AGT CAC 3'; Ex.1.4 reverse, 5'- AAA ATG GTC AGA GAA ACC TTT ATC TGT ATC- 3'; Ex.2.9 forward, 5'- TCA GGA TTC CTA CAG GAA GCA AGT AGT A- 3' and Ex.2.11 reverse, 5'- ATA CAC AAA GAA AGC CCT CCC CA -3'. After 10 min. denaturation, 1.5 units of AmpliTaq DNA polymerase (Perkin Elmer) was added under the hot start conditions. Amplification was performed for 35-40 cycles at 94°C for 30 sec, 60°C for 1 min., 72°C for 1 min. Final extension at 72°C was performed for 3 min. PCR products (4 μl) were analyzed on a 2% agarose gel to verify the presence of the PCR product of the expected size. For DNA isolated from paraffin sections, only Exon 1 of the K-*ras* gene was amplified. PCR products were stored at -20°C.

Step 2: LDR

LDRs were carried out in a 20 μl mixture containing 20 mM Tris-HCl, pH 7.6; 10 mM MgCl2; 100 mM KCl; 10 mM DTT; 1 mM NAD^+; 25 nM of the allele-specific primers and indicated amounts of the PCR products obtained in the pre-amplification step. The reaction mixture was heated for 1.5 min. at 94°C prior to adding 25 fmol of wild-type *Tth* DNA ligase or *Tth*-K294R mutant ligase. The ligase we used was purified in house and quantified by either protein assays or SDS-PAGE. LDR mixtures were cycled for 20 times at 94°C for 15 sec and at 65-K294R mutant ligase. The ligase we used was purified in house and quantified by either protein assays or SDS-PAGE. LDR mixtures were cycled for 20 times at 94°C for 15 sec and at 65°C for 4 min. Reactions were stopped by chilling the tubes in an ethanol-dry ice bath, and by adding 0.5 μl of 0.5 mM EDTA. Aliquots of 2.5 μl of the reaction products were mixed with 2.5 μl of loading buffer (83% formamide, 8.3 mM EDTA, and 0.17% Blue Dextran) and 0.5 μl Rox-1000, or TAMRA 350 molecular weight marker, denatured at 94°C for 2 min., chilled rapidly on ice prior to loading on an 8 M urea-10% polyacrylamide gel, and subjected to electrophoresis using an ABI 373 DNA sequencer. Fluorescent products were visualized and quantified using the ABI Gene Scan 672 software. The amount of the ligation products was calculated using the calibration curve obtained with the known amounts of DNA (1 fmol = 600 peak area units).

Example: Detection of Point Mutations in K-ras Oncogene

To analyze point mutations in K-ras oncogene we used DNA from the cell lines of the wild-type (HT29) and known mutant K-*ras* genotypes (SW620, G12V; SW1116, G12A; LS180, G12D; DLD1, G13D). The LDR primers for detecting mutations in K-ras codons 12, 13, and 61 are presented in Table 1. Multiplex LDRs using sets of three or twenty-six primers were used to analyze one and nineteen possible single-base mutations at K-*ras* codons 12, 13 and 61, respectively. In the three primer set experiment, 50 nM of the two allele-specific primers for G12D and G12V and 100 nM of the common primer were used. For the twenty-six primer set, 25 nM (500 fmol) of all nineteen allele-discriminating primers for codons 12, 13, and 61 and 50-75 nM of all seven common primers were used in a single reaction. PCR amplicon containing the G12V mutation was diluted from 1:200 to 1: 4000-fold by the wild-type DNA. LDR was performed using 5 nM of *Tth* ligase (Luo *et al*., 1996).

Point mutations in the K-*ras* gene occur early in the development of colorectal neoplasms and are found in 35-50% of colorectal adenomas and cancers. Most of these K-*ras* mutations are localized in codon 12, and to a lesser degree in codons 13 and 61. The K-*ras* genotyping presents the dual challenge of distinguishing multiple mutations in the neighboring codons 12 and 13 from wild-type sequence, as well as the H-, N-*ras* genes, which are highly homologous to K-*ras*. Probes specific for each mutation can interfere with one another during hybridization or polymerase extension. To approach the above challenges, a primary gene-specific PCR assay was developed. Independent K-*ras*-specific PCR primers were designed for exon 1, containing codons 12 and 13, and for exon 2, containing codon 61. For LDR, the discriminating

Table 1. LDR primer sets for detecting mutations in K-*ras* codons 12, 13, and 61.

#	K-*ras* LDR primers	Tm (°C)	Primer Size, nt	LDR Product size, bp	Sequence (5'- 3')
1	Fam-K-ras c12.2D	67.6	24	44	Fam-AAA ACT TGT GGT AGT TGG AGC TGA
2	Tet-K-ras c12.2A	67.6	25	45	Tet-CAA AAC TTG TGG TAG TTG GAG CTG C
3	Fam-K-ras c12.2V	67.6	26	46	Fam-ACA AAA ACT TGT GGT AGT TGG AGC TGT
4	K-ras c12 Com-2	69.8	20[a]		pTGG CGT AGG CAA GAG TGC CT-Bk
5	Tet-K-ras c12.1S	66.6	26	47	Tet-ATA TAA ACT TGT GGT AGT TGG AGC TA
6	Fam-K-ras c12.1R	66.6	27	48	Fam-AAT ATA AAC TTG TGG TAG TTG GAG CTC
7	Tet-K-ras c12.1C	66.6	28	49	Tet-CAA TAT AAA CTT GTG GTA GTT GGA GCT T
8	K-ras c12 Com-1	69.6	21[a]		pGTG GCG TAG GCA AGA GTG CCA A-Bk
9	Fam-K-ras c13.4D	67.2	21	51	Fam-TGT GGT AGT TGG AGC TGG TGA
10	Tet-K-ras c13.4A	67.2	22	52	Tet-ATG TGG TAG TTG GAG CTG GTG C
11	Fam-K-ras c13.4V	67.2	23	53	Fam-AAT GTG GTA GTT GGA GCT GGT GT
12	K-ras c13 Com-4	66.8	30[a]		pCGT AGG CAA GAG TGC CTT GAC $(A)_9$-Bk
13	Tet-K-ras c13.3S	66.6	22	54	Tet-CTT GTG GTA GTT GGA GCT GGT A

14	Fam-K-ras c13.3R	66.6	23	55	Fam-ACT TGT GGT AGT TGG AGC TGG TC
15	Tet-K-ras c13.3C	66.6	24	56	Tet-AAC TTG TGG TAG TTG GAG CTG GTT
16	K-ras c13 Com-3	69.7	32[a]		pGCG TAG GCA AGA GTG CCT TGA $(A)_{11}$-Bk
17	Tet-K-ras c61.7HT	68.2	25	59	Tet-AGA TAT TCT CGA CAC AGC AGG TCA T
18	Fam-K-ras c61.7HC	68.2	26	60	Fam-AAG ATA TTC TCG ACA CAG CAG GTC AC
19	K-ras c61 Com-7	68.9	36[a]		pGAG GAG TAC AGT GCA ATG AGG GAC $(A)_{10}$-Bk
20	Tet-K-ras c61.6R	66.2	23	61	Tet-GAT ATT CTC GAC ACA GCA GGT CG
21	Fam-K-ras c61.6L	66.2	24	62	Fam-AGA TAT TCT CGA CAC AGC AGG TCT
22	Tet-K-ras c61.6P	66.2	25	63	Tet-AAG ATA TTC TCG ACA CAG CAG GTC C
23	K-ras c61 Com-6	69.1	38[a]		pAGA GGA GTA CAG TGC AAT GAG GGA $(A)_{14}$-Bk
24	Fam-K-ras c61.5K	67.4	23	64	Fam-GGA TAT TCT CGA CAC AGC AGG TA
25	Tet-K-ras c61.5E	67.4	24	65	Tet-AGG ATA TTC TCG ACA CAG CAG GTG
26	K-ras c61 Com-5	69.0	41[a]		pAAG AGG AGT ACA GTG CAA TGA GGG CAA $(A)_{14}$-Bk

[a] Primers 4, 8, 12, 16, 19, 23 and 26 are 5' phosphorylated and contain a 3'-C3 spacer, Bk (Glen Research, Sterling, VA) which prevents possible DNA polymerase extension of the LDR primer. DNA polymerase can be carried over from the PCR pre-amplification step.

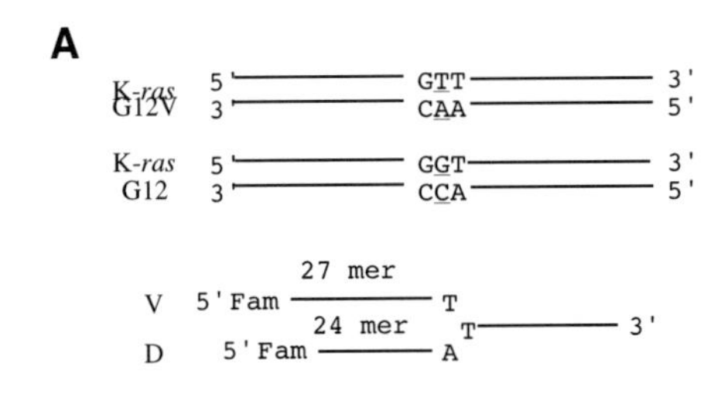

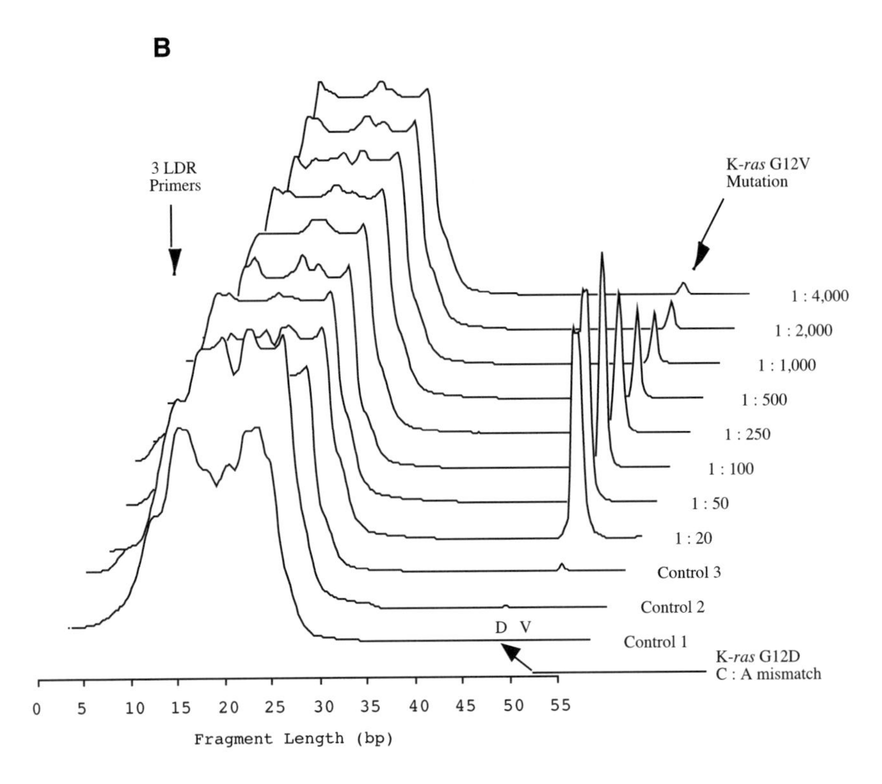

Figure 4. Quantitative detection of G12V mutation of the K-*ras* gene in an excess of wild DNA. A: Schematic diagram of mutant G12V and normal G12 double-stranded templates with the two discriminating primers (V, valine; D, aspartic acid), and one common primer used in the LDR. Sequence discriminating primers are labeled with Fam and have different length, thus the LDR products can be discriminated by length on high resolution sequencing gel. See Table 1 for primer sequences.
B: Electropherograms of LDR products obtained in the experiment with dilution of the mutant DNA with wild DNA. Dilutions are indicated on the right. The high peaks on the right (46 bp) show the signal generated by the V target while the peak D (44 bp), barely visible in control lane #2 indicates the background signal, a result of misligation. Minor V peak in control lane #3 is due to the spillover signal from the adjacent lane. The broader peaks on the left are unligated primers. Controls 1, 2, 3 are three independently performed control experiments. Reproduced from Khanna *et al.*, 1999 with permission.

oligonucleotides contained the mutant base at their 3' end and were fluorescently-labeled. Up to three discriminating oligonucleotides competed for ligation with an adjacent phosphorylated common oligonucleotide, which contained a poly A tail and C3-spacer blocking the 3' end.

The sensitivity of PCR/LDR was determined by analyzing samples containing various dilutions of mutant DNA by wild-type DNA. Mutant and wild-type DNA samples were PCR amplified independently, and then mixed, allowing for a range of mutant-to-wild-type ratios to be tested. Using a primer set that targeted mutations G12V -G12D, the maximal sensitivity achieved was 1 part of mutant DNA in 4,000 parts of wild-type templates with a signal-to-noise ratio greater than 9:1 (Figure 4). In the higher level multiplex LDR, a total of 26 primers were used to test all 19 possible single-base mutations at K-*ras* codons 12, 13 and 61. In this case, the sensitivity of G12V mutation detection was 1 mutant: 500 wild-type templates, with a signal-to-noise ratio of 3:1.

Then multiplex PCR/LDR was performed on 144 primary colon carcinomas previously analyzed by direct sequencing of PCR amplicons obtained from DNA from micro-dissected paraffin sections. In these experiments the same 19 possible point mutations in codons 12, 13, and 61 were probed with 26 K-ras specific primers (Figure 5A). Analysis was done in a blinded fashion using coded tumor samples, and the full spectrum of point mutations in codons 12 and 13 was successfully detected (Figure 5B). Genotyping by PCR/LDR and direct sequencing showed concordance in 134 of 144 tumors. The ten discordant cases were reanalyzed by PCR/LDR using a single primer pair targeted to the specific mutation(s). In eight cases, single primer PCR/LDR analysis confirmed the presence of a K-*ras* mutation in the tumor sample. We assume that the concentration of some of the mutants may be too low to be identified by direct sequencing. In one of these eight cases we identified two mutations via PCR/LDR in the same paraffin section. For the two other discordant cases identified as mutants by direct sequencing and wild-type by multiplex PCR/LDR, neither mutation could be confirmed by single primer PCR/LDR. One of these cases contained an unexpected double point mutation

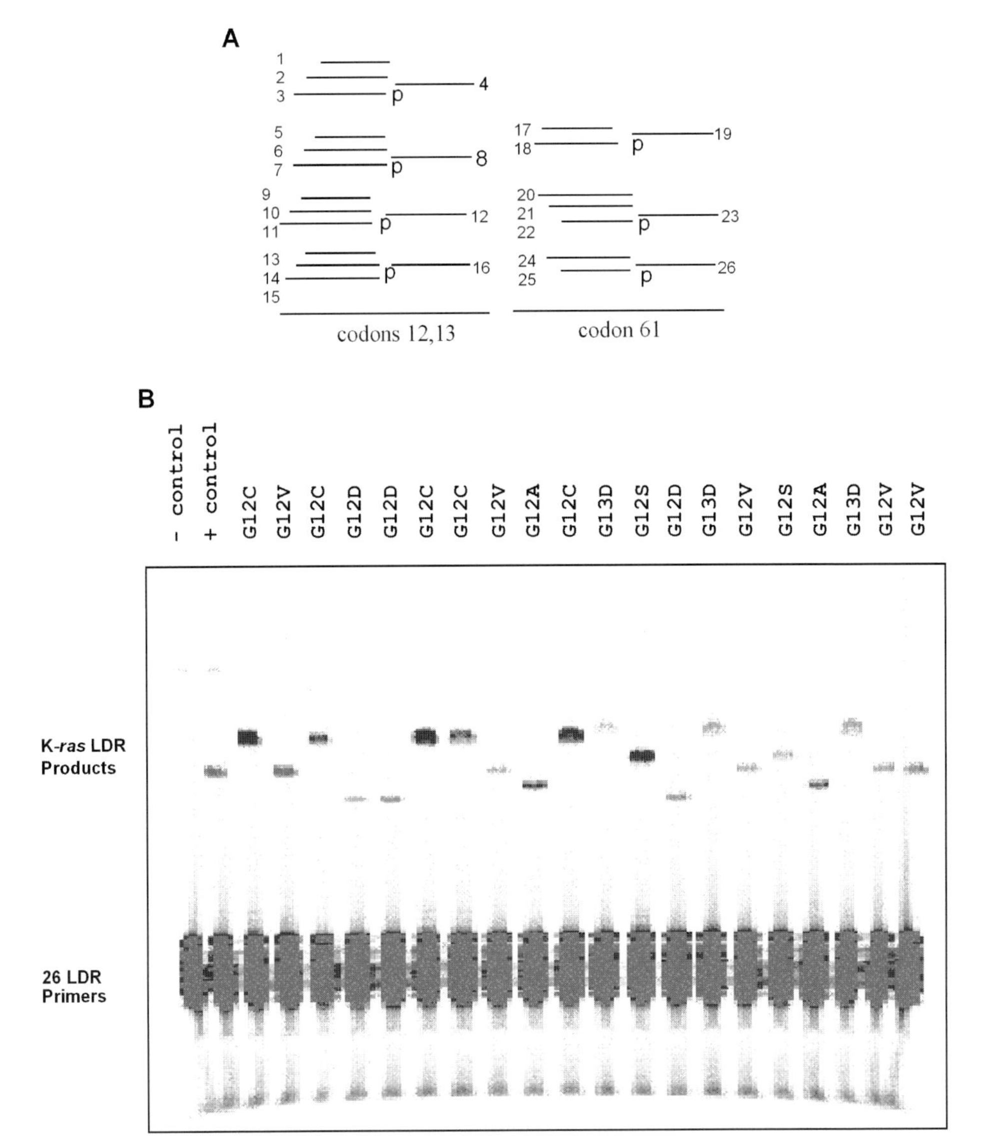

Figure 5. Example of identification of different K-*ras* mutation in 20 paraffin-embedded microdissected colon tumor samples (reproduced from Khanna *et al.*, 1999 with permission). DNA extracted from microdissected samples was PCR pre-amplified and subjected to a multiplex LDR using the twenty-six primer set and the Tth-K294R mutant DNA ligase.
A: Schematic outline of the primer positions to analyze single-base mutations in codons 12, 13, and 61 of the K-*ras* gene, p-5' phosphate. See Table 1 for the primer sequences.
B: Example of a polyacrylamide gel with identification of single-base mutations in K-*ras* gene. Each LDR product reveals the presence of the specific K-*ras* mutation indicated at the top of the figure. Specific mobility and a type of fluorescence label allowed for identification of each mutation. The picture presented is obtained after running the samples on ABI sequencing instrument.

(Asp-12, Cys-12) which was subsequently confirmed by cloning and sequencing of the PCR product. Our results show that because of the high sensitivity, multiplex PCR/LDR is able to identify K-ras mutations in non-microdissected tumor tissues as well (data not shown).

Discussion

As we show here, LCR faithfully detected as few as 200 initial target molecules, as well as discriminated both β^A and β^S alleles directly from genomic DNA. Moreover, LCR was able to discriminate T-T, G-T, C-T, or C-A 3'-terminal mismatches, in contrast to allele-specific PCR (Barany, 1991a; Kwok *et al.*, 1990). The major drawback of LCR is the potential for false positives due to target-independent ligation of the four ligation primers. This negative feature is eliminated in LDR, where only one pair of ligation primers is applied. However, because LDR achieves only linear amplification, a PCR pre-amplification step is often needed, such as in the PCR/LDR approach. It is also possible to perform ligation reaction first and then amplify the ligation signal by PCR, such as in ligation-mediated PCR (LMPCR) (Pfeifer and Riggs, 1996; Pfeifer *et al.*, 1989; Pfeifer and Tommasi, 2000; Rodriguez *et al.*, 2000; Schouten *et al.*, 2002). In a recent report, LDR was coupled with single pair fluorescence resonance energy transfer (LDR-spFRET) to achieve real-time detection of low abundant point mutations in K-ras gene (Wabuyele *et al.*, 2003)

Several PCR based techniques can detect single-base mutations present in a minority of human tumor cells, such as allele-specific PCR (AS-PCR), primer mediated RFLP or electrophoresis-based methods. However, these amplification methods have several drawbacks:
(1) They generate false-positive signals due to low match/mismatch discrimination; (2) they can not precisely identify the full spectrum of mutations at a given gene locus; (3) they are difficult to multiplex; (4) they require extensive optimization.

PCR/LDR overcomes many of the above limitations by separating the amplification (PCR) and mutations discrimination (LDR) steps. A distinguishing feature of PCR/LDR is that mis-ligations do not undergo subsequent amplification, therefore reducing the chance of false positive reactions. Any low-level polymerase errors remain unselected, and thus contribute only a minimum to background noise. Since PCR/LDR can be set up for the presence of any base mutation (*e.g.* in K-*ras* codon 12), a profile of background noise due to misligations is generated, which may serve as internal control allowing distinguish true positive signal (Figure 4, controls). As we showed here, thermostable ligases provide a tremendous sensitivity of one mutation in 500 wild-type sequences in the 19 - plex set. For comparison, T4 DNA ligase was able to identify three mutations in K-*ras* codon 12 with only 1% sensitivity under conditions requiring blocking oligonucleotides and high salt conditions to suppress misligations on wild-type templates (Jen *et al.*, 1994; Powell *et al.*, 1993; Redston *et al.*, 1995). In summary, PCR/LDR offers several advantages in cancer mutation detection. It (1) allows for large-scale multiplexing (Pfeifer and Riggs, 1996; Schouten *et al.*, 2002); (2) provides quantitative detection of mutations on a high background of normal sequences; (3) allows for detection of closely-clustered mutations; and (4) is amenable to automation.

Ligase-based amplification techniques are also explored for detection of RNA and protein molecules (Nilsson *et al.*, 2001; Zhang *et al.*, 1998). Given the high fidelity of thermostable DNA ligases and robustness of ligase-based amplification, it is likely that ligase-based techniques will be one of the leaders in genome analysis and molecular diagnosis.

Acknowledgements

I thank Dr. Francis Barany for kind help and Jeremy Parker for critically reading the manuscript. The preparation of the manuscript is supported by CSREES/USDA (SC-1700153, technical contribution number 4888) and NIH (GM 067744).

References

Abravaya, K., Carrino, J.J., Muldoon, S., and Lee, H.H. (1995). Detection of point mutations with a modified ligase chain reaction (Gap- LCR). Nucleic Acids Res. *23*, 675-82.

Aravind, L., and Koonin, E.V. (1999). Gleaning non-trivial structural, functional and evolutionary information about proteins by iterative database searches. J. Mol. Biol. *287*, 1023-40.

Ausina, V., Gamboa, F., Gazapo, E., Manterola, J.M., Lonca, J., Matas, L., Manzano, J.R., Rodrigo, C., Cardona, P.J., and Padilla, E. (1997). Evaluation of the semiautomated Abbott LCx Mycobacterium tuberculosis assay for direct detection of Mycobacterium tuberculosis in respiratory specimens. J. Clin. Microbiol. *35*, 1996-2002.

Barany, F. (1991a). Genetic disease detection and DNA amplification using cloned thermostable ligase. Proc. Natl. Acad. Sci. USA. *88*, 189-93.

Barany, F. (1991b). The ligase chain reaction in a PCR world [published erratum appears in PCR Methods Appl 1991 Nov; 1(2):149]. Pcr Methods Appl, *1*, 5-16.

Barany, F., and Gelfand, D.H. (1991). Cloning, overexpression and nucleotide sequence of a thermostable DNA ligase-encoding gene. Gene *109*, 1-11.

Cao, W. (2001). DNA ligases and ligase-based technologies. Clin. Appl. Immunol. Rev. *2*, 33-43.

Cao, W. (2002). DNA ligases: structure, function and mechanism. Curr. Organic Chem. *6*, 827-839.

Cherepanov, A.V., and de Vries, S. (2002). Dynamic mechanism of nick recognition by DNA ligase. Eur. J. Biochem. *269*, 5993-9.

Doherty, A.J., and Suh, S.W. (2000). Structural and mechanistic conservation in DNA ligases. Nucleic Acids Res. *28*, 4051-8.

Engler, M.J., and Richardson, C.C. (1982). The enzymes in, Boyer, P. D. (ed.). (New York: Academic Press), pp. 3-29.

Favis, R., Day, J.P., Gerry, N.P., Phelan, C., Narod, S., and Barany, F. (2000). Universal DNA array detection of small insertions and deletions in BRCA1 and BRCA2. Nat. Biotechnol. *18*, 561-4.

Fredriksson, S., Gullberg, M., Jarvius, J., Olsson, C., Pietras, K., Gustafsdottir, S.M., Ostman, A., and Landegren, U. (2002). Protein detection using proximity-dependent DNA ligation assays. Nat. Biotechnol. *20*, 473-7.

Gerry, N.P., Witowski, N.E., Day, J., Hammer, R.P., Barany, G., and Barany, F. (1999). Universal DNA microarray method for multiplex detection of low abundance point mutations. J. Mol. Biol, *292*, 251-62.

Huang, J., Kirk, B., Favis, R., Soussi, T., Paty, P., Cao, W., and Barany, F. (2002). An endonuclease/ligase based mutation scanning method especially suited for analysis of neoplastic tissue. Oncogene *21*, 1909-21.

Iannone, M.A., Taylor, J.D., Chen, J., Li, M.S., Rivers, P., Slentz-Kesler, K.A., and Weiner, M.P. (2000). Multiplexed single nucleotide polymorphism genotyping by oligonucleotide ligation and flow cytometry. Cytometry *39*, 131-40.

Jarvius, J., Nilsson, M., and Landegren, U. (2003). Oligonucleotide ligation assay. Meth. Mol. Biol. *212*, 215-28.

Jen, J., Powell, S.M., Papadopoulos, N., Smith, K.J., Hamilton, S.R., Vogelstein, B., and Kinzler, K.W. (1994). Molecular determinants of dysplasia in colorectal lesions. Cancer Res. *54*, 5523-6.

Kehl, S.C., Georgakas, K., Swain, G.R., Sedmak, G., Gradus, S., Singh, A., and Foldy, S. (1998). Evaluation of the abbott LCx assay for detection of Neisseria gonorrhoeae in endocervical swab specimens from females. J. Clin. Microbiol. *36*, 3549-51.

Khanna, M., Park, P., Zirvi, M., Cao, W., Picon, A., Day, J., Paty, P., and Barany, F. (1999). Multiplex PCR/LDR for detection of K-ras mutations in primary colon tumors. Oncogene *18*, 27-38.

Korn, S.H., Moerkerk, P.T., and de Goeij, A.F. (1993). K-ras point mutations in routinely processed tissues: non-radioactive screening by single strand conformational polymorphism analysis. J. Clin. Pathol. *46*, 621-3.

Kwok, S., Kellogg, D.E., McKinney, N., Spasic, D., Goda, L., Levenson, C., and Sninsky, J.J. (1990). Effects of primer-template mismatches on the polymerase chain reaction: human immunodeficiency virus type 1 model studies. Nucleic Acids Res. *18*, 999-1005.

Landegren, U., Kaiser, R., Sanders, J., and Hood, L. (1988). A ligase-mediated gene detection technique. Science *241*, 1077-80.

Lehman, I.R. (1974). DNA ligase: structure, mechanism, and function. Science *186*, 790-797.

Lieber, M.R. (1999). The biochemistry and biological significance of nonhomologous DNA end joining: an essential repair process in multicellular eukaryotes. Genes Cells *4*, 77-85.

Luo, J., Bergstrom, D.E., and Barany, F. (1996). Improving the fidelity of Thermus thermophilus DNA ligase. Nucleic Acids Res. *24*, 3071-8.

Martin, I.V., and MacNeill, S.A. (2002). ATP-dependent DNA ligases. Genome Biol. *3*, REVIEWS3005.

Nilsson, M., Antson, D.O., Barbany, G., and Landegren, U. (2001). RNA-templated DNA ligation for transcript analysis. Nucleic Acids Res. *29*, 578-81.

Nilsson, M., Barbany, G., Antson, D.O., Gertow, K., and Landegren, U. (2000). Enhanced detection and distinction of RNA by enzymatic probe ligation. Nat. Biotechnol. *18*, 791-3.

Pfeifer, G.P., and Riggs, A.D. (1996). Genomic sequencing by ligation-mediated PCR. Mol. Biotechnol. *5*, 281-8.

Pfeifer, G.P., Steigerwald, S.D., Mueller, P.R., Wold, B., and Riggs, A.D. (1989). Genomic sequencing and methylation analysis by ligation mediated PCR. Science *246*, 810-3.

Pfeifer, G.P., and Tommasi, S. (2000). *In vivo* footprinting using UV light and ligation-mediated PCR. Methods Mol. Biol. *130*, 13-27.

Powell, S.M., Petersen, G.M., Krush, A.J., Booker, S., Jen, J., Giardiello, F.M., Hamilton, S.R., Vogelstein, B., and Kinzler, K.W. (1993). Molecular diagnosis of familial adenomatous polyposis. N. Engl. J. Med. *329*, 1982-7.

Pritchard, C.E., and Southern, E.M. (1997). Effects of base mismatches on joining of short oligodeoxynucleotides by DNA ligases. Nucleic Acids Res. *25*, 3403-7.

Redston, M.S., Papadopoulos, N., Caldas, C., Kinzler, K.W., and Kern, S.E. (1995). Common occurrence of APC and K-ras gene mutations in the spectrum of colitis-associated neoplasias. Gastroenterology *108*, 383-92.

Rodriguez, H., Akman, S.A., Holmquist, G.P., Wilson, G.L., Driggers, W.J., and LeDoux, S.P. (2000). Mapping oxidative DNA damage using ligation-mediated polymerase chain reaction technology. Methods *22*, 148-56.

Ruiz-Serrano, M.J., Albadalejo, J., Martinez-Sanchez, L., and Bouza, E. (1998). LCx: a diagnostic alternative for the early detection of Mycobacterium tuberculosis complex. Diagn. Microbiol. Infect. Dis. *32*, 259-64.

Sato, Y., Mukai, K., Matsuno, Y., Furuya, S., Kagami, Y., Miwa, M., and Shimosato, Y. (1990). The AMeX method: a multipurpose tissue-processing and paraffin-embedding method. II. Extraction of spooled DNA and its application to Southern blot hybridization analysis. Am. J. Pathol. *136*, 267-71.

Schouten, J.P., McElgunn, C.J., Waaijer, R., Zwijnenburg, D., Diepvens, F., and Pals, G. (2002). Relative quantification of 40 nucleic acid sequences by multiplex ligation-dependent probe amplification. Nucleic Acids Res. *30*, e57.

Shuman, S. (1995). Vaccinia virus DNA ligase: specificity, fidelity, and inhibition. Biochemistry *34*, 16138-47.

Shuman, S., and Schwer, B. (1995). RNA capping enzyme and DNA ligase: a superfamily of covalent nucleotidyl transferases. Mol. Microbiol. *17*, 405-10.

Sriskanda, V., Schwer, B., Ho, C.K., and Shuman, S. (1999). Mutational analysis of Escherichia coli DNA ligase identifies amino acids required for nick-ligation *in vitro* and for *in vivo* complementation of the growth of yeast cells deleted for CDC9 and LIG4. Nucleic Acids Res. *27*, 3953-3963.

Timson, D.J., Singleton, M.R., and Wigley, D.B. (2000). DNA ligases in the repair and replication of DNA. Mutat. Res. *460*, 301-18.

Tomkinson, A.E., Chen, L., Dong, Z., Leppard, J.B., Levin, D.S., Mackey, Z.B., and Motycka, T.A. (2001). Completion of base excision repair by mammalian DNA ligases. Prog. Nucleic Acid Res. Mol. Biol. *68*, 151-64.

Tomkinson, A.E., Tappe, N.J., and Friedberg, E.C. (1992). DNA ligase I from Saccharomyces cerevisiae: physical and biochemical characterization of the CDC9 gene product. Biochemistry *31*, 11762-71.

Tong, J., Barany, F., and Cao, W. (2000). Ligation Reaction Specificities of an NAD+-dependent DNA Ligase from Hyperthermophile Aquifex aeolicus. Nucleic Acids Res. *28*, 1447-1454.

Tong, J., Cao, W., and Barany, F. (1999). Biochemical properties of a high fidelity DNA ligase from Thermus species AK16D. Nucleic Acids Res. *27*, 788-94.

Wabuyele, M.B., Farquar, H., Stryjewski, W., Hammer, R.P., Soper, S.A., Cheng, Y.W., and Barany, F. (2003). Approaching Real-Time Molecular Diagnostics: Single-Pair Fluorescence Resonance Energy Transfer (spFRET) Detection for the Analysis of Low Abundant Point Mutations in K-ras Oncogenes. J. Am. Chem. Soc. *125*, 6937-45.

Wilkinson, A., Day, J., and Bowater, R. (2001). Bacterial DNA ligases. Mol. Microbiol. *40*, 1241-8.

Wilson, T.E., Topper, L.M., and Palmbos, P.L. (2003). Non-homologous end-joining: bacteria join the chromosome breakdance. Trends Biochem. Sci. *28*, 62-6.

Wu, D.Y., and Wallace, R.B. (1989a). The ligation amplification reaction (LAR)--amplification of specific DNA sequences using sequential rounds of template-dependent ligation. Genomics *4*, 560-9.

Wu, D.Y., and Wallace, R.B. (1989b). Specificity of the nick-closing activity of bacteriophage T4 DNA ligase. Gene *76*, 245-54.

Yeakley, J.M., Fan, J.B., Doucet, D., Luo, L., Wickham, E., Ye, Z., Chee, M.S., and Fu, X.D. (2002). Profiling alternative splicing on fiber-optic arrays. Nat. Biotechnol. *20*, 353-8.

Young, D.C., Craft, S., Day, M.C., Davis, B., Hartwell, E., and Tong, S. (2000). Comparison of Abbott LCx Chlamydia trachomatis assay with Gen-Probe PACE2 and culture. Infect. Dis. Obstet. Gynecol. *8*, 112-5.

Yu, H., Merchant, B., Scheffel, C., Yandava, C., Drazen, J.M., and Huff, J. (2001). Automated detection of single nucleotide polymorphism in beta-2 adrenergic receptor gene using LCx(R). Clin. Chim. Acta. *308*, 17-24.

Zanchetta, N., Nardi, G., Tocalli, L., Drago, L., Bossi, C., Pulvirenti, F.R., Galli, C., and Gismondo, M.R. (2000). Evaluation of the abbott LCx HIV-1 RNA quantitative, a new assay for quantitative determination of human immunodeficiency virus type 1 RNA. J. Clin. Microbiol. *38*, 3882-6.

Zhang, D.Y., Brandwein, M., Hsuih, T.C., and Li, H. (1998). Amplification of target-specific, ligation-dependent circular probe. Gene *211*, 277-85.

Zhang, Y., Kaur, M., Price, B.D., Tetradis, S., and Makrigiorgos, G.M. (2002). An amplification and ligation-based method to scan for unknown mutations in DNA. Hum. Mutat. *20*, 139-47.

Section 2

Thermocycling Methods of DNA Amplification

2.1

High Multiplexity PCR Based on PCR Suppression

Natalia E. Broude, Adriaan W. van Heusden and Richard Finkers

Abstract

We present a protocol for efficient amplification of a large number of DNA targets using a single-tube polymerase chain reaction (PCR) based on PCR suppression. This method allows amplification of each DNA target with only one target-specific primer whereas the other primer is common for all targets. Thus, this approach requires $n+1$ primers for n targets instead of $2n$ in conventional PCR and substantially reduces the cost and complexity of the assay. The method has been successfully applied for amplification of 30 SNP-related DNA targets from human genomic DNA and for genotyping plant genomes.

Background

Analyses of complex genomes for mutations require a pre-amplification step to increase the concentrations of the DNA targets to an analyzable level. Among different amplification techniques, PCR is still the most widely used. Multiplex variants of PCR allow simultaneous synthesis

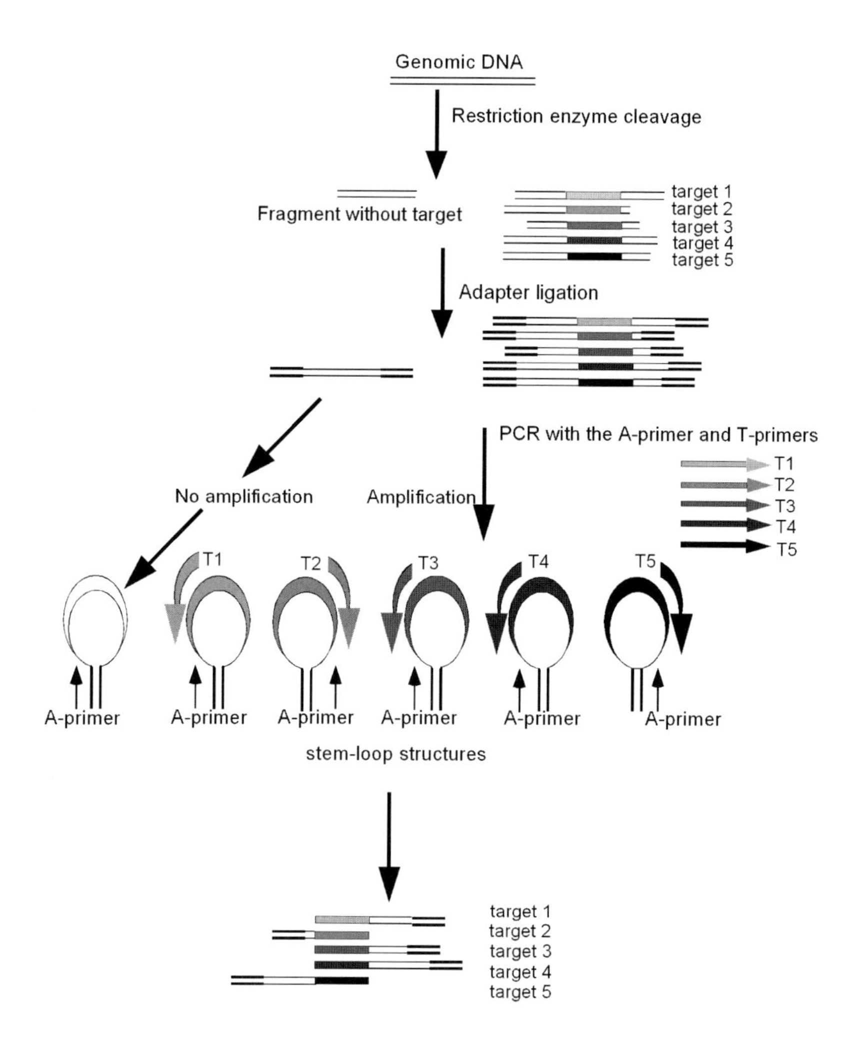

Figure 1. Outline of multiplex PCR based on suppression principle. A-primer, adapter-primer; T-primer, target-specific primer. Genomic DNA is digested with a restriction enzyme and ligated with a specially designed adapter with high GC-content. As a result, long self-complementary ends flank each single-stranded DNA fragment. Self-complementary termini of DNA fragments form duplexes during each PCR step, so that the fragments adopt large stem-loop structures. The formation of duplex structures at the fragment ends inhibits PCR with the A-primer (PCR suppression). PCR is efficient with the A- and T-primers. T-primer finds its target in the single-stranded regions of the stem-loop structures and DNA polymerase extends it with low efficiency. However, in the next PCR cycles this DNA product is efficiently amplified by a T-primer and A-primer, because it does not have self-complementary ends and is not a subject to the suppression effect. Thus, only DNA fragments with the target are efficiently amplified on the background of all other fragments without a target.

of several targets in one tube and thus offer a possibility to increase throughput and more efficient use of the DNA samples (Chamberlain *et al.*, 1988; Edwards and Gibbs, 1994). Most methods developed for multiplex PCR are limited to amplification of 5-10 different amplicons due to primer-primer interactions, which substantially reduce the amplification efficiency.

We have developed an alternative approach to multiplex PCR based on PCR suppression (PS) (Broude *et al.*, 2001a, b). In the PS-based PCR (see Figure 1), genomic DNA is digested with a restriction enzyme and ligated with a specially designed adapter (Siebert *et al.*, 1995, Lukyanov *et al.*, 1995). The ligation of adapter results in formation of large stem-loop structures by each DNA fragment due to the presence of self-complementary sequences at the fragment's ends. The stable duplexes at the fragment ends make PCR with the adapter-primer (A-primer) alone relatively inefficient. This effect is called PCR suppression (PS). Instead, the PCR is efficient with two primers, namely the A-primer and T-primer (target-primer). The T-primer finds its complement in the single-stranded regions of the stem-loop structures, anneals to the target site and DNA polymerase extends it with low efficiency. However, in the next cycles this DNA product is efficiently amplified by the T- and A-primers, because it does not have self-complementary ends and is not a subject to the PS-effect anymore. Consequently, only DNA fragments with the target binding site are efficiently amplified on the background of all other fragments without a target.

If used in the multiplex PCR, the PS approach offers substantial advantages. Because the A-primer is the same for all targets, only $n + 1$ primers are needed instead of $2n$ primers in conventional PCR to amplify n targets. Hence the method employs significantly simpler primer design, provides an increased multiplexing level and reduces costs due to the twice lower number of primers.

Protocol

Materials

Non-phosphorylated oligonucleotides serving as PCR primers or adapters, unlabeled or fluorescently labeled, were purchased from Integrated DNA Technologies. Genomic DNA was isolated using different Qiagen kits depending on the source of DNA according to the manufacturer's protocol (Qiagen). Different thermostable DNA polymerases can be used for amplification. We used *Taq* DNA polymerase from Promega, AmpliTaq and AmpliTaq Gold from Applied Biosystems, KlenTaq from Ab peptides and the long-range DNA polymerase from the Expand Long DNA kit (Boehringer).

Step 1: Primer Design

The criteria for primer design in PS PCR are the same as in conventional PCR (Wu *et al.*1991; Rychlik, 1995): the primers should not be self-complementary or mutually complementary, they should not adopt unconventional structures and their melting temperatures should be kept in a narrow temperature window. Because, only one target-specific primer is needed, its uniqueness is crucial. Usually, the length of the primers in the PS-based PCR is about 30-35 nt (see Table 1), which ensures uniqueness of sequence and allows high annealing temperatures during PCR (65-68°C). The choice between a direct or reverse T-primer in PS PCR depends on the location of the closest restriction site used to digest DNA and on the particular sequence upstream and downstream of the target. Usually, there is a choice between several alternative primer sequences and only empirical approach allows choosing the best variant. As a starting point to design primers we recommend Primer3 software (www-genome.wi.mit.edu/cgi-bin/primer/primer3), which we used during this study. Examples of T-primers used to genotype plant DNA are shown in Table 1. Note that in our practice we did not drop any of the targets because of the absence of a restriction site in close vicinity or inability to find an appropriate site for the primer. To minimize possible primer-primer interactions we used a simple rule of thumb: the last 3-4 nucleotides at the 3'- end were designed to contain the same nucleotides (*e.g.* G, C, A) (Broude *et al.*, 2001b).

Table 1. Primers used to genotype plant genomes.

#	Target site	Primer (5'-3')	Primer length, (nt)	Primer T_m [1]	Product length (bp)[2]
1	API21	TCTCCCTTCT GTACTTGGCG ACCCAAGAAT TGGACG	36	65.2	138-142
2	API73	TGAAGAAATT TCCGGAAGTT CTTGGATGCG	30	57.2	217
3	API76	CGCTTACGCT GTTAGTATAC CGCATGCTTC G	31	61.7	110-114
4	OJ39	ACGATGCCAG GACCAGGCCC TACGTTATCA AG	32	65.6	152-156
5	Omega-3 fatty acid desaturase	GAGAGGAGGG CTTACAACAC TTGATCGTGA CTATGG	36	63.5	255
6	JSW19	CATATTAATC ATGGCTCAAT AGTATGTCGC ACTCGC	36	61.7	269-291
7	JSW21	TGATGTTCAC CTTTTAGTTG TGGACTTCAA TGAGG	35	60.7	259
8	TG176	AGCCGGTTGA CCTGCAATTA GTAGGCACAG TG	32	63.6	315-337
9	TG505	CATTTGGCTG CCGAAAATAG AGATCAAGCT ACTAC	35	63.1	216
10	Adapter-primer 1	TGTAGCGTGA AGACGACAGA A	21	57.3	
11	Adapter-primer 2	GAAAGGGCGT GGTGCGGACG CGG	23	70.3	

[1] The Tm's of the primers were calculated using a nearest-neighbor thermodynamic parameter set (Generunner 3.0).
[2] The expected lengths of the DNA fragments in the PS PCR. This is the length from the restriction site to the T-primer plus the length of the adapter.

Step 2: Preparation of Adapter-Ligated DNA

Genomic DNA (500 ng) is digested with *Rsa*I restriction enzyme (20 U) (New England Biolabs) (or any other convenient restriction endonuclease) in 50 µl at 37°C for 90 min, fresh enzyme is added (10 U) and incubation is continued for another 90 min. If the restriction enzyme generates sticky ends, the overhang is blunted with dNTPs using DNA polymerase. The digested DNA is purified by phenol extraction and ethanol precipitation. DNA precipitate is dissolved in 20 µl of sterile water and ligated with an excess adapter (2 µM) at 16°C overnight. The ligation is performed in 25 µl final volume, containing 1x ligation buffer (50 mM Tris-HCl, pH 7.5, 10 mM $MgCl_2$, 10 mM dithiotreitol, 1 mM ATP, 25 mg/ml bovine serum albumin) and 5 U of T4 DNA ligase (Life Technologies). The adapter used by us consists of two pre-annealed complementary oligonucleotides: 5'-TGTAGC GTGAAGACGACAGAAAGGGCGTGGTGCGGACGCGGG and 5'-CCCGCGTCCGC. The complementary oligonucleotides are of different length to ensure the right polarity of ligation to blunt-ended genomic DNA fragments. The recessed ends of the ligated adapters are automatically filled in during the first round of the subsequent PCR reaction in the presence of dNTPs and DNA polymerase. The ligation is terminated by incubation at 75°C for 10 min. and DNA is purified from excess adapter with a QIAquick column (Qiagen). Ligated DNA is aliquoted and kept at -20°C. In the experiments with plant DNAs the last purification step was omitted.

Step 3: DNA Amplification

Adapter-ligated DNA (3-5 ng) is amplified by PCR in a 25 µl reaction volume containing 1.6 x PCR buffer (buffer depends on the DNA polymerase), 2.5 mM $MgCl_2$, 250 µM of each dNTP, 5-10 pmol of each primer and 2.5 U of thermostable DNA polymerase. The PCR mixtures, containing all components but primers, are denatured at 95°C for 3-10 min. The primer mixture containing all T-primers and A-primer 1 (Table 1) is added at 95°C and 38 cycles of PCR (94°C, 10 sec; 68°C, 15 sec; 72°C, 1 min) are performed. Our results showed that the mixtures of different DNA polymerases performed better than single DNA polymerases. We tried several combinations of DNA polymerases. In the 30 plex PCR three DNA polymerases (*Taq*: AmpliTaq: long range DNA polymerase mix) were used at the ratio 1:1:3, respectively. In experiments with plant DNAs amplification was performed with a 1:1 mix of AmpliTaq Gold (Applied Biosystems) and Hotstart DNA polymerase (Qiagen). In this case no hot-start was necessary and all primers were included in the reaction mix. When the number of targets is >15, the extension time should be prolonged (to 2-3 min) to allow efficient amplification of all targets. In case of many targets, two-step PCR may produce better results that the one-step amplification (see Example 1). In this case, the first step PCR is performed for 22-23 cycles. Then, the PCR products are diluted 1:1000 and re-amplified with the same target-primers but with the A-primer 2 instead of A-primer 1 (Table 1) for 23-25 cycles.

Step 4: Analysis of PCR Products

The PCR products are analyzed by electrophoresis in a 2% agarose gel. Alternatively, the A-primer common for all targets can be fluorescein-labeled and PCR products can be analyzed by high resolution denaturing PAGE using a fluorescence-based automated DNA sequencer. PCR products (2-3 µl) are denatured for 3 min at 90°C in a "stop" solution (Pharmacia-Biotech) containing 6 mg/ml of dextran blue in deionized formamide, loaded on a 6% denaturing polyacrylamide gel and analyzed using an ALF sequencer. The results are visualized using the Fragment Manager software provided with the instrument. A fluorescein-labeled 50-bp ladder (Pharmacia-Biotech) is used as a size marker. Alternatively PCR products can be analyzed using an ABI377 sequencer (Applied Biosystems). In this case, Genescan-500 Rox size standards are used as a size marker and the results are visualized using the Genotyper 3.6 software (Applied Biosystems).

Example 1

We have demonstrated the potential of PS-based PCR by amplifying 30 different DNA targets from human genomic DNA (Broude *et al.*, 2001b). The DNA targets included mostly single nucleotide polymorphisms (SNPs) plus several anonymous sequences from human chromosome 7 (Broude *et al.*, 2001b). All T-primers were first tested in single PCRs with the common A-primer. If a single product of expected size was generated, the T-primer was included into the multiplex reaction. If the T-primer produced several bands, it was either not used in subsequent multiplex experiments, or was modified (usually extended 1-2 nt) to generate single specific product of the expected size.

The primers were 29-35 nt long, with the T_m varying from 61°C to 71°C and the sizes of the amplicons varying from 78 to 580 bp. It must be emphasized that the amplicon lengths depend on two major parameters. If the target sequence has a specific point of interest, like SNPs, the amplicon length depends on the distance between the SNP and the closest restriction site. However, in many applications it is necessary to amplify just gene-specific fragments which lengths can be arbitrary. In this case the amplicon lengths should be considered depending on the downstream analyses of the PCR products. The necessity of different size targets is mostly determined by the use of the gel-based analysis methods. If the analytical procedure avoids gel-based separation methods [*e.g.* DNA microarray-based analysis of the PCR products (Gerry *et al.*, 1999), or molecular beacon-based real-time detection (Nazarenko *et al.*, 2002)], it is possible to design primers generating PCR products of very similar size, which should result in more uniform amplification of different targets.

Since we analyzed the PCR products by gel electrophoresis, we designed primers generating amplicons of different sizes. Therefore, to equalize the efficiencies of amplification of different size fragments we changed several parameters. Three major parameters have been found to be most effective in optimizing multiplex amplification. First, different mixtures of DNA polymerases perform multiplex PCR with different efficiencies (Broude *et a*l., 2001a, b); second, T-primer concentrations can be adjusted for more uniform amplification of targets of different sizes; and third, the extension time should be increased if amplification of many targets is performed.

Figure 2 shows a representative pattern of a 30-plex PCR product separated on denaturing polyacrylamide gel. All targeted DNA fragments were detected on the display (Broude *et al.*, 2001b).

Example 2

For genotyping plant genomes (tomato and onion) for specific candidate genes we followed five strategic steps:

1. Targets were selected from available EST sequences with known (putative) functions and corresponding T-primers were designed.
2. All T-primers were first tested in single PCRs with the common A-primer and adapter-ligated chromosomal DNA. DNA was digested with three different restriction enzymes to increase the chances of finding polymorphisms.
3. All chosen targets were tested for polymorphisms in the parental genotypes of the mapping populations.
4. Then single reactions were combined in a multiplex format. The T-primer concentrations were adjusted for efficient amplification of all targets.
5. Genotyping of all individuals of the segregating population was performed under optimized reaction conditions.

For tomato genotyping we used a Recombinant Inbred Line mapping population of 98 lines originating from an interspecific cross between *Lycopersicon esculentum* and *Lycopersicon pimpinellifolium*. The core of the molecular linkage map for this mapping population is based on AFLP™ markers (Vos *et al.*, 1995) anchored with a few Restriction Fragment Length Polymorphism (RFLP) markers on each chromosome. We were especially interested in genes putatively involved in salt tolerance. Candidate gene sequences were obtained in our own expression studies or through data mining. To show the robustness of the method, two T-primers were designed based on the sequence of an RFLP probe. No discrepancies were found using the multiplex PCR compared to the RFLP mapping. The main problem we encountered to make large multiplex reactions was to find enough polymorphisms between the parental genotypes of this cross (*L. pimpinellifolium*

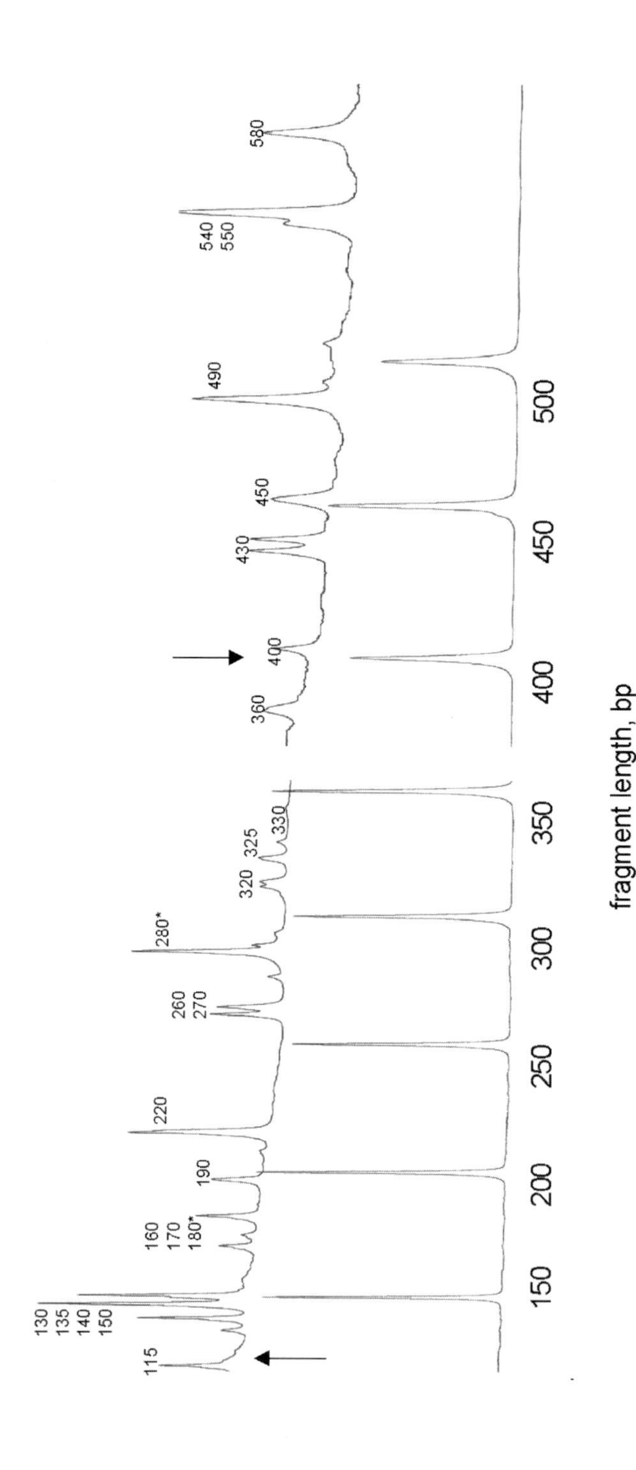

Figure 2. PAGE fluorescent image of 30-plex PCR. Human genomic DNA was digested with *Rsa* I restriction enzyme and ligated with the PS-driving adapters. Suppression PCR was performed in two steps under optimized conditions (see text for the details). Numbers above the peaks indicate the fragment lengths in bp. Three fragments, 78, 87 and 92 bp-long, are beyond the displayed area. Asterisks mark the bands containing several overlapping DNA fragments. Arrows show two unidentified products. A 50-bp fluorescent DNA ladder was used as a size marker.

is the closest wild relative to *L. esculentum*). In total, we tried 42 candidate genes with three different restriction enzymes and found polymorphisms only in eight cases. Of these eight polymorphisms, three were obtained with adapter-ligated DNA digested with the restriction enzyme *Alu*I. These three were successfully amplified in a multiplex PCR containing also two monomorphic targets (Figure 3A) and these markers were successfully placed on the genetic linkage map. One additional fragment was amplified in this multiplex PCR, which was not observed during amplification of the individual single reaction (marked by an arrow).

For genotyping in *Allium* we used an F_2 population based on the interspecific cross between *Allium cepa* and *Allium roylei* (van Heusden *et al.*, 2000). Molecular analysis of onion and other *Allium* species is hampered by an enormously large and complex nuclear genome making it one of the largest genomes among cultivated plants (16 times larger then tomato). The GC-content of onion DNA is the lowest for any angiosperm (32%). Many duplications and repetitive sequences are also known to be present in the *Allium* genome. Candidate genes were obtained from EST-libraries of onion and selection was especially aimed at candidate genes with a possible role in carbohydrate and sulphur metabolism. For eleven targets, individual primers were designed; six of those gave one or more amplification products and of those six five were polymorphic. Four of these five targets were successfully amplified in a multiplex PCR with DNA digested with the restriction enzyme *Rsa*I. (Figure 3A).

In both plant families the results show that the PS-based multiplex PCR technique can be successfully applied for genotyping. The main problem we encountered with analysis of tomato candidate gene markers was the lack of polymorphisms between the parental genotypes for the selected candidate genes, which did not allow us to design high-level multiplex reactions. However, all targets chosen for amplification were successfully co-amplified in one reaction. The major problem for the genotyping of *Allium* genome was the T-primer design since more then 50% of the tested primers were failing in amplification of a right product. This may be partly explained by the composition and structure of the *Allium* genome. More critical T-primers design is needed for

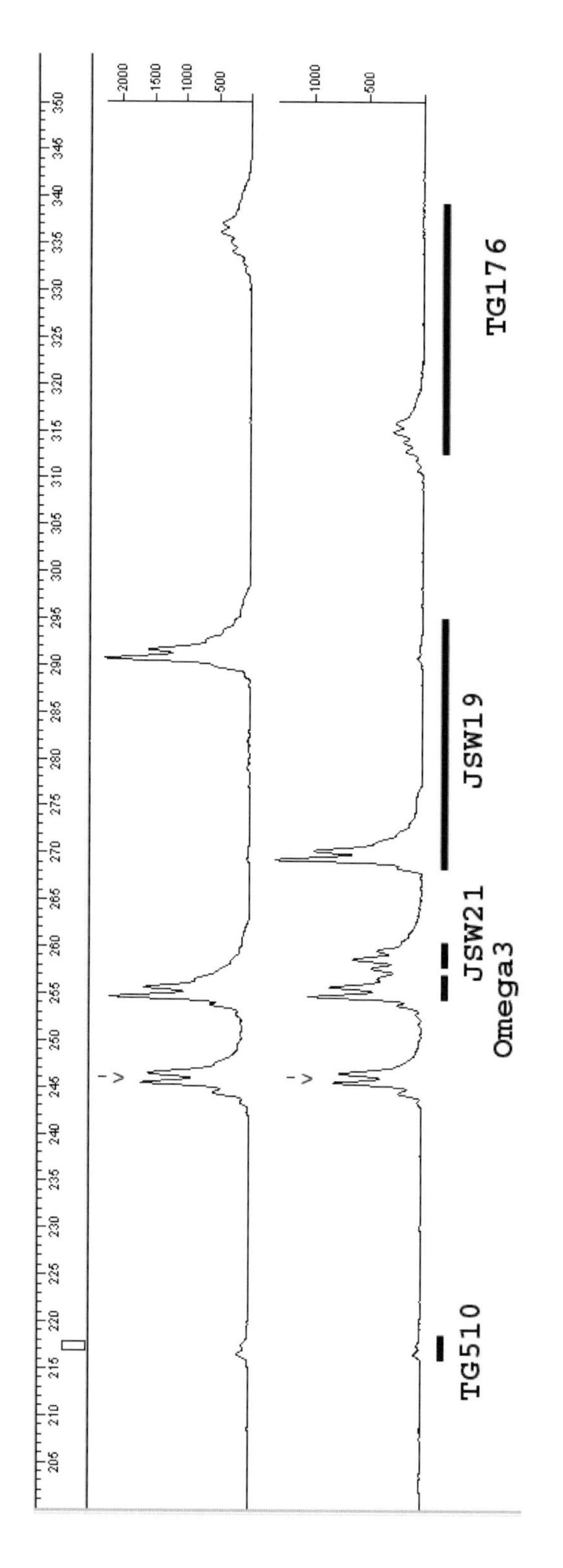
A
TG176
JSW19
JSW21
Omega3
TG510

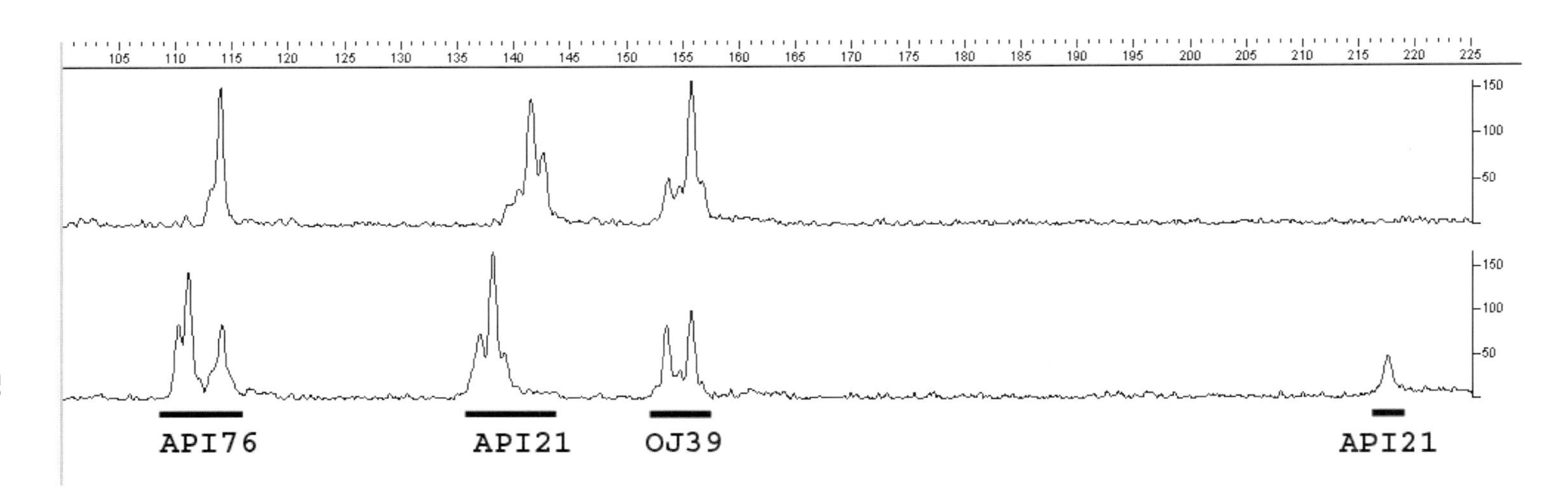

Figure 3. Genotyping of plant genomes using PS-based multiplex PCR.
A: Multiplex PCR of five tomato markers on the parents of the mapping population *L. esculentum* x *L. pimpinellifolium*. Tomato genomic DNA was digested with *Alu*I restriction enzyme and ligated with the PS-driving oligonucleotide adapters. Names and size ranges of each (candidate gene/ RFLP) marker are below the fragments. One extra band appeared during multiplex amplification which was not observed in the individual PCR reactions (245 bp fragment, marked by an arrow). The ruler above the picture shows the fragment size in bp. Relative fluorescence (arbitrary units) is shown on the right.
B: Multiplex PCR of four onion candidate gene markers on the parents of the mapping population *A. cepa* x *A. roylei*. Genomic DNA was digested with *Rsa*I restriction enzyme and ligated with PS-driven oligonucleotide adapters. Names and size range or each candidate gene marker are indicated below the fragments. For other designations see (A).

successful amplification of each target. However, both examples show that once enough polymorphisms are found and the T-primers are chosen, which work in individual PS PCR, it is feasible to combine them in one tube for simultaneous amplification of many DNA targets. This is especially important in large mapping populations which will make candidate gene/marker determination more efficient.

Discussion

The presented examples demonstrate the successful use of the PS-based multiplex PCR for genotyping complex genomes. The uniform amplification of all targets from plant genomes was achieved after simple optimization of one-step PCR. It should be emphasized that no nested PCR was necessary for amplification. Given high complexity of the plant genomes and presence of large amounts of repetitive DNA, these results demonstrate high potential of the PS-based multiplex approach. In human genome, it was necessary to apply two-step PCR to efficiently amplify 30 different targets. However, it was shown that single-step protocol works well when lower number of targets (up to 15) need to be amplified (Broude *et al.*, 2001a).

However, several issues still need more studies. The major yet unsolved problem with the PS-based approach is the choice of the T-primers, which are efficient in a single suppression-based PCR. Several basic questions should be experimentally addressed before the choice of the T-primers will be a technical question. For example, influence of the size of the single-stranded loops and position of the primers within the loops should be experimentally tested to better predict successful T-primers.

It should be mentioned however, that multiplex PCR is more tolerant to the mismatch priming that single target PCR. Indeed, it was shown that some of the T-primers generated extra bands in single PCR indicative of mismatch priming (Broude *et al.*, 2001b). These extra bands, however, were reduced to background in high level multiplex PCRs. The more primers participate in the reaction and find their perfect match, the less favorable is mismatch priming. Additionally, even if some mismatch products were generated, their relative concentrations in a 30-plex PCR product would be 30 fold less than in a single PCR.

In conclusion, we demonstrated that by following the principles well-established for conventional and some modified versions of PCR (*e.g.* long-range PCR, multiplex PCR) amplification of several tens of DNA targets can be achieved. In the future, combination of the PS-based multiplex PCR with high-throughput analyses methods, like microarray hybridization or real-time detection with fluorescent probes, will further extend the potential of this approach.

References

Broude, N.E., Zhang L., Woodward, K., Englert, D., and Cantor, C.R. (2001). Multiplex allele-specific target amplification based on PCR suppression. Proc. Natl. Acad. Sci. USA. *98*, 206-211.

Broude, N.E., Driscoll K, and Cantor, C.R. (2001). High level multiplex DNA amplification. Antisense Nucleic Acid Drug Dev. *11*, 327-322.

Chamberlain, J.S., Gibbs, R.A., Ranier, J.E., Nguyen, P.N., and Caskey, C.T. (1988). Deletion screening of the Duchenne muscular dystrophy locus via multiplex DNA amplification. Nucleic Acids Res. *16*, 11141-11156.

Edwards, M.C.and Gibbs R.A. (1994). Multiplex PCR: advantages, development, and applications. PCR Methods Appl. *3*, S65-75.

Gerry, N.P., Witowski, N.E., Day, J., Hammer, R.P., Barany, G., and Barany, F. (1999) Universal DNA microarray method for multiplex detection of low abundance point mutations. J. Mol. Biol. *292*, 251-62.

Henegariu, O., Heerema, N.A., Dlouhy, S.R., Vance, G.H., and Vogt, P.H. (1997). Multiplex PCR: critical parameters and step-by-step protocol. Biotechniques *23*, 504-511.

Lukyanov, K.A., Launer, G.A., Tarabykin, V.S., Zaraisky, A.G., and Lukyanov, S.A. (1995). Inverted terminal repeats permit the average length of amplified DNA fragments to be regulated during preparation of cDNA libraries by polymerase chain reaction. Anal. Biochem. *229*, 198-202.

Nazarenko, I., Lowe, B., Darfler, M., Ikonomi, P., Schuster, D., and Rashtchian A. (2002). Multiplex quantitative PCR using self-quenched primers labeled with a single fluorophore. Nucleic Acids Res. *30*, e37

Rychlik, W. (1995). Selection of primers for polymerase chain reaction. Mol. Biotechnol. *3*, 129-134.

Siebert, P.D., Chenchik, A., Kellogg, D.E., Lukyanov, K.A., and Lukyanov, S.A. (1995). An improved PCR method for walking in uncloned genomic DNA. Nucleic Acids Res. *23*, 1087-1088.

van Heusden, A.W., van Ooijen, J.W., Vrielink, R., Verbeek, W.H.J., Wietsma, W.A., and Kik., C. (2000). Genetic mapping in an interspecific cross in *Allium* with amplified fragment length polymorphism (AFLP^{TM}) markers. Theoretical Appl. Genetics *100*, 118-126.

van Heusden, A.W., Koornneef, M., Voorrips, R.E., Brüggemann, W., Pet, G., Vrielink-van Ginkel, R., Chen, X., and Lindhout, P. (1999). Three QTLs from *Lycopersicon peruvianum* confer a high level of resistance to *Clavibacter michiganensis ssp michiganensis*. Theoretical Appl. Genetics *99*, 1068-1074.

Vos, P., Hogers, R., Bleeker, M., Reijans, M., van de Lee, T., Hornes, M., Frijters, A., Pot, J., Peleman, J., Kuiper, M., *et al.* (1995). AFLP: a new technique for DNA fingerprinting. Nucleic Acids Res. *23*, 4407-4414.

Wu, D.Y, Ugozzoli, L., Pal, B.K., Qian, J., Wallace, R.B. (1991). The effect of temperature and oligonucleotide primer length on the specificity and efficiency of amplification by the polymerase chain reaction. DNA Cell Biol. *10*, 233-238.

2.2

On-Chip PCR: DNA Amplification and Analysis on Oligonucleotide Microarrays

Martin Huber, Christian Harwanegg, Manfred W. Mueller and Wolfgang M. Schmidt

Abstract

We describe a protocol for DNA microarray-based amplification, a method which we call on-chip PCR, suitable both for amplification and simultaneous characterization of a DNA sample. In contrast to conventional PCR, the reaction is performed directly on a flat surface of a glass chip and reaction products are visualized by fluorescence scanning of the chip rather than electrophoretic separation. The underlying principle of sequence detection is based on the well-known process of semi-nested PCR. On-chip PCR includes liquid phase PCR with two sequence-specific primers, which is performed on the top of a glass chip with covalently bound nested, allele-specific PCR primers. PCR products generated in the liquid phase are then re-amplified in a semi-nested PCR directly on the chip surface. Thus in on-chip PCR amplification and sequence detection are combined within a single step. The oligonucleotide microarray is designed such that positive signals reveal the presence and nature of the target DNA of interest. In the first example, we show how on-chip PCR can be employed to identify single nucleotide polymorphisms (SNPs) in human genomic DNA. In the second, we describe the use of on-chip PCR for detection and identification of pathogens. The results demonstrate the beneficial combination of conventional PCR amplification with microarray technology for the development of simple and rapid DNA diagnostic systems.

Background

The accurate and sensitive analysis of DNA sequences is a prerequisite in DNA diagnostics. With respect to throughput, DNA microarray (also called "DNA chip") technology promises to allow massive parallel analysis of a high number of diagnostic parameters (Graber *et al.*, 1998; Wang *et al.*, 2000). In addition to widely used microarray techniques based on pure hybridisation (Southern 1996), a number of enzyme-based approaches have been developed in combination with microarray detection. Among these methods mini-sequencing (Nikiforov *et al.*, 1994; Head *et al.*, 1997; Pastinen *et al.*, 1997; Dubiley *et al.*, 1999), allele-specific primer elongation (Shumaker *et al.*, 1996; Erdogan *et al.*, 2001; Pastinen *et al.*, 2000) and solid-phase PCR (Adessi *et al.*, 2000; Strizhkov *et al.*, 2000) employ specificity provided by DNA polymerases. These techniques, however, consist of several separate steps: amplification, purification of reaction products, and often preparation of single-stranded DNA template prior to the actual microarray-based analysis. These steps increase time and cost of such assays, ultimately limiting their potential in DNA diagnostic applications.

To overcome these limitations we developed an on-chip one-step PCR technique (Huber *et al.*, 2001; Huber *et al.*, 2002; Mitterer *et al.*, 2004) consisting of DNA sample amplification in the liquid phase and a sequence-specific semi-nested solid-phase PCR on a surface of coated glass slides (see Figure 1). During thermal cycling a PCR product is generated in the fluid phase, serving as a template for solid-phase

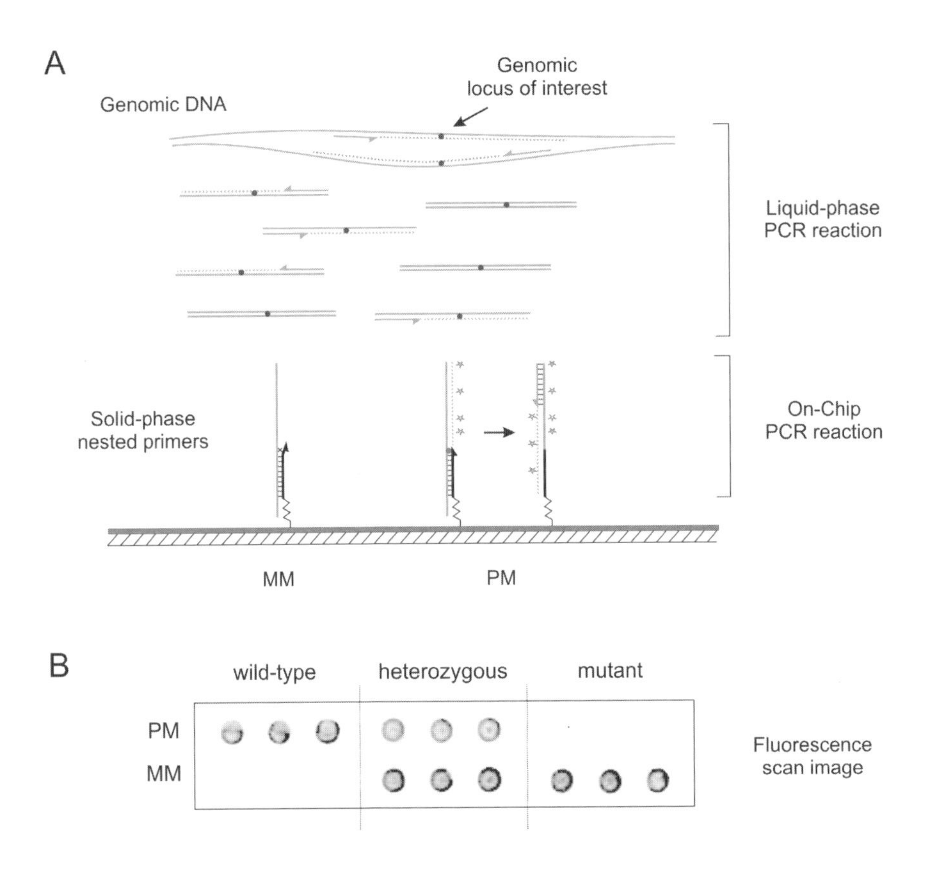

Figure 1. Schematic outline of on-chip PCR.
A: Experimental strategy. Genomic DNA is used as a template for simultaneous PCR amplification in the liquid phase and allele-specific amplification in solid-phase on a glass microarray of genes of interest (indicated by dots). The reaction mixture contains in solution DNA primers for gene-specific amplification and all other components of PCR. Nested primers, perfectly matched (PM) and mismatched (MM), are tethered to the glass slide. DNA polymerase extends the perfectly matched primers during the cycling reaction and generates fluorescently labelled extension products, indicated by asterisks. Mismatched primers are not extended by the DNA polymerase. Genomic DNA is shown as curved lines and PCR products as straight lines. Primers in the liquid phase are indicated by grey arrows. The solid-phase primers are shown as black arrows.
B: Allele genotyping after on-chip PCR by fluorescence scanning. As reaction products remain covalently bound to the glass surface throughout the on-chip PCR, they can be detected *via* standard fluorescence scanning of the glass chip. Genotype information is deduced from reading the fluorescence intensities of spots containing the respective allele-specific primers.

primer extension. Elongated solid-phase immobilized products are then subject to second-strand synthesis, initiating a solid-phase PCR that is driven by the immobilized nested primer and the second primer in solution. This semi-nested solid-phase PCR is a reaction with low amplification efficiency (estimated at 20-30% of PCR in solution) but constitutes an integral part of the process sufficient for target DNA detection. The glass chip contains up to several hundreds of covalently attached specific oligonucleotides, designed for interrogating multiple single nucleotide positions within the genomic regions of interest. After PCR, amplification products remain covalently bound to the glass chip and can be visualized and analysed due to incorporation of fluorescent dyes into amplicons during PCR. Data interpretation is facilitated by computer automated identification of positive signals, fluorescence intensity extraction and an algorithm-based, unsupervised genotype assignment. Technically on-chip PCR is a one-step method that provides accurate sequence information through a high number of sequences and/or allele-specific variants. The technical challenges of on-chip PCR are buried in the - actually unfavourable - reaction format and require a number of modifications in terms of surface coating and additives to the reaction mixtures, which will be the subject of the following protocol section.

Protocol

Materials

For the SNP analysis, genomic DNA can be extracted from EDTA-treated blood using the QIAamp DNA Blood Midi Kit (Qiagen, Hilden, Germany) protocol. In our experiments genomic DNA was isolated from cervical carcinoma cell lines HeLa S3 (ATCC CCL-2.2), CaSki (ATCC CRL-1550), and SiHa (ATCC HTB-35). These DNAs were used as positive controls for human papillomavirus DNAs (HPV-16 in CaSki and SiHa; HPV-18 in HeLa S3) (Meissner, 1999). For the HPV detection, cervical swaps, stored at - 80 °C in sample transfer medium (Digene, Gaithersburg, Maryland), were vortexed, briefly centrifuged at 12,000 rpm, and 200 μl of the supernatant was used for DNA extraction following essentially the QIAprep Mini protocol (Qiagen).

DNA oligonucleotides used as PCR primers can be synthesised with an Expedite 8909 Nucleic Acid Synthesiser (Perceptive Biosystems, Foster City, CA) using standard phosphoramidite chemistry (Proligo, Hamburg, Germany). Oligonucleotides for microarray attachment should be synthesised with a 5' terminal $(CH_2)_6$-NH_2 modification (Cruachem Ltd, Glasgow, UK) and purified by chromatography on a BioCAD Sprint system (PerSeptive Biosystems, Foster City, CA). The 5'- monomethyltritoxy group of the amino linker can be used as a hydrophobic tag for purification of full-length DNA oligos.

For SNP detection we analysed previously characterized polymorphic nucleotide positions in the human *MBL2* gene coding for the mannose-binding lectin: allelic variants in codon 52 (CGT -> TGT), codon 54 (GGC -> GAC), and codon 57 (GGA -> GAA) (Sumiya *et al.*, 1991; Madsen *et al.*, 1995) as well as two SNPs in the promoter region of this gene: nucleotide position –550 (G -> C) and –221 (G -> C). The *MBL2* gene sequence can be retrieved from EMBL or GenBank (Accession number AL583855). The mutations studied are referenced in the Human Gene Mutation Database (http://www.hgmd.org) (Krawczak and Cooper, 1997). In this study, alleles with the reference (wild-type) nucleotide are denoted A1, whereas alleles containing the changed nucleotide (mutation or polymorphism) are referred to as A2.

Human papillomavirus genome sequences were obtained from entries in the Los Alamos National Laboratory database (http://hpv-web.lanl.gov/stdgen/virus/hpv). All papillomavirus type-specific and subtype-specific primer sequences were located in the E6 and E7 gene region of the corresponding viral genomes.

Step 1: Primer Design

Considering the experimental setup of an on-chip PCR-based sequence detection and analysis system, primer design is a major challenge. The primary goal is to design hundreds of primer sequences that perform equally well at the same melting temperature and cycling conditions. In addition, there are two other important considerations for the primer design. First, primer sequences for target amplification in the fluid phase must meet all requirements for primers in conventional multiplex PCR; they have to be carefully checked for potential dimer formation with other primer sequences in the reaction mixture. A step-wise, empirical optimisation might be required in some cases. A helpful discussion of designing multiplex PCR protocols has been published by others (Henegariu *et al.*, 1997). Second, nested primer sequences for the solid-phase PCR should be selected for maximum allelic discrimination, which sometimes has to be achieved empirically, *e.g.* by selection of best performing primers from a larger set of possible sequences. In general, 3' mismatches that are reported to be less refractory, like G-T or C-A mismatches (Ayyadevara *et al.*, 2000; O'Meara *et al.*, 2002), should be avoided, as well as high GC content at the 3' termini. Here, one should exploit the advantage of possibility of using one of two possible (sense and antisense) primer sequences in order to bypass unfavourable sequence context. In a multiplex set-up, fragment length should be controlled and limited to 100-250 bp to prevent preferential amplification of products with smaller size. Advantageously, fragment length of all PCR amplicons can be similar because the analysis of reaction products is performed on solid-phase rather than by electrophoretic separation. The location of nested primers within the primary amplification product controls the length of the nested PCR product and therefore influences the amount of label that can be incorporated by the DNA polymerase during the solid-phase PCR reaction. Therefore, an equal length of the solid-phase products may contribute to uniform signal intensities.

The web based primer selection program *Primer3* (http://www-genome.wi.mit.edu/cgi-bin/primer/primer3_www.cgi) is our recommended choice for selecting primer sequences (Rozen and Skaletsky, 2000).

Step 2: Preparation of Polymer-Coated Glass Slides

The nature and quality of surface coating of glass slides is a critical issue in on-chip PCR. In our initial studies we screened a multitude of commercially available microarray glass slides and found that only slides containing a polymer-type surface coating were suitable for on-chip PCR. The polymer layer increases the spatial distance between the glass-surface and the PCR primers which decreases steric hindrance for template DNA and polymerase at the primer binding-region (Vasiliskov *et al.*, 1999; Huber *et al.*, 2001). Further, we found that attachment chemistry creating stable, covalent amide bonds between the polymer and the 5' amino group of PCR primers, is compatible with high temperature thermal cycling during PCR. The following protocol introduces a rather simple procedure for generating such polymer-coated slides suitable for on-chip PCR.

In brief, the glass slides are treated with trimethoxysilane derivatized polyethyleneimine (PEI) and subsequently activated with an amine-reactive cross-linker. Standard glass slides (Melvin Brand, Sigma-Aldrich) are cleaned in HCl:methanol (1:1) for 24 hours at room temperature, washed thoroughly with de-ionised water and dried under an air stream. The slides are then incubated in a solution containing 3 % (v/v) trimethoxysilylpropyl modified polyethyleneimine (Gelest, Tullytown, PA) in 95 % ethanol for 1 hour with vigorous agitation at ambient temperature. Afterwards the slides are washed in 95 % ethanol, dried under air and cured at 80 °C for 1 hour. Surface activation of the slides is achieved by treating with the homo-bifunctional cross-linker ethylene glycol-bis(succinic acid N-hydroxysuccinimide) ester (EGS) (Pierce, Rockford, IL).

Step 3: Attachment of Oligonucleotide Primers to the Glass Chips

For coupling to the glass slides the amino modified oligonucleotides are dissolved in 150 mM sodium phosphate buffer pH 8.5 at 24°C, supplemented at a concentration of 20 μM with 0.1 % (w/v) SDS and spotted using an Affymetrix 417™ Arrayer (Affymetrix, Santa Clara, CA) equipped with 125 μm pins. In order to keep an average spot diameter of 190 -200 μm it is important to control the temperature at 20 °C and relative humidity at 50-60% during the spotting process. In addition to the Affymetrix spotter, which is based on the so-called Pin-and-Ring™ technology, we also used a spotting robot that employed the commonly used Split-Pin technology (TeleChem SMP 3 Stealth

Pins). We advise to carefully adjust spotting buffer, additives like SDS, and spotting conditions like temperature and humidity according to the type of spotting technology used. Manufacturers of microarraying machines and substrates typically provide protocols for an optimal operating range regarding to the above mentioned parameters.

The arrayed slides are incubated in a NaCl-saturated humid chamber at 25 °C for at least 12 hours. Spotted glass slides can be stored at least 3 months until used in on-chip PCR. Prior to usage, the slides are immersed in 150 mM ethanolamine, 100 mM Tris (pH 9.0) at 55 °C for 20 minutes, in order to block unreacted succinimide esters. The slides are then washed thoroughly with deionised water and dried again.

Step 4: On-Chip PCR

The technical challenge lies in performing PCR in a small volume (10 μl) that is spread in a very thin layer over an area of 22 x 22 mm between a glass slide and a glass cover slip. For satisfactory efficiency, the enzymatic reaction performed on a flat surface requires a number of additives to the reaction mixture. First, the reaction is performed under higher salt (300 mM KCl, 50 mM Tris) and magnesium concentration (3 mM) compared to conventional PCR and contains bovine serum albumin (BSA) as a blocking agent. Second, the reaction contains a self-seal reagent (MJ Research), which polymerises upon contact with air at high temperature and thereby seals the reaction mixture at the edges to prevent evaporation during PCR.

A 10 μl PCR mix contains 2 x HotStar Taq PCR-buffer, 50 μM each dNTP, 20 μM Cy3-dCTP (Amersham Pharmacia Biotech Europe, Freiburg, Germany), 2.5 μg/μl BSA, 25 % (v/v) Self-Seal Reagent (MJ Research, Waltham, MA), 3 units HotStar™ Taq DNA Polymerase (Qiagen, Hilden, Germany) and liquid phase-primers at 0.025 - 0.4 μM. Prior to thermal cycling, human genomic DNA (20 ng) in sterile water is added. After pipetting the reaction mix onto the oligonucleotide array, a clean glass cover slip (22 x 22 mm) is mounted carefully using forceps. Glass slides were put into a PTC 200 *In Situ* slide thermocycler (MJ Research) and cycling is carried out according to the following scheme: 80°C for 10 min, 95°C for 5 min, 30-50 cycles at 95°C for 30 sec, 60-64°C for 1 min and 70-72°C for 1 min. After cycling the slides are placed in 0.1 x SSC, 0.1 % SDS for 10 min with gentle agitation. After removing the cover slips the slides are washed again for 10 min in 0.1 x SSC, 0.1 % SDS, then washed with de-ionised water and dried under an air stream. We found that denaturing washes after the amplification procedure were not necessary suggesting that no significant non-specific hybridisation occurred (Huber *et al.*, 2001).

Step 5: On-Chip PCR Data Analysis

Incorporation of Cy3-dCTP during the DNA amplification reaction is measured using an Affymetrix 428™ Array Scanner (Affymetrix, Santa Clara, CA). Fluorescence intensities (medians after local background subtraction) are calculated using Genepix 3.0 software (Axon Instruments, Foster City, CA).

Data analysis and interpretation of the genotype is performed using a scripting extension of the Genepix software programmed in Visual Basic Script Version 5.5 (Microsoft Corporation, Redmond, WA). The script facilitates automated data analysis by extracting median intensity (minus background) data from the GenePix program's result tab and automatically calculates discrimination ratios from average values derived from the replicates according to the equation: $DF = (FI_{A1}-FI_{A2})/(FI_{A1}+FI_{A2})$, where DF is the discrimination factor, and FI is the average fluorescence intensity of the spots corresponding to the two possible allelic variants A1 and A2. The program then automatically assigns genotype information to the DF values according to the following scheme: $DF \geq 0.4$: A1/A1 genotype (wild-type), $DF \leq -0.4$: A2/A2 genotype (mutant) and $-0.25 \leq DF \leq 0.25$: A1/A2 genotype (heterozygous). DF values not matching to this scheme are flagged for visual inspection by the user. This procedure of genotype prediction is very robust and does not require any kind of data transformations that would be necessary in cases where either of the spots is undetectable.

Examples

Example 1: Detection of *MBL2* Genotypes by On-Chip PCR

Figure 2 shows an example of allelic discrimination using the on-chip PCR protocol. In this example, the method was used to measure allelic variants in the human *MBL2* gene encoding the mannose-binding lectin, an important component of innate immunity (Hoffmann *et al.*, 1999). The assessment of allelic variants in the *MBL2* gene is of great clinical importance in newborns with recurrent infections or immune suppressed patients at high risk for a variety of infections (Kilpatrick, 2002).

Six PCR primers for multiplex liquid phase PCR were designed to amplify three fragments of the *MBL2* gene: 157 bp long fragment (spanning a part of exon 1 containing codons 52, 54 and 57), and two promoter fragments, one 156 bp long containing nucleotide position –221 and another fragment, 157 bp long, containing nucleotide position – 550.

The microarray layout was designed for simultaneous testing with several allele-specific primers (Figure 2a) and additionally employed a high confidence rate by arraying 12 spots (in form of 4 triplicates) for

Figure 2. Detection of SNPs in the human *MBL2* gene by on-chip PCR.
A: Genomic DNA was subjected to on-chip PCR with primers specific for 5 different polymorphic loci in the *MBL2* gene using an oligonucleotide microarray comprising a set of three different allele-specific primer pairs for each locus. Perfect match primers (PM) were specific for the normal reference *MBL2* sequence, whereas mismatch primers (MM) were designated for detection of the respective variant nucleotides. The solid-phase primers interrogated the variants with either their 3' first, PM / MM, second, PM (+1) / MM (+1), or third nucleotide residue, PM (+2) / MM (+2).
B: Microarray layout with mutant and control (normal) primers spotted in triplicates. *MBL2* polymorphisms covered codons 52, 54, and 57 in exon 1; and nucleotides –221 and –550 in promoter (Pro). Fluorescent spotting controls (guide dots) served as markers for orientation, while poly$(dT)_{15}$ probes (with, "T15-NH2", and without, "T15", terminal amino link) were included as quality controls for surface chemistry. The scheme shows the layout of one of the 4 identical sub-arrays.
C: Fluorescence scan images of on-chip PCR performed with two different DNA

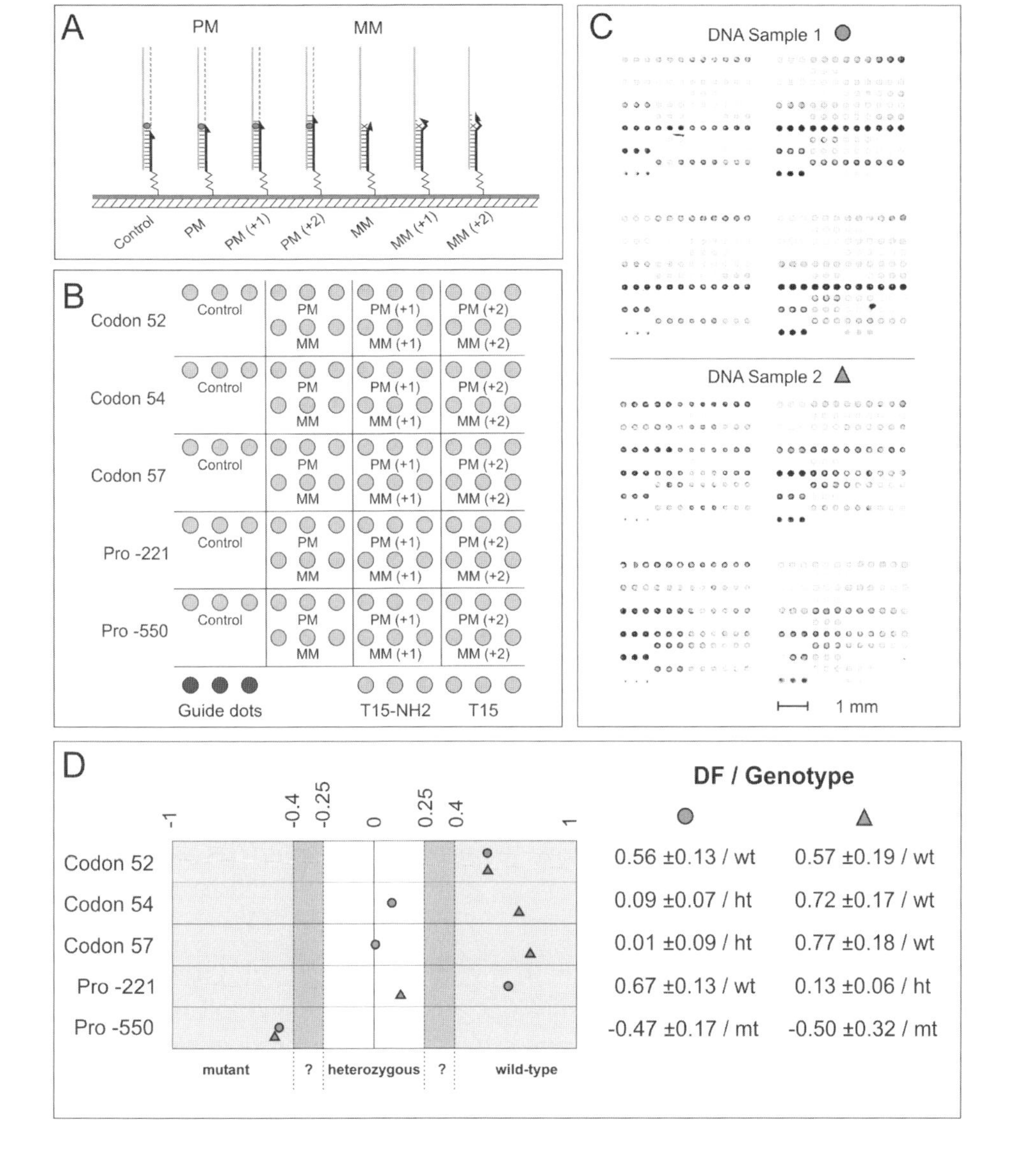

samples analysed for *MBL2* polymorphisms (see layout in B). The images are exemplary and show typical on-chip PCR results. Note that due to the replicate layout a precise and robust genotyping was achieved in spite of recurrent technical problems, such as dust particles covering spots (compare DNA sample 1 upper left and lower right array), scratches (compare DNA sample 2 lower right array) or air bubbles caught beneath the cover slip.
D: Graphical representation of discrimination factor (DF) values calculated from spot fluorescence intensities and genotype assignment inferred from the average DF values from the 4 identical sub-arrays. WT, wild-type, HT, heterozygote, MT, mutant. The DF values are mean values with the standard deviations indicated.

each solid-phase oligonucleotide. The solid-phase primers interrogated the allelic variants with either the first, second, or third nucleotide residue from their 3'-terminus. The array consisted of 4 identical sub-arrays, each containing 3 allele-specific primer pairs (pairs are indicated by PM and MM) together with positive controls and guide dots (see Figure 2b). This “high-replicate” array pattern provided auxiliary security in the event of possible technical troubles and greatly facilitated automated analysis steps like grid placement and spot finding. As shown in Figure 2c, on-chip PCR successfully identified the allelic variants of the five different polymorphic loci within the *MBL2* gene. The correct genotype was derived unambiguously from average fluorescent intensity ratios calculated from spots that correspond to the 3 primer positions tested for each gene (Figure 2d). We evaluated the reliability of the method in more than 30 assays and compared data to the results obtained by sequencing of the 3 different PCR products involved in the analysis. The on-chip PCR did not produce any false results in more than 150 genotypes assigned. The careful selection of the allele-specific primers together with the replicate layout of the microarray resulted in unequivocal discrimination values for the allelic variants studied.

Example 2: Detection of Human Papillomaviruses by On-Chip PCR

Figure 3 shows an example of identification and sub-typing of viral pathogens using the on-chip PCR protocol. The method was used to identify human papillomavirus (HPV) types in cervical samples and human cervical carcinoma cell lines containing HPV-16 and HPV-18 (Meissner, 1999). The detection of HPV types is of great clinical importance in screening programs aimed at reducing the incidence of cervical cancer. This frequent type of cancer is associated in over 99% of all cases with persistent infection with at least one of 15 high-risk HPV types (Walboomers *et al.*, 1999; Munoz *et al.*, 2003). Exact identification of the HPV type (and subtypes) can significantly help in patient care and management because it is known that variants of HPV-16 and HPV-18 are preferentially associated with neoplasia and therefore impose a higher risk for cervical cancer (Villa *et al.*, 2000).

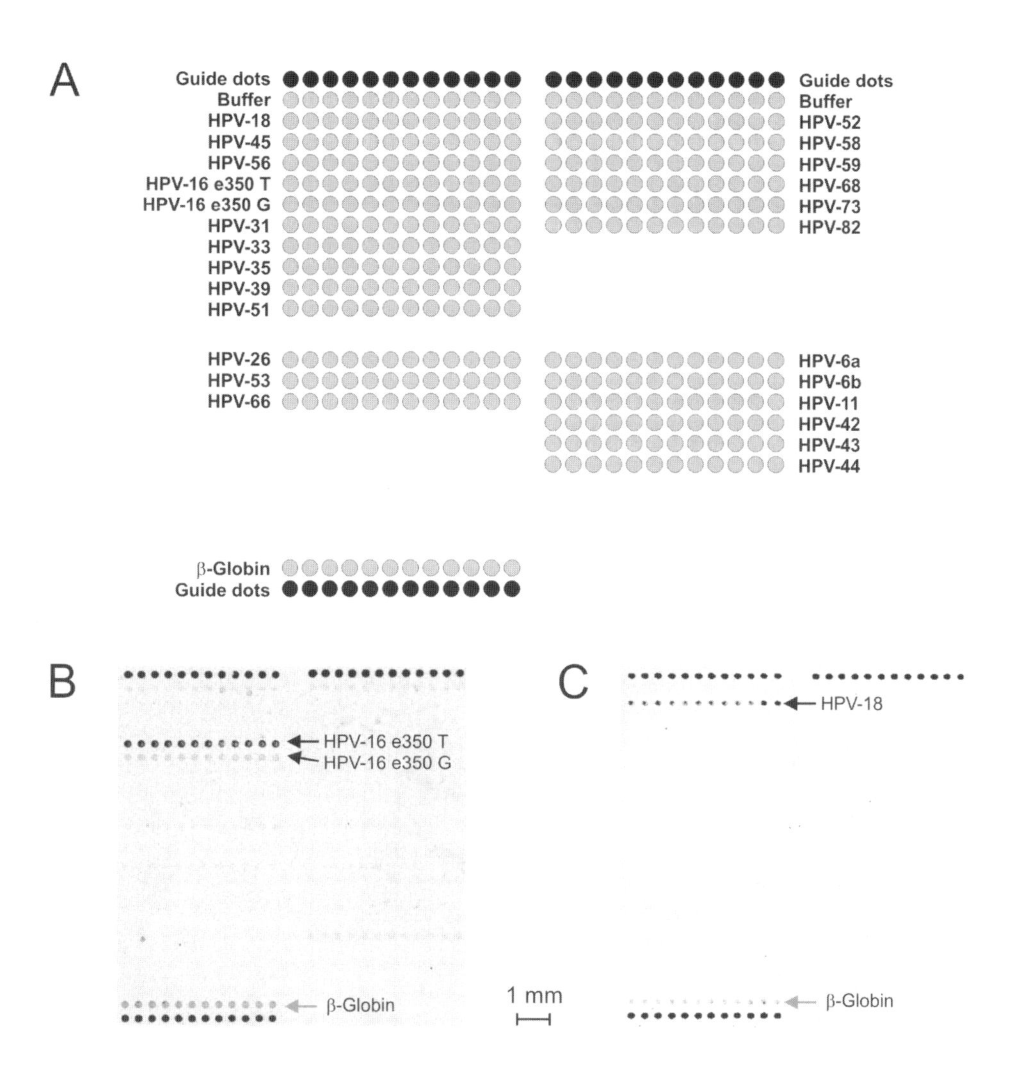

Figure 3. Detection of human papillomavirus by on-chip PCR.
A: DNA isolated from cervical carcinoma cell lines and cervical swabs was subjected to on-chip PCR using an oligonucleotide microarray containing covalently attached primers specific for 25 different HPV types and variant subtypes. The layout of the arrayed solid-phase PCR primers is shown. The HPV type-specific primers and human β-globin control were spotted with 11 horizontal replicates.
B, C: Fluorescence scan images. The DNA sample shown in (B), isolated from a clinical cervical swab sample, contained the HPV-16 variant E6 350T, whereas the DNA isolated from the cervical carcinoma cell line HeLa shown in (C) contained HPV type 18.

On-chip PCR is well suited for this purpose because it provides high sensitivity and specificity required for virus identification. The reaction mix contained 38 PCR primers, intended to amplify 23 different HPV types, belonging to the high-risk, probable high-risk and low-risk groups (Munoz *et al.*, 2003). All fragments span a part of the E6 and/or E7 region of the respective viral genomes. This region is maintained during late stages of infection (*i.e.* integration of the viral DNA into the host cell genome) and consists of known intratype and intertype sequence variation (Villa *et al.*, 2000). PCR fragments were from 132 bp to 224 bp long. Some highly homologous HPV types were amplified with consensus primers specific for 2 types, such as HPV-18 and HPV-45, or HPV-33 and HPV-58. The primers were designed also for the human β-globin gene, which served as an amplification control.

The microarray layout (depicted in Figure 3b) consisted of 25 type- and subtype (variant) specific solid-phase primers, spotted with 11 horizontal replicates, allowing the correct identification of HPV. As shown for a cervical carcinoma cell line and a cervical swab sample, on-chip PCR successfully detected the presence of a HPV infection. In both cases, the on-chip PCR result was confirmed by conventional PCR and sequencing. The examples shown are part of preliminary studies including more than 200 cervical samples. We are currently evaluating the clinical performance of on-chip PCR in HPV testing in a routine laboratory setup.

Discussion

The results clearly demonstrate the robustness and accuracy of the on-chip PCR suggesting its applicability as a novel DNA analysis tool. When compared to the widely employed microarray hybridisation method, the on-chip PCR has several important advantages. The initial design of the arrayed oligonucleotides is easier because it allows a choice between 3-4 possible primers for the sense and antisense strand, depending on the position of the target nucleotide relative to the 3' end. This circumvents time-consuming pilot experiments for the design of allele-specific oligonucleotides to ensure optimal signal intensities and discrimination in microarray hybridisation. Next, high density arrays, like Affymetrix SNP GeneChip, require a high density tiling array setup and complex mathematical algorithms for computer-guided analysis (Chee *et al.*, 1996). In contrast, we found that a simple ratio-formulating algorithm can accurately assign genotypes in an automated fashion after on-chip PCR. The automated analysis is further supported by the multiple test setup integrating the simultaneous probing of sense and antisense strands, which would be difficult to implement in a microarray hybridisation procedure that normally requires single-stranded targets for accurate measurement. As a further advantage, there is no need for washing steps or any kind of stringency inducing post-processing after amplification in order to increase specificity.

On-chip PCR requires very little hands-on-time since it integrates amplification and genotyping into a single assay. Virtually only a single pipetting step is needed to launch the reaction from a pre-prepared master mix; this not only avoids laborious sample preparation steps, like single-stranded target DNA synthesis, but also represents an important benefit with respect to potential sample contamination risk inherent to all PCR applications. The method completely overcomes the need for changing reaction tubes or wells and allows obtaining sequence information directly from a genomic DNA sample in less than three hours with only 15 minutes hands-on-time.

On-chip PCR is well suited for detection of conserved genes such as bacterial 23S rRNA genes by using consensus primers (Mitterer *et al.*, 2004). Further, on-chip PCR permits multiplexing levels similar to conventional PCR thus allowing the simultaneous analysis of a high number of sequence loci (Huber *et al.*, 2002). Although high multiplex levels in PCR have been described by others (Wang *et al.*, 1998), we failed to generate multiplex reactions higher then 15-20 genes in a single solid-phase PCR assay. To overcome this limitation of on-chip PCR, a multitude of individual multiplex PCR assays can be performed in separate micro-fabricated reaction chambers which would extend the number of analyzed genes. Note that successful PCR in cavities of silicon devices has been demonstrated by others (Cheng *et al.*, 1996; Shoffner *et al.*, 1996). As another possibility, PCR approaches addressed to extend the number of individual amplifications could be integrated into on-chip PCR as well. Thus, previously described strategies such as multiplex PCR based on PCR suppression (Broude *et al.*, 2001) or the degenerate oligonucleotide primer based PCR (DOP-PCR) (Jordan *et al.*, 2002) could – although not tested extensively in our studies – significantly enhance PCR multiplexing level. Whole-genome amplification using Phi 29 DNA polymerase (Dean *et al.*, 2001; Hosono *et al.*, 2003) could be combined with on-chip PCR to considerably improve the procedure by decreasing PCR cycle numbers, especially when testing clinical samples where template DNA is limited. However, all these modifications would require an additional template preparation or amplification step and would therefore negatively affect the advantageous one-step format of on-chip PCR.

The described method in its present format can be reliably employed for the detection of any kind of variations in DNA sequences. Our current studies address technical improvements and evaluate alternative reaction formats trying to establish on-chip PCR-based integrated devices for use in DNA diagnostics.

Acknowledgements

We thank Dr. O. Bodamer (Department of Pediatrics, University of Vienna) and Iris Schmidt (Institute of Virology, Veterinary University of Vienna) for supplying us with DNA samples used in this work.

References

Adessi, C., Matton, G., Ayala, G., Turcatti, G., Mermod, J.J., Mayer, P., and Kawashima, E. (2000). Solid phase DNA amplification: characterisation of primer attachment and amplification mechanisms. Nucleic Acids Res. *28*, e87.

Ayyadevara, S., Thaden, J.J., and Shmookler Reis, R.J. (2000). Discrimination of primer 3'-nucleotide mismatch by Taq DNA polymerase during polymerase chain reaction. Anal. Biochem. *284*, 11-18.

Broude, N.E., Zhang, L., Woodward, K., Englert, D., and Cantor, C.R. (2001). Multiplex allele-specific target amplification based on PCR suppression. Proc. Natl. Acad. Sci. USA. *98*, 206-211.

Chee, M., Yang, R., Hubbell, E., Berno, A., Huang, X.C., Stern, D., Winkler, J., Lockhart, D.J., Morris, M.S., and Fodor, S.P.A. (1996). Accessing genetic information with high density DNA arrays. Science *274*, 610-614.

Cheng, J., Shoffner, M.A., Hvichia, G.E., Kricka, L.J., and Wilding, P. (1996). Chip PCR. II. Investigation of different PCR amplification systems in microfabricated silicon-glass chips. Nucleic Acids Res. *24*, 380-385.

Dean, F.B., Nelson, J.R., Giesler, T.L., and Lasken, R.S. (2001). Rapid amplification of plasmid and phage DNA using Phi 29 DNA polymerase and multiply-primed rolling circle amplification. Genome Res. *11*, 1095-1099.

Dubiley, S., Kirillov, E., and Mirzabekov, A. (1999). Polymorphism analysis and gene detection by minisequencing on an array of gel-immobilized primers. Nucleic Acids Res. *27*, e19.

Erdogan, F., Kirchner, R., Mann, W., Ropers, H.H., and Nuber, U.A. (2001). Detection of mitochondrial single nucleotide polymorphisms using a primer elongation reaction on oligonucleotide microarrays. Nucleic Acids Res. *29*, e36.

Graber, J.H., O'Donnel, M.J., Smith, C.L., and Cantor, C.R. (1998). Advances in DNA diagnostics. Curr. Opin. Biotechnol. *9*, 14-18.

Head, S.R., Rogers, Y.H., Parikh, K., Lan, G., Anderson, S., Goelet, P., and Boyce-Jacino, M.T. (1997). Nested genetic bit analysis (N-GBA) for mutation detection in the p53 tumor suppressor gene. Nucleic Acids Res. *25*, 5065-5071.

Henegariu, O., Heerema, N.A., Dlouhy, S.R., Vance, G.H., and Vogt, P.H. (1997). Multiplex PCR: critical parameters and step-by-step protocol. Biotechniques *23*, 504-511.

Hoffmann, J.A., Kafatos, F.C., Janeway, C.A., and Ezekowitz, R.A. (1999). Phylogenetic perspectives in innate immunity. Science *284*, 1313-1318.

Hosono, S., Faruqi, A.F., Dean, F.B., Du, Y., Sun, Z., Wu, X., Du, J., Kingsmore, S.F., Egholm, M., and Lasken, R.S. (2003). Unbiased whole-genome amplification directly from clinical samples. Genome Res. *13*, 954-964.

Huber, M., Losert, D., Hiller, R., Harwanegg, C., Mueller, M.W., and Schmidt, W.M. (2001). Detection of single base alterations in genomic DNA by solid-phase PCR on oligonucleotide microarrays. Anal. Biochem. *299*, 24-30.

Huber, M., Mündlein, A., Dornstauder, E., Schneeberger, Ch., Tempfer, C.B., Mueller, M.W., and Schmidt, W.M. (2002). Accessing single nucleotide polymorphisms in genomic DNA by direct multiplex PCR amplification on oligonucleotide microarrays. Anal. Biochem. *303*, 25-33.

Jordan, B., Charest, A., Dowd, J.F., Blumenstiel, J.P., Yeh, R.F., Osman, A., Housman, D.E., and Landers, J.E. (2002). Genome complexity reduction for SNP genotyping analysis. Proc. Natl. Acad. Sci. USA. *99*, 2942-2947.

Kilpatrick, D.C. (2002). Mannan-binding lectin: clinical significance and applications. Biochim. Biophys. Acta *1572*, 401-413.

Krawczak, M. and Cooper, D.N. (1997). The human gene mutation database. Trends Genet. *13*, 121-122.

Madsen, H.O., Garred, P., Thiel, S., Kurtzhals, J.A., Lamm, L.U., Ryder, L.P., and Svejgaard, A.J. (1995). Interplay between promoter and structural gene variants control basal serum level of mannan-binding protein. Immunol. *155*, 3013-3020.

Meissner, J.D. (1999). Nucleotide sequences and further characterization of human papillomavirus DNA present in the CaSki, SiHa and HeLa cervical carcinoma cell lines. J. Gen. Virol. *80*, 1725-1733.

Mitterer, G., Huber, M., Leidinger, E., Kirisits, C., Lubitz, W., Mueller, M.W., and Schmidt, W.M. (2004). Microarray-based identification of bacteria in clinical samples by solid-phase PCR amplification of 23S ribosomal DNA sequences. J. Clin. Microbiol. *42*, 1048-1057.

Munoz, N., Bosch, F.X., de Sanjosé, S., Herrero, R., Castellsagué, X., Shah, K.V., Snijders, P.J.F., Meijer C.J.L.M., and the International Agency for Research on Cancer Multicenter Cervical Cancer Study Group. (2003). Epidemiologic classification of human papillomavirus types associated with cervical cancer. New Engl. J. Med. *348*, 518-527.

Nikiforov, T.T., Rendle, R.B., Goelet, P., Rogers, Y.H., Kotewicz, M.L., Anderson, S., Trainor, G.L., and Knapp, M.R. (1994). Genetic bit analysis: a solid phase method for typing single nucleotide polymorphisms. Nucleic Acids Res. *22*, 4167-4175.

O'Meara, D., Ahmadian, A., Odeberg, J., and Lundeberg, J. (2002). SNP typing by apyrase-mediated allele-specific primer extension on DNA microarrays. Nucleic Acids Res. *30*, e75.

Pastinen, T., Kurg, A., Metspalu, A., Peltonen, L., and Syvanen, A.C. (1997). Minisequencing: a specific tool for DNA analysis and diagnostics on oligonucleotide arrays. Genome Res. *7*, 606-614.

Pastinen, T., Raitio, M., Lindroos, K., Tainola, P., Peltonen, L., and Syvanen, A.C. (2000). A system for specific, high-throughput genotyping by allele-specific primer extension on microarrays. Genome Res. *10*, 1031-1042.

Rozen, S. and Skaletsky, H.J. (2000). Primer3 on the WWW for general users and for biologist programmers. In: Bioinformatics Methods and Protocols: Methods in Molecular Biology. S. Krawetz, S. Misener S. eds Humana Press, Totowa, NJ, p. 365-386.

Shoffner, M.A., Cheng, J., Hvichia, G.E., Kricka, L.J., and Wilding, P. (1996). Chip PCR. I. Surface passivation of microfabricated silicon-glass chips for PCR. Nucleic Acids Res. *24*, 375-379.

Shumaker, J.M., Metspalu, A., and Caskey, C.T. (1996). Mutation detection by solid phase primer extension. Hum. Mutat. *7*, 346-354.

Southern, E.M. (1996). DNA chips: analysing sequence hybridisation to oligonucleotides on a large scale. Trends Genet. *12*, 110-115.

Strizhkov, B.N., Drobyshev, A.L., Mikhailovich, V.M., and Mirzabekov, A.D. (2000). PCR amplification on a microarray of gel-immobilized oligonucleotides: detection of bacterial toxin- and drug-resistant genes and their mutations. Biotechniques *29*, 844-850.

Sumiya M., Super M., Tabona P., Levinsky R.J., Arai T., Turner M.W. and Summerfield, J.A. 1991. Molecular basis of opsonic defect in immunodeficient children. Lancet *337*, 1569-1570.

Vasiliskov, A.V., Timofeev, E.N., Surzhikov, S.A., Drobyshev, A.L., Shick, V.V., and Mirzabekov, A.D. (1999). Fabrication of microarray of gel-immobilized compounds on a chip by copolymerization. Biotechniques *27*, 592-594.

Villa, L.L., Sichero, L., Rahal, P., Caballero, O., Ferenczy, A., Rohan, T., and Franco, E.L. (2000). Molecular variants of human papillomavirus types 16 and 18 preferentially associated with cervical neoplasia. J. Gen. Virol. *81*, 2959-2968.

Walboomers, J.M., Jacobs, M.V., Manos, M.M., Bosch, F.X., Kummer, J.A., Shah, K.V., Snijders, P.J., Peto, J., Meijer, C.J., and Munoz, N. J. (1999). Human papillomavirus is a necessary cause of invasive cervical cancer worldwide. Pathol. *189*, 1-3.

Wang, D.G., Fan, J.B., Siao, C.J., Berno, A., Young, P., Sapolsky, R., Ghandour, G., Perkins, N., Winchester, E., Spencer, J., *et al.* (1998). Large-scale identification, mapping, and genotyping of single-nucleotide polymorphisms in the human genome. Science *280*, 1077-1082.

Wang, J. (2000). From DNA biosensors to gene chips. Nucleic Acids Res. *28*, 3011-3016.

2.3

Analysis of Somatic Mutations via Long-Distance Single Molecule PCR

Yevgenya Kraytsberg, Ekaterina Nekhaeva, Connie Chang, Konstantin Ebralidse and Konstantin Khrapko

Abstract

This chapter advocates the use of single molecule PCR (smPCR) as a tool in mutational analysis. smPCR is compared to the widely used vector-mediated cloning of PCR products. We argue that smPCR is capable of circumventing the problems inherent in post-PCR cloning, namely, the PCR-derived errors, allelic preference, and the template jumping artifact. These arguments are substantiated by examples of smPCR applications: an estimate of the error rate of smPCR, a study of linkage between multiple mutations within a single cell, and a measurement of frequencies of deletions in mitochondrial DNA in an aged human tissue.

Background: The Problems of Clone-by-Clone Analysis of PCR Products

Studies of somatic mutations entail analysis of complex mixtures of wild type and various mutant DNA molecules. One efficient way of analyzing such mixtures is based on cloning of a large number of individual DNA molecules followed by analysis of individual clones for mutations. The distribution of mutations in the original sample is then approximated by the distribution of mutant clones (Benzer and Freese, 1958). At present time, analysis of clones is done mostly by direct sequencing. The clone-by-clone approach is feasible only if the expected frequency of mutations in the mixture is sufficiently high, so that not too much sequencing is wasted on low-informative non-mutant clones. Human mitochondrial DNA (mtDNA) appears to contain a substantial number of somatic mutations, *i.e.* a few per 10^4 nucleotides, though estimates vary by more than an order of magnitude. This translates into a few mutants detected per 20 typical sequencing reactions, which is acceptable given low current sequencing costs. Thus, clone-by-clone analysis has been widely used in mitochondrial mutational analysis (Monnat and Loeb, 1985a), (Bodenteich *et al.*, 1991), (Kovalenko *et al.*, 1996), (Jazin *et al.*, 1996), (Michikawa *et al.*, 1999), (Simon *et al.*, 2001), (Lin *et al.*, 2002), (Khaidakov *et al.*, 2003), (Del Bo *et al.*, 2003), (Kamiya and Aoki, 2003). This technique should be applicable to nuclear DNA mutations in hypermutable regions or in cases where some kind of enrichment of mutants is feasible.

Although clones for mutational analysis may be generated directly from purified mtDNA (Monnat and Loeb, 1985a), (Bodenteich *et al.*, 1991), in recent studies researchers preferred to PCR amplify DNA prior to cloning. PCR simplifies the subsequent cloning procedure, besides sometimes it is unavoidable if the amount of tissue/DNA is limiting. Pre-cloning PCR is mandatory if nuclear DNA is to be studied. Despite being useful and sometimes unavoidable, the cloning of PCR products has, in addition to being rather laborious, the following significant disadvantages inherent to the PCR procedure.

1. PCR-derived Mutations

Most of the studies using cloning of PCR products report relatively high PCR error rates that limit the sensitivity of the approach. For example, in a recent study of mutational burden in mtDNA of the brain (Lin *et al.*, 2002), the frequency of detected mutations in the aged brain was $3x10^{-4}$, and as much as one third of them was due to PCR errors. In the young brain, the frequency of mutations could not even be estimated, because apparently it was below the frequency of errors.

2. Template Jumping

PCR amplification may result in "template jumping artifact" (Paabo *et al.*, 1990), *i.e.* condition where DNA polymerase switches template while synthesizing a new DNA strand. This artifact is important if there is more than one mutation per molecule within the DNA region under consideration. In this case template switching may separate mutations which originally were on the same DNA molecule or, conversely, join originally separated mutations, just as recombination does. This may create combinations of mutations that may not have existed in the original sample, but will be indistinguishable from the genuine ones upon cloning.

3. Allelic Preference

PCR amplification may distort the distribution of mutations if templates bearing different mutations are amplified with different efficiencies. This effect, called allelic preference, is particularly important when large deletions are considered, because DNA fragments of different length may have drastically different PCR amplification efficiencies.

In this chapter, we demonstrate that the above-mentioned problems can be avoided by using single molecule PCR (smPCR) of native (non-amplified) DNA instead of the conventional vector-mediated cloning of PCR products. smPCR was originally introduced by Jeffreys (Jeffreys *et al.*, 1988) and has since been used, among other applications, in studies of minisatellite variability (Jeffreys *et al.*, 1990), to determine the linkage phase of human haplotypes (Ruano *et al.*, 1990), to create expression libraries (Lukyanov *et al.*, 1996), for quantification of specific mutations (Vogelstein and Kinzler, 1999), to study recombination (Yauk *et al.*, 2003) and alternative splicing (Zhu *et al.*, 2003). We will now consider a smPCR protocol with particular emphasis on amplification of long (up to 16 kb) fragments for the purposes of mutational analysis, and illustrate how smPCR can help to avoid the problems inherent in the cloning of PCR products.

Protocol

Step 1: DNA Isolation

In a typical smPCR protocol, template DNA is diluted so that, on average, there are about 0.3 amplifiable DNA templates per PCR reaction, *i.e.* approximately one third of PCRs yields a product. As a result, most "positive" reactions have been started with only one template molecule and the PCR products are essentially clones. The high extent of dilution implies that the method of DNA isolation is not important for the success of the procedure, as long as it preserves amplifiable DNA molecules of the appropriate length. We therefore reduce DNA isolation procedure to the lysis of cells or tissue by gentle shaking for a few hours at 37°C in 10 mM EDTA pH 8, 0.5% SDS, 2 mg/ml proteinase K (at least 20 microliters per milligram of protoplasm). DNA does not require any further purification and can be stored frozen at -70°C for years.

Step 2: smPCR Conditions

smPCR is performed in 96-well PCR plates (or pieces thereof) in volumes as small as a three microliters per well. The use of overlaying mineral oil is essential. We found TaKaRa LaTaq amplification system (following the manufacturer's protocol) to be the best choice for this application. The use of high T_m primers 30-40 nucleotides long with G/C content of at least 50% allows to combine the annealing and elongation steps. PCR machines with high ramping rates (*e.g.* Primus from MWG Biothech) are preferred. Long-range smPCR requires the use of two pairs of (nested) primers to ensure robustly high yield of amplification; thus PCR is performed in two steps. Additional reaction mixture for the second step amplification (with the second set of primers and additional amount of polymerase) can be added directly over the oil layer of the first step reaction. The second step mixture (typically twice the volume of the first step reaction) passively sinks and mixes in upon heating of the plate. Obviously, the single-template PCR is highly sensitive to contamination. It is therefore suggested that all pre-PCR steps (including tissue processing, cellular DNA and PCR mixture preparation) are performed in a clean facility completely isolated from the room where PCR itself and post-PCR operations (*i.e.* gel electrophoresis, PCR purification, etc.) are conducted.

Step 3. Estimation of the Number of Template Copies

To determine the optimal conditions of smPCR, the sample is serially diluted over several orders of magnitude and dilutions are subjected to PCR amplification for a total of about 50 cycles (see comments below) in two steps (*e.g.* 28+22 cycles). It is difficult to predict the optimal total number of cycles, as it depends on the length of the amplifiable DNA, the type of sequence, etc. Empirical approach, as described below, is probably most availing. Upon PCR, samples are run on an agarose gel to determine the presence and the length of the product. Normally, one observes PCR products of the expected length up to a certain dilution and no product at higher dilutions. In theory, the "last positive dilution" should have contained one or just a few DNA templates. The optimal template concentration lies somewhere within the interval between the "last positive" and "first negative" dilution.

To more precisely pinpoint the optimal concentration, multiple (30 or more) PCR reactions are run at an intermediate DNA concentration from within that interval, and positive wells are counted. Normally, the numbers of positive and negative wells should be comparable at this point, otherwise concentration should be increased (if there are no positives), or decreased (if there are no negatives) ~30-fold. The fraction of positive wells approximates the average number of templates per well, though a correction based on Poisson distribution (Table 1) should be used at higher fractions of positives. DNA concentration is then readjusted to about 0.3 templates per well and used to perform a large number of smPCRs (depending on the statistical requirements of a particular experiment). It is helpful to bear in mind that 95% confidence limits on the number of wells are about twice the square root of that number. Individual reactions can be used directly for restriction analysis, or, upon removal of primers, for direct sequencing.

Optimization of PCR Conditions and the Number of Cycles

It is imperative that the last positive dilution produces large amount of high quality PCR product. This guarantees that single templates are efficiently amplified under the chosen PCR conditions. PCR reaction should have reached or be close to a plateau, but the product should not be too "overamplified". If a high molecular weight smear or a product failing to enter the gel is dominating the reaction, then the number of cycles is too high. On the other hand, if PCR product of the last positive dilution forms a weak band barely exceeding the background,

Table 1. Poisson distribution for quantitation of the concentration of templates in smPCR*.

Fraction of positives	0.1	0.2	0.3	0.4	0.5	0.6	0.7	0.8	0.9	0.95
Templates per well	0.11	0.22	0.36	0.51	0.69	0.92	1.20	1.61	2.30	3.00
Fraction of multiplets	0.01	0.02	0.05	0.09	0.15	0.23	0.34	0.48	0.67	0.80

* Example of use: if 24 of 30 reactions in a smPCR experiment are "positive" (*i.e.* contain a product of the appropriate length), then one should refer to the column with the "fraction of positives" of 24/30 = 0.8. The corresponding number of 1.61 templates per well implies that the reaction mixture used in the experiment actually contained a total of 1.61 x 30 = 48 template molecules (not 24, as one might have concluded without Poisson correction). The expected fraction of multiplets of 0.48 implies that about 30 x 0.48=~14 wells in this experiment are expected to be "multiplets", *i.e.* to contain mixtures of products originating from two or more different initial templates.

then either the number of cycles is too low or PCR conditions need to be otherwise optimized. The result of smPCR in each well should be a clear yes or no. The absence of a clear difference between positive and negative wells (*i.e.* signal/background ratio well above 10) may indicate that the "last positive dilution" contains many more than just a few templates. Alternatively, the "first negative dilution" may be not really negative, but rather is only perceived as such because PCR is unable to sufficiently increase the concentration of the product and make it clearly visible.

Example 1. Single Molecule PCR as a Means to Avoid PCR Errors

To estimate the error rate of smPCR, we retrospectively analyzed sequences of human mtDNA from a 29- and a 39- year old subjects amplified using smPCR. About 100 smPCR products yielded a total of about 60 kb of high quality sequence, which included potentially mutagenic homopolymeric sequences. Mutations were searched for by multiple alignment of sequences using the online alignment tool at http://searchlauncher.bcm.tmc.edu/multi-align/multi-align.html. One smPCR product carrying a mutation was found in this 60-kb dataset. It should be noted that this single mutation is not necessarily a PCR error but may rather be an *in vivo* lesion that, at this frequency, might well be plausible in a 29- year old person. Thus this likely is an over-estimate of the error rate. In other words, the error rate of the procedure is on the order of 1.6×10^{-5} errors/nucleotide or less, which is lower than the rates reported for cloning of PCR products, *i.e.* 10^{-4} (Lin *et al.*, 2002), 6.3×10^{-5} (Del Bo *et al.*, 2003), and 5.5×10^{-5} (Kamiya and Aoki, 2003). Note that these latter numbers may be underestimates, because these studies did not ensure that control DNA was subject to a sufficient number of PCR duplications.

The error rate of the conventional cloning of PCR products is higher than that of smPCR because in the former, every *in vitro* polymerase error become indistinguishable from genuine *in vivo* mutations upon cloning since each of the errors affects 100% molecules of a clone, just as any genuine *in vivo* mutation does. In contrast, in smPCR, any *in vitro* polymerase error, even if introduced in the very first cycle of smPCR, will only affect the daughters of the nascent DNA strand in which the error happened, *i.e.* a half (single-stranded initial template molecule), or a quarter (double-stranded template) of all molecules to be synthesized in the corresponding PCR. Individual mutational events that happen at subsequent cycles will affect exponentially smaller fractions of molecules. Only genuine *in vivo* mutations will affect 100% of the corresponding smPCR products. Thus, if one intentionally disregards all mutations that constitute about 50% or less of a smPCR clone, then smPCR procedure would be essentially PCR error-resistant, because all PCR-derived mutations will be efficiently filtered out. Note that such filtering is performed by default by any sequence analysis software: a nucleotide position containing less than 50% admixture of a second (mutant) nucleotide is usually interpreted as a non-mutant or at most as an unreadable one. The error rate of smPCR reported above includes this kind of filtering.

Example 2. smPCR Averts Template Jumping in Linkage Analysis of Multiple mtDNA Mutations

Here we describe an example of smPCR-based linkage analysis within a population of mtDNA of a single cell containing a mixture of two mutations. Methodologically, this example is interesting in that linkage analysis is obviously highly sensitive to the template jumping. Such an analysis is entirely dependent on the ability of smPCR to avert this kind of artifact.

In this example, mtDNA from muscle of a person with mild symptoms of a mitochondrial disorder, which contained a high fraction (30-40%) of at least two A to G mutations at positions 189 and 3243 of the mitochondrial genome was used. Single cell analysis showed that most cells contained a mutation at either a relatively high fraction or lacked it completely, consistent with our earlier studies (Nekhaeva *et al.*, 2002). A few cells, however, contained a mixture of both mutations and of the wild type, which implied that heterologous sequences were present side by side within a single cell, and thus potentially offered a possibility to study intracellular interactions of mitochondrial genomes.

Table 2. Distribution of haplotypes of individual mtDNA molecules amplified from a single muscle fiber carrying multiple mutations.

Haplotype	189A/3243A	189G/3243A	189A/3243G	189G/3243G
Number of molecules	16 (43%)	0 (0%)	6 (16%)	15 (41%)

One of such cells containing about 40% and 60% of the mutations 189G and 3243G, respectively, was selected for smPCR analysis to determine the precise distribution of mutations within individual molecules. There could be a few possibilities (and their combinations). The two mutations could have been located on separate molecules, so that neither molecule contains both mutations (60% molecules are 189A/3243G, and 40% molecules are 189G/3243A). Alternatively, mutations could be linked with each other, so that most mutations reside on the same molecules, and an excess of 3243a>g resides on separate molecules (40% 189g/3243g, 20% 189a/3243g and 40% 189a/3243a). In another scenario, mutations could have been distributed randomly so that all four possible combinations are distributed accordingly to statistical expectations.

DNA from a section of a single muscle fiber was isolated as described (Khrapko *et al.*, 1999), and subjected to 4 kb-long smPCR. Thirty six individual molecules were amplified and analyzed using digestion with mutation-specific restriction endonucleases. The results of the analysis are presented in Table 2. Note that the distribution of genotypes is markedly asymmetric. One of the four possible combinations of mutations (189g/3243a) is completely absent. This asymmetry may reflect the history of the introduction of mutations in this cell. For example, the 3243g mutation could have been the first to occur. After mtDNA molecules carrying this mutation expanded to a significant fraction of the cell, one of them could have acquired an additional 189g mutation. The double mutant molecule then could have been successful in expansion, ultimately creating the observed distribution of mutations. In this scenario, no exchange of genetic material (recombination) between mtDNA molecules is assumed. Alternatively, the observed distribution may reflect selection against the 189g/3243a genome in a cell with relatively frequent recombination between heterologous molecules. Other scenarios are also possible. Analysis of a larger number of cells is necessary to determine the mechanism responsible for the asymmetry.

Whatever the interpretation of the data, from the methodological point of view, this experiment demonstrates the possibility of very subtle type of molecular analysis using smPCR. One is not only able to separate individual mutations on their carrier molecules, but also to preserve the native linkage relationship (linkage phase) between the mutations. Note that jumping PCR would have erased the asymmetry observed in this example. Indeed, we observed intensive template jumping resulting in erasing of linkage phase between distant mutations in an experiment involving cloning of PCR products (unpublished), which forced us to turn to smPCR in the studies of the linkage phase.

On the face of it, template jumping is not expected to be of a concern in smPCR, because even if jumping does occur between identical molecules of a clone, that does not result in a new molecular structure. This argument however, holds only for the later stages of smPCR, when most of DNA present is monoclonal. The situation is more complex at the earlier stages of PCR. The single molecularity of smPCR refers to a single *amplifiable* molecule (which means a continuous DNA molecule with no impassable adducts/modifications). Additional "non-amplifiable", broken/modified molecules may be present and may even represent a majority of DNA molecules during the initial PCR cycles. Jumping between the amplifiable template and these "non-amplifiable" templates or between different "non-amplifiable" templates could result in amplifiable recombinant molecules (Pääbo *et al.*, 1990). Fortunately, the conditions of extreme dilution in smPCR apparently preclude any jumping from occurring during the first PCR cycles. Jumping is a bimolecular reaction with respect to template concentration and thus is efficiently inhibited by template dilution. In our hands, no template jumping was observed between markers located 2 kb apart in 150 smPCR clones derived from a restriction digested sample with at least 100-fold ratio of "non-amplifiable" to amplifiable templates. Even though template jumping during smPCR appears rather unlikely, appropriate controls are due, especially if the quality of DNA is low and high proportion of "non-amplifiable" templates is anticipated.

As a first control, the number of positive wells in smPCR should be proportional to the concentration of the input DNA, consistent with monomolecularity of the non-jumping PCR. As a further control, a clearly distinguishable heterologous reference template may be added to the sample and the smPCR products can be checked for composite molecules involving the reference template.

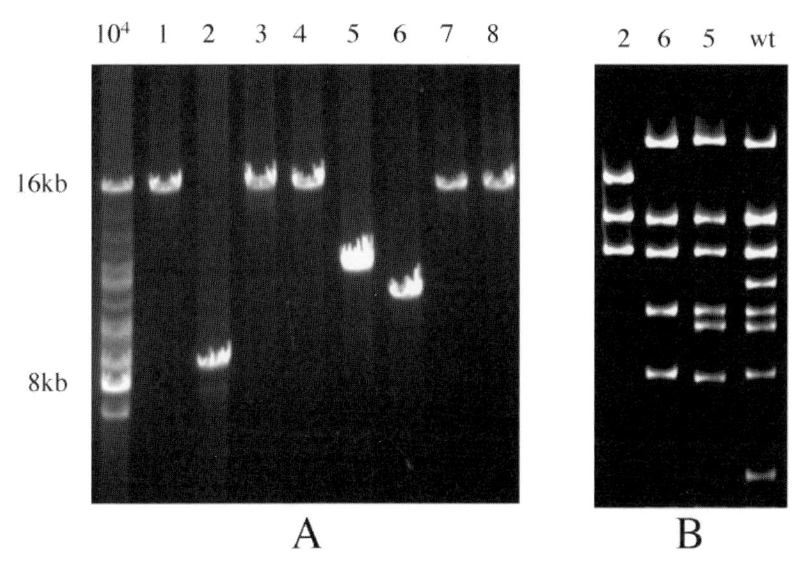

Figure 1. Quantitation and analysis of mtDNA with deletions by smPCR.
A: Comparison of conventional PCR and smPCR. The leftmost lane ("10^4") shows a conventional PCR from about 10^4 molecules of s. nigra mtDNA. In addition to the band corresponding to the full-length wild type mtDNA (16kb) there is a complex mixture of various DNA fragments with deletions. It is tempting to conclude from the relative band intensity that fragments with deletions (bands below the wild type), especially the shorter ones, greatly outnumber the wild type in this sample. The results of several smPCRs of the same DNA sample presented in lanes 1 to 8 (deliberately overloaded) demonstrate that such a conclusion is incorrect. The 8 positive reactions from a typical 24-reaction smPCR experiment are shown. Most reactions contain pure wild type DNA fragment, which thus appears to be the predominant species in the DNA sample. Even products with deletions are mostly long DNA fragments. This implies that the bias in favor of the short mutant mtDNA characteristic of conventional PCR has been eliminated in smPCR.
B: Restriction analysis of the smPCR products. Lane numbers correspond to those in (A); wt is the restriction pattern of the full-length wild type mtDNA. The distinct restriction pattern in each lane proves that the smPCR products with deletions are pure clones; the absence of particular bands present in the wild type pattern can be used to locate the breakpoints of the deletions.

Example 3. Quantification of Deletions in mtDNA: smPCR Eliminates Allelic Preference

Deletions in mtDNA are associated with mtDNA diseases and are likely to play a role in the aging process. Quantification of deletions in mtDNA is a long-standing problem, which has generated voluminous literature. The problem is that whenever one tries to use PCR to amplify a naturally occurring complex mixture of wild type and deleted species of various lengths, the differences in amplification efficiencies result in over-representation of the shortest species, which are amplified with much higher efficiency than the long species and, most notably, the full-length wild type mtDNA (Kajander *et al.*, 1999a). smPCR offers a straightforward way to solve this problem by separating potential competitors into different wells and subjecting them to saturating number of PCR cycles. In addition, this approach automatically provides individual deleted DNA fragments suitable for further analysis.

To estimate the fraction of deleted mtDNA in human substantia nigra, DNA was isolated from an autopsy frozen section of the substantia nigra of an elderly person (81 years old). Substantia nigra is known to accumulate a large proportion of mitochondrial genomes with deletions with age (Soong *et al.*, 1992). DNA was subjected to both conventional PCR (starting with about 10^4 template molecules) and smPCR. The results of this experiment are presented in Figure 1. The data show that smPCR is capable of both averting allelic preference typical of conventional PCR (Figure 1A), and providing mtDNA fragments with deletions in a pure form suitable for further analysis (Figure 1B).

Concluding Remarks

Single molecule PCR offers a simple and efficient approach for the molecule-by-molecule mutational analysis capable of circumventing the various PCR-related artifacts. It is therefore surprising that it is not widely used. We believe that the low popularity of the approach is related to the common belief that single molecule PCR is a technically difficult and unreliable procedure. In our experience, smPCR is a straightforward and robust approach even when long DNA fragments are involved.

Acknowledgements

This work has been supported in part by NIH grants ES 11343 and AG 19787.

References

Benzer, S., and Freese, E. (1958). Induction of Specific mutations with 5-Bromouracil. Proc. Natl. Acad. Sci. USA. *44*, 112.

Bodenteich, A., Mitchell, L. G., and Merril, C. R. (1991). A lifetime of retinal light exposure does not appear to increase mitochondrial mutations. Gene *108*, 305-310.

Del Bo, R., Crimi, M., Sciacco, M., Malferrari, G., Bordoni, A., Napoli, L., Prelle, A., Biunno, I., Moggio, M., Bresolin, N., *et al.* (2003). High mutational burden in the mtDNA control region from aged muscles: a single-fiber study. Neurobiol. Aging *24*, 829-838.

Jazin, E. E., Cavelier, L., Eriksson, I., Oreland, L., and Gyllensten, U. (1996). Human brain contains high levels of heteroplasmy in the noncoding regions of mitochondrial DNA. Proc. Natl. Acad. Sci. USA. *93*, 12382-12387.

Jeffreys, A. J., Neumann, R., and Wilson, V. (1990). Repeat unit sequence variation in minisatellites: a novel source of DNA polymorphism for studying variation and mutation by single molecule analysis. Cell *60*, 473-485.

Jeffreys, A. J., Wilson, V., Neumann, R., and Keyte, J. (1988). Amplification of human minisatellites by the polymerase chain reaction: towards DNA fingerprinting of single cells. Nucleic Acids Res. *16*, 10953-10971.

Kajander, O. A., Kunnas, T. A., Perola, M., Lehtinen, S. K., Karhunen, P. J., and Jacobs, H. T. (1999a). Long-extension PCR to detect deleted mitochondrial DNA molecules is compromized by technical artefacts. Biochem. Biophys. Res. Commun. *254*, 507-514.

Kamiya, J., and Aoki, Y. (2003). Associations between hyperglycaemia and somatic transversion mutations in mitochondrial DNA of people with diabetes mellitus. Diabetologia *46*, 1559-1566.

Khaidakov, M., Heflich, R. H., Manjanatha, M. G., Myers, M. B., and Aidoo, A. (2003). Accumulation of point mutations in mitochondrial DNA of aging mice. Mutat. Res. *526*, 1-7.

Khrapko, K., Bodyak, N., Thilly, W. G., van Orsouw, N. J., Zhang, X., Coller, H. A., Perls, T. T., Upton, M., Vijg, J., and Wei, J. Y. (1999). Cell-by-cell scanning of whole mitochondrial genomes in aged human heart reveals a significant fraction of

myocytes with clonally expanded deletions. Nucleic Acids Res *27*, 2434-2441.
Kovalenko, S. A., Tanaka, M., Yoneda, M., Iakovlev, A. F., and Ozawa, T. (1996). Accumulation of somatic nucleotide substitutions in mitochondrial DNA associated with the 3243 A-to-G tRNA(leu)(UUR) mutation in encephalomyopathy and cardiomyopathy. Biochem. Biophys. Res. Commun. *222*, 201-207.
Lin, M. T., Simon, D. K., Ahn, C. H., Kim, L. M., and Beal, M. F. (2002). High aggregate burden of somatic mtDNA point mutations in aging and Alzheimer's disease brain. Hum. Mol. Genet. *11*, 133-145.
Lukyanov, K. A., Matz, M. V., Bogdanova, E. A., Gurskaya, N. G., and Lukyanov, S. A. (1996). Molecule by molecule PCR amplification of complex DNA mixtures for direct sequencing: an approach to *in vitro* cloning. Nucleic Acids Res. *24*, 2194-2195.
Michikawa, Y., Mazzucchelli, F., Bresolin, N., Scarlato, G., and Attardi, G. (1999). Aging-dependent large accumulation of point mutations in the human mtDNA control region for replication. Science *286*, 774-779.
Monnat, R., Jr., and Loeb, L. A. (1985a). Nucleotide sequence preservation of human mitochondrial DNA. Proc. Natl. Acad. Sci. USA. *82*, 2895-2899.
Nekhaeva, E., Bodyak, N. D., Kraytsberg, Y., McGrath, S. B., Van Orsouw, N. J., Pluzhnikov, A., Wei, J. Y., Vijg, J., and Khrapko, K. (2002). Clonally expanded mtDNA point mutations are abundant in individual cells of human tissues. Proc. Natl. Acad. Sci. USA. *99*, 5521-5526.
Paabo, S., Irwin, D. M., and Wilson, A. C. (1990). DNA damage promotes jumping between templates during enzymatic amplification. J. Biol. Chem. *265*, 4718-4721.
Pääbo, S., Irwin, D. M., and Wilson, A. C. (1990). DNA damage promotes jumping between templates during enzymatic amplification. J. Biol. Chem. *265*, 4718-4721.
Ruano, G., Kidd, K. K., and Stephens, J. C. (1990). Haplotype of multiple polymorphisms resolved by enzymatic amplification of single DNA molecules. Proc. Natl. Acad. Sci. USA. *87*, 6296-6300.
Simon, D. K., Lin, M. T., Ahn, C. H., Liu, G. J., Gibson, G. E., Beal, M. F., and Johns, D. R. (2001). Low mutational burden of individual acquired mitochondrial dna mutations in brain. Genomics *73*, 113-116.
Soong, N. W., Hinton, D. R., Cortopassi, G., and Arnheim, N. (1992). Mosaicism for a specific somatic mitochondrial DNA mutation in adult human brain. Nat. Genet. *2*, 318-323.
Vogelstein, B., and Kinzler, K. W. (1999). Digital PCR. Proc. Natl. Acad. Sci. USA. *96*, 9236-9241.
Yauk, C. L., Bois, P. R., and Jeffreys, A. J. (2003). High-resolution sperm typing of meiotic recombination in the mouse MHC Ebeta gene. Embo J. *22*, 1389-1397.
Zhu, J., Shendure, J., Mitra, R. D., and Church, G. M. (2003). Single molecule profiling of alternative pre-mRNA splicing. Science *301*, 836-838.

2.4

Digital PCR Analysis of Allelic Status in Clinical Specimens

Wei Zhou, Tanisha Williams, Cecile Colpaert, Aki Morikawa and Diansheng Zhong

Abstract

Digital PCR is a quantitative method for the analysis of known genetic alterations existing in a fraction of particular cells. To this end, genomic DNA is isolated and diluted to such extent that single molecules can be individually amplified in separate asymmetric PCRs. The presence or absence of genetic alterations is then determined by fluorescent probes. This method can be expediently used in a range of biotechnological and clinical applications. For example, it might be employed for detection of various mutations during carcinogenesis, such as point mutations, gene amplifications or changes in gene expression. Here, we describe a protocol for allelic imbalance quantitation in primary tumors using digital PCR. It is called digital SNP analysis, which stands for digital PCR based single-nucleotide polymorphism analysis.

Background

Cancer is a genetic disease that is characterized by accumulation of multiple genetic alterations in oncogenes and tumor suppressor genes during tumorigenesis (Vogelstein and Kinzler, 1998). Identification of genetic mutations responsible for each tumor type is essential not only for the basic understanding of the disease but also for the development of successful treatments. Compared to normal DNA, the tumor genome is a dynamic, unstable entity constantly undergoing changes due to defects in the DNA repair or mitotic checkpoint machinery (Lengauer *et al.*, 1997, 1998). Although much of the genetic change may be nonspecific alterations, some of them are causative because they lead to the activation of oncogenes, such as *ras* and *myc*, or to the inactivation of tumor suppressor genes, which include *RB1*, *p53*, *WT1*, *APC*, *BRCA1* and *BRCA2*.

Many tumor suppressor genes are identified by loss of heterozygosity (LOH) studies. In primary tumors, however, the analysis of LOH is complicated by a number of technical problems that limit its reproducibility and interpretation. In particular, primary tumors are typically the mixtures of neoplastic and normal cells, so that DNA from the normal non-neoplastic cells can mask LOH.

Digital SNP analysis is a novel method for allelic imbalance studies in primary tumors derived from digital PCR (Vogelstein and Kinzler, 1999). Digital PCR is based on the following principles. First, single-molecule PCR amplification is possible when genomic DNA is appropriately diluted (Monckton and Jeffreys, 1991; Lukyanov *et al.*, 1996; Mitra and Church, 1999; Mitra *et al.*, 2003). This means that PCR of multiple highly-diluted aliquots of the same sample of genomic DNA will separately amplify individual DNA molecules representing different alleles of the heterozygous genome. Thus, even though the starting DNA sample contains a mixture of allele-specific molecules, most of the resultant PCR products are individually homogenous (part of them still will be assorted) but each of them belongs to one of the two different types and can, therefore, be easily distinguished with variety of techniques. This method was originally used for the detection of a mutant *ras* oncogene which was present in only 3.9% of c-Ki-Ras alleles in the stool samples from a colorectal cancer patient (Vogelstein and Kinzler, 1999).

We have additionally incorporated several useful features into digital PCR, thus making it more suitable for allelic imbalance analysis in archived clinical specimens (Zhou, 2003).

1. Single nucleotide polymorphic (SNP) markers are used instead of microsatellites. These markers are single-base alterations thus not susceptible to size-based amplification bias usually associated with the PCR amplification of microsatellites. In addition, small PCR products (≤100 bp) can be designed for each polymorphic marker, and the amplification of small products is robust in paraffin samples in spite of DNA degradation.
2. The detection of different alleles of each SNP locus is carried out with the SNP-specific molecular beacons (MBs). MBs are fluorogenic DNA probes featuring the stem-loop structures tagged by fluorophore and quencher (Tyagi and Kramer, 1996; Tyagi *et al.*, 1998; Marras *et al.*, 1999) and brightly fluorescing upon hybridization (see Figure 1). Use of MBs makes it possible to readily detect SNPs in the close-tube homogeneous format without employment of radioisotopes and/or gel electrophoresis. In addition, the stem-loop structure of MBs allows to get a much better match/mismatch discrimination (as compared to conventional linear oligonucleotide probes), which is important in the SNP analysis (Bonnet *et al.*, 1999; Broude, 2002).
3. Since SNPs are generally bi-allelic (Brookes, 1999), allelic status of a sample is quantitatively determined by simultaneously using a pair of allele-specific MBs with two different fluorophores and directly counting the number of selectively ‘colored’ individual alleles cloned by digital asymmetric PCR. Normal tissues and tumors without allelic loss will have approximately equal counts for both alleles. Tumor samples with distinct allelic loss will have different counts for each allele even if they are contaminated by regular cells.

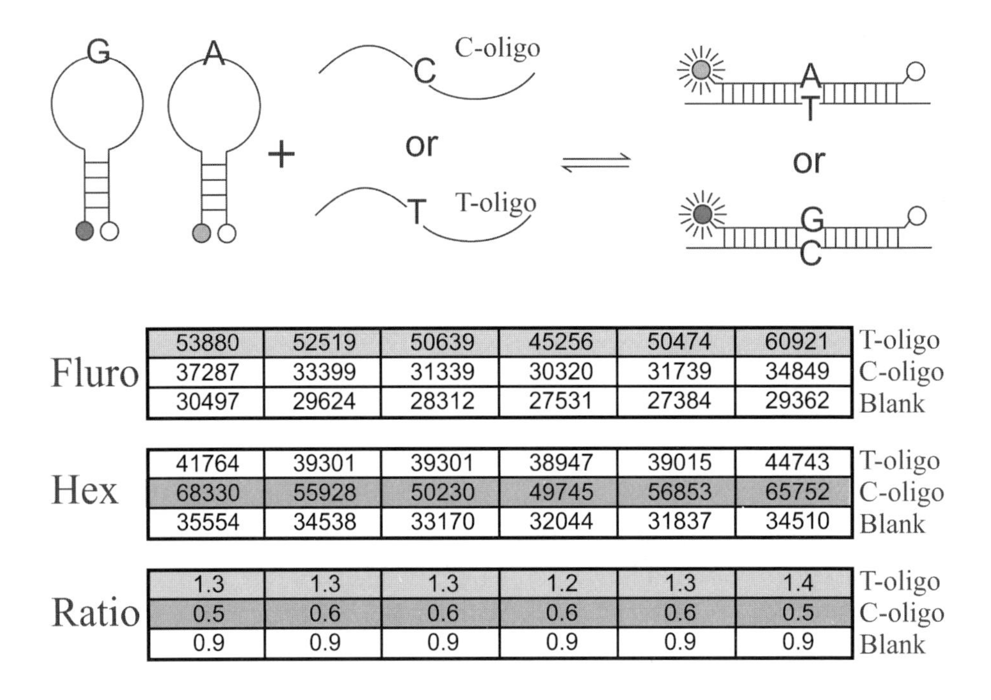

Fluro	53880	52519	50639	45256	50474	60921	T-oligo
	37287	33399	31339	30320	31739	34849	C-oligo
	30497	29624	28312	27531	27384	29362	Blank
Hex	41764	39301	39301	38947	39015	44743	T-oligo
	68330	55928	50230	49745	56853	65752	C-oligo
	35554	34538	33170	32044	31837	34510	Blank
Ratio	1.3	1.3	1.3	1.2	1.3	1.4	T-oligo
	0.5	0.6	0.6	0.6	0.6	0.5	C-oligo
	0.9	0.9	0.9	0.9	0.9	0.9	Blank

Figure 1. Evaluation of SNP-specific molecular beacons. SNP-specific MBs and two target oligonucleotides (T-oligo and C-oligo) were designed for detecting polymorphic SNP allele. SNP-A MB is labeled with fluorescein and SNP-G MB is labeled with Hex. Reaction mixtures contained both MBs and one of the target oligonucleotides (T-oligo or C-oligo) or blank controls. Reactions were done in replicates of six, and fluorescent signals were measured for both fluorescein and Hex. Signal ratios (r) were calculated by dividing the fluorescent signal by the Hex signal (for T-oligo $r \geq 1.2$; for C-oligo $r \leq 0.6$). Consistency of r values for blank samples serves as internal control.

4. The presence or absence of allelic imbalance in a sample is unambiguously established by a statistical approach, sequential probability ratio test (SPRT; Wald, 1947). This test is able to ascertain whether a specific tumor specimen exhibits the allelic loss with up to 50% normal tissue contamination.

In this chapter we describe a detailed protocol for this method, called digital SNP analysis, and show its advantages over other techniques of allelic status checkup.

Protocol

Materials

Genomic DNA was isolated from human specimens with a QIAquick Kit (Qiagen). 10x PCR buffer [200 mM Tris-HCl (pH 8.4), 500 mM KCl, 50 mM $MgCl_2$], 10 mM dNTPs and recombinant Taq DNA polymerases were purchased from Invitrogen. DMSO was purchased from Sigma.

SNPs and PCR Primers

SNPs from a defined genomic region can be identified from the NCBI dbSNP database (www.ncbi.nlm.nih.gov/SNP). Since SNPs are located in the middle of the amplicon, both alleles can be amplified by the same PCR primers. As most archived clinical specimens may be degraded to a various extent, choice of primers that will yield the PCR amplicons ≤100 bp is recommended. In the example presented here, we amplifed the 60-87 bp DNA fragments corresponding to seven different SNP regions on human chromosome 8p (see Table 1).

SNP-Specific Molecular Beacons

Figure 1 schematically depicts the detection of each of the two SNP alleles by MBs (all MBs are listed in Table 1). Sequences of the SNP-specific MBs are identical except for the polymorphic nucleobase base representing the SNP (for example, an A for the paternal T allele and a G for the maternal C allele). One MB is labeled with fluorescein and the other is labeled with hexachloro-6-carboxyfluorescein (Hex); Dabcyl is used as quencher. The stem portion of MBs should be 4 to 5 bp in length, and the stem 5'-CACG/5'-CGTG is usually used in our MBs. The loop portion of MBs can be between 17 to 23 bp in length, and the melting temperature (T_m) of this sequence, when complexed with its complement, should be ~50°C. To maintain a consistent stem length, the first and the last nucleotides in the loop should not be complementary to each other. Design of MBs can be simplified

Table 1. SNPs and Molecular Beacons.

	SNPs	Location	Forward Primer	Reverse Primer
	CH. 8p			
1	WIAF-1085	8p23.1	CACTGAATGCTCTGCCATGA	AACCTGTCCTTGTGGGTGAT
2	WIAF-1622	8p21.2	GAGGGACCCGGCCTTCCTGG	GAGCGGCACCTCGCCACACA
3	WIAF-1044	8p23.2	GTGTTTTTATCAATGATGCCG	GGAAGTGTTTGATGGTAAGTTTCTTT
4	WIAF-1596	8p21.3	CCCACTTGGGTCTCTTTCAA	AAAAGGCTTTAGAACAGGAACG
5	WIAF-3833	8p22	GGAGAGGGGGCTGCTTAG	GGTCTGATGTCAGTGAACGG
6	TSC0032628	8p21.3-8p22	GGCCCCAGTCGGCATTAAAT	GGACTAGCCGCACCCAAAG
7	TSC0028254	8p21.3-8p22	GAAATCAGGTTAGTTTAGTCTACAGC	TTGCCTGAAGCTACCTCCAT

	Flouorescein Beacon	Hex Beacon	PCR product (bp)
1	CACGATGAGCCACAAGCAGCACGTG	CACGATGAGCCGCAAGCAGCCGTG	60
2	CACGGCCGTGAACACCCTGTCGTG	CACGGCCGTGAGCACCCTGTCGTG	61
3	CACGGGTCACTGCTATACAAAGATTCGTG	CACGGGTCACTGCTCTACAAAGATTCGTG	69
4	CACGCAAGTGAATGTTCCTTTCGTCGTG	CACGCAAGTGAATTTTCCTTTCGTCGTG	56
5	CACGTTGCTCTTCAATTCCGTTCCGTG	CACGTTGCTCTTCAGTTCCGTTCCGTG	72
6	CACGGAACAACTGTTGAATGATGGCGTG	CACGGAACAACTGCTGAATGATGGCGTG	87
7	CACGCGATGACTACTCTCTCTTGCCGTG	CACGCGATGATTACTCTCTCTTGCCGTG	76

by using the Beacon Designer software from PREMIER Biosoft International (www.premierbiosoft.com).

Step 1: Evaluating the Specificity of MBs and Their Fluorescent Response to Hybridization

It is crucial that SNP-specific MBs hybridize only to their perfectly matched DNA targets. Specificity of MBs can be tested by hybridizing them with synthetic target oligonucleotides. For example, to test the specificity of MBs for WIAF-1622 SNPs (Table 1), two 25 nt oligonucleotides were synthesized (T-oligo, $^{5'}$GCCACACAGGGT GTTCACGGCCATG$^{3'}$, and C-oligo, $^{5'}$GCCACACAGGGTGCTCA CGGCCATG$^{3'}$). These oligonucleotides are identical except for the polymorphic nucleobase representing SNP (underlined). Hybridization is carried out in replicates of six using a multi-well PCR plate. First, molecular beacons are mixed together and added to the wells containing either T-oligo, or C-oligo, or blank controls. The final reaction volume per well is 3 µl, and the reaction mixture contains 20 mM Tris-HCl (pH 8.4), 50 mM KCl, 6% DMSO, 5 mM $MgCl_2$, 0.15 mM of each MBs and 0.2 mM of target oligonucleotide. The PCR plate is heated to 95°C for 2 min and cooled to 65°C for 5 min. The fluorescence in each well is then determined at room temperature using BIO-TEK FL600 fluorescence plate reader for both fluorescein (490 nm excitation, 520 emission) and Hex (520 nm excitation, 590 nm emission).

Figure 1 provides a typical readout from a control experiment. Background signals are measured in the blank controls and are usually ≤35,000 a.u. (arbitrary units; relative instrumental fluorescence units). Fluorescent signals for fluorescein and Hex are usually ≥50,000 a.u. in the presence of target sequence and are close to background signals in the presence of non-target sequence. The MB responses over background can be identified using a standard cutoff value (*i.e.* average of the blank controls plus three standard deviations). Signal ratios (fluorescent signal from fluorescein divided by fluorescent signal from Hex) are normally very consistent. In case of WIAF-1622 SNPs, they are ≥1.2 for T-oligonucleotides and ≤0.6 for C-oligonucleotidess. These key allele-discriminating parameters should be determined in similar model experiments with known alleles (oligonucleotides) for all MB/ SNP combinations and used subsequently for their differentiation in 'blind' experiments.

Step 2: Choice of a Proper SNP Marker

Allelic imbalance (AI) analysis requires the presence of polymorphic alleles in the tested normal sample. Therefore, prior genotyping of normal tissues by the MB-based digital PCR is necessary to determine

Table 2. Genotyping of normal genomic DNA. PCR primers were designed for the amplification of SNP alleles. Asymmetric PCR reactions were carried out in replicates of six on genomic DNA isolated from normal tissues (Sample 1-3) along with control genomic DNA templates containing A (A-ctrl) allele, G (G-ctrl) allele, or blank control (Blank). Sample 1 and 2 contains only A alleles or only G alleles, respectively. Therefore, this SNP marker is not suitable for allelic imbalance analysis in these two samples. Allelic imbalance analysis can be carried out with this SNP marker for Sample 3 because it contains both A and G alleles in the normal genomic DNA.

Fluor	23041	21714	24647	22239	18403	18406	sample 1
	46904	42539	35023	33396	37830	38377	sample 2
	23389	41804	37827	32969	31559	33341	sample 3
	40144	39900	39420	43745	42967	41664	A-ctrl
	21692	22584	25638	15689	18201	16092	G-ctrl
	13706	14057	14154	14292	16507	15265	Blank
Hex	48005	43351	52888	45283	37660	36329	sample 1
	34782	32410	28080	26942	29804	32197	sample 2
	48869	38990	49363	37253	35815	39262	sample 3
	30390	30271	29707	32764	33112	31928	A-ctrl
	43696	45869	53038	31095	35044	36735	G-ctrl
	18531	18955	18891	19062	21930	22046	Blank
Ratio	0.5	0.5	0.5	0.5	0.5	0.5	sample 1
	1.3	1.3	1.2	1.2	1.3	1.2	sample 2
	0.5	1.1	0.8	0.9	0.9	0.8	sample 3
	1.3	1.3	1.3	1.3	1.3	1.3	A-ctrl
	0.5	0.5	0.5	0.5	0.5	0.4	G-ctrl
	0.7	0.7	0.7	0.7	0.8	0.7	Blank

whether the individual sample (patient) is heterozygous for a specific SNP: only in that case this particular marker will be suitable for the analysis of a chosen sample. Results of a typical test involving the WIAF-1622 SNP are given in Table 2. They demonstrate that this SNP locus can be used for sample 3 but not for samples 1 and 2. Both latter samples require different SNPs to be analyzed for potential allelic imbalance.

PCR Conditions

Since detection of SNP alleles by MBs requires single-stranded target DNA, an asymmetric PCR is performed to generate single-stranded amplicons from genomic DNA. The concentration of genomic DNA needs to be determined in advance, so that on average less than one DNA molecule will be added per well from the properly diluted sample. PCR is carried out in 3 μl reaction volume with 0.3 μM forward primer and 1.2 μM reverse primer in the reaction mixture containing 20 mM Tris-HCl (pH 8.4), 50 mM KCl, 6%DMSO, 5 mM $MgCl_2$, 1mM dNTPs, 0.15 mM of each MBs and 0.12 U recombinant Taq DNA polymerase (Invitrogen). To increase the amount of single-stranded PCR amplicons, the following cycling conditions were used: step 1 - 94°C, 1 min; step 2 - 4 cycles at 94°C for 15 sec, 64°C for 15 sec, 70°C for 15 s; step 3 - 4 cycles at 94°C for 15 sec, 61°C for 15 sec, 70°C for 15 sec; step 4 - 4 cycles at 94°C for 15 sec, 58°C for 15 sec, 70°C for 15 sec; step 5 - 60 cycles at 94°C for 15 sec, 55°C for 15 sec, 70°C for 15 sec; step 6 - 1 cycle at 94°C for 1 min, 70°C for 5 min. The forward primer is usually exhausted after 45 cycles in step 5, and the remaining 15 cycles in this step are used to generate single-stranded PCR products.

Step 3: Quantitation of Allelic Imbalance (AI) in Tumors

DNA is isolated from the microdissected tumor and autologous normal tissue and amplified in separate PCRs. The PCR mixtures are robotically distributed to the wells of a 384-well plate and each well contains 3 μl of final reaction mix. 180 wells each are used for

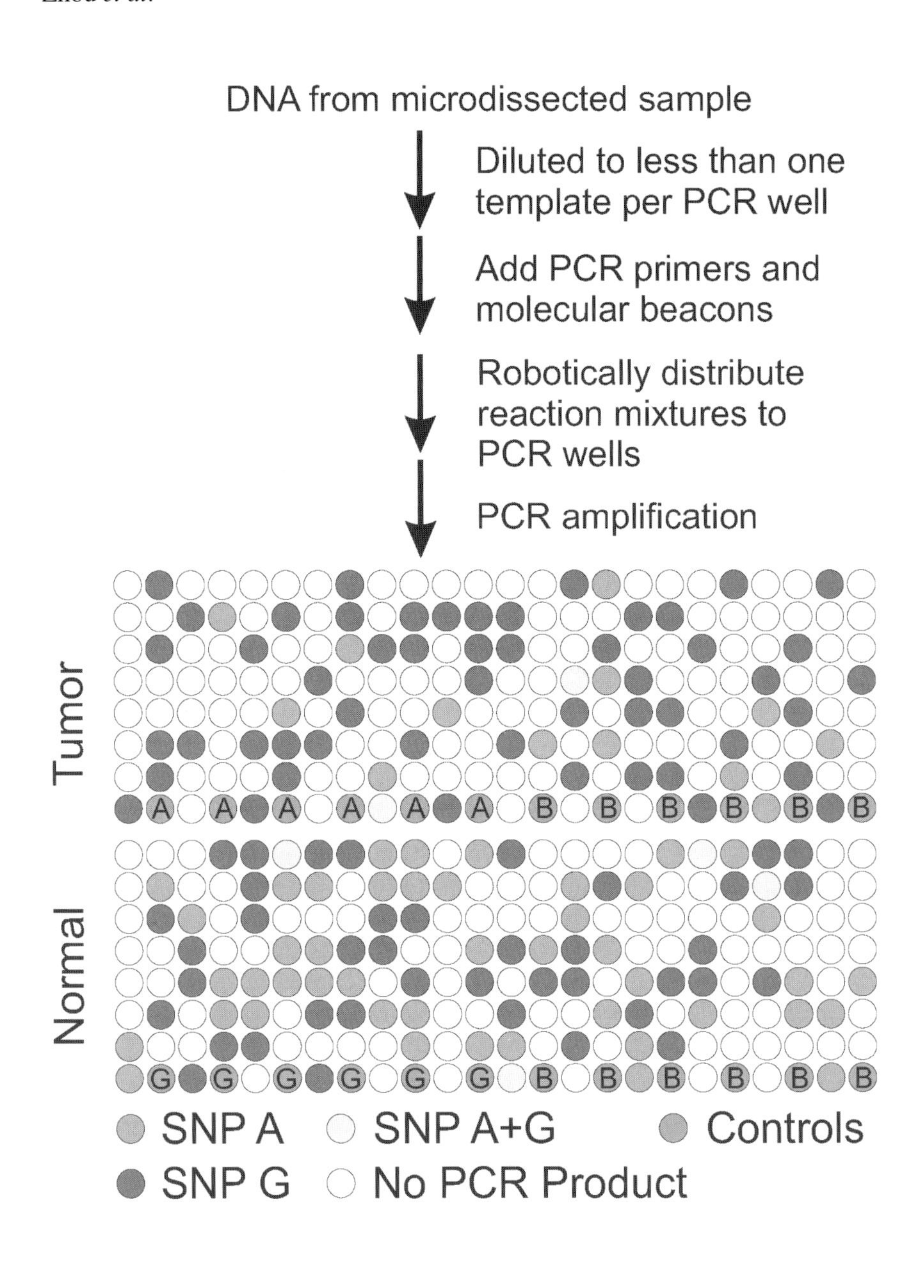

Figure 2. Schematic representation of digital PCR analysis of allelic status. Genomic DNA was isolated from tumor and autologous normal breast tissues and diluted to the extent that single molecules can be individually amplified in separate PCRs. Green and Red wells contain PCR products from only one of the parental alleles. Wells with background fluorescence are colorless and yellow wells contain both alleles (green + red = yellow). Grey wells contain control samples used for confirming the expect value of fluorescence (A for SNP A, G for SNP G, B for blank control). **The colour version of this figure is located in the colour plate section at the back of this book.**

the analysis of tumor or normal DNA; the rest 24 wells are used as blank controls. Then asymmetric digital PCR is performed as in step 2. As a result, the majority of PCR products in each well will represent either maternal or paternal allele only whereas a minor part of them will contain both alleles. After PCR, the fluorescence in each well is determined by fluorometry, and the number of alleles of each type is directly estimated as shown in Figure 1. In a typical case presented in Figure 2, 65 and 84 single alleles were counted in the DNA samples isolated from the tumor and autologous normal breast tissues respectively. The tumor-derived sample contained 13 A-type alleles (green wells) and 52 G-type alleles (red wells). In contrast, the normal tissue yields 40 A-type alleles and 44 G-type alleles. Therefore it can be concluded, that in this case tumor tissue has AI compared to normal tissue containing balanced number of alleles. Statistical analysis of these data finally verifies this conclusion.

Data Analysis and Interpretation

The likelihood technique is used to evaluate the strength of the evidence for AI in each tumor sample. The specific methodology used here is the Sequential Probability Ratio Test or SPRT (Wald, 1947), a method which allows two probabilistic hypotheses to be compared as data accumulate, and allows on average a smaller amount of testing for a given level of confidence than any other method. Hypothesis 1 is that a tumor-derived sample is in allelic balance, *i.e.* the alleles have the same proportion in the tumor cells as they do in normal cells. Hypothesis 2 is that one of the two alleles is lost in every tumor cell. Given the conservative assumption that at least 50% of the DNA from the microdissected samples originated from neoplastic rather than non-neoplastic cells, the hypothesis of allelic imbalance corresponds to the probabilistic hypothesis that in the second case the observed proportion of alleles would be ≥66.7%. A SPRT is therefore constructed to choose between the hypotheses $P = 50\%$ (no allelic imbalance) and $P = 66.7\%$ (allelic imbalance), with a threshold likelihood ratio of 8. This ratio corresponds roughly to a two-sided 95% confidence interval on the allelic proportion that does not include 50%.

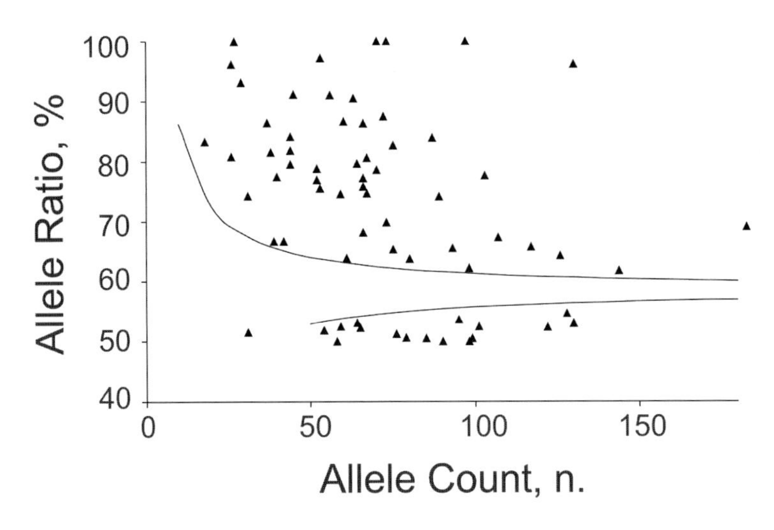

Figure 3. Results of chromosome 8p allelic status determination in tumor samples. Chromosome 8p allelic status in 68 tumor samples was determined by analyzing eight SNP alleles. Sequential Probability Ratio Test (SPRT) was applied to analyze the data. The total number of alleles that were counted in a tumor sample is plotted on the x-axis, and the observed proportion of the two alleles is plotted on the y-axis. All points above the upper boundary (curve 1 described by Eq. 1) were interpreted to have allelic imbalance, while those below the lower boundary (curve 2 described by Eq. 2) were interpreted to be in allelic balance. Each point on the graph represents the result from an individual tumor sample. DNA from adjacent normal tissues was purified from the same slides used to isolate DNA from the tumors and analyzed in identical fashion. In each of the 68 cases, the data points from the normal tissues (not shown) were below curve 2, indicating allelic balance.

The two curves representing these assumptions are plotted along with a scatterplot of the data obtained from the tumor and normal cells (see Figure 3). The equation for the upper boundary of the SPRT, which had to reflect the fact that the proportion for any observed imbalance would always be greater than 50%, is:

$$P_1 = \ln 16/n\ln 2 + \ln 1.5/\ln 2 \cong 4/n + 0.585, \qquad (1)$$

where n is the total number of alleles counted (Royall, 1997; Wald, 1947). The equation for the lower boundary of the SPRT is:

$$P_2 = \ln 1.5/\ln 2 - \ln 16/n\ln 2 \cong 0.585 - 4/n. \qquad (2)$$

For each case, the total number of alleles counted in each sample is plotted on the abscissa and the ratio of wells containing the allele with the higher counts to the total number of wells containing either allele is plotted on the ordinate (Figure 3). Tumors represented by points above curve 1 (Eq. 1) were interpreted to have allelic loss, meaning that the likelihood ratio for $P = 66.7\%$ versus $P = 50\%$ exceeded 8.0. Those below the curve 2 (Eq. 2) were categorized as having no allelic loss.

Example: Evaluation of Chromosome 8p Allelic Imbalance in Early-Stage Breast Cancer

Since chromosome 8p is commonly lost in breast cancer (Anbazhagan *et al.*, 1998; Seitz *et al.*, 1997; Utada *et al.*, 2000), we chose to evaluate the frequency of chromosome 8p allelic imbalance in breast cancer. We have designed eight SNP-specific molecular beacon pairs for SNP markers on chromosome 8p (Table 1) and carried out allelic imbalance analysis for 68 early-stage, node-negative breast primary tumors (Colpaert *et al.*, 2001). We focused analysis on DNA from patients with early stage disease without lymph node involvement or distant metastasis at the time of the surgery. To simplify data interpretation, we decided to include breast cancer patients with only local treatment but not systemic therapies.

Allele analysis was carried out on all patients with SNP markers on 8p. Each SNP marker is usually informative for 25-35% of cases. Sixty eight cases were informative for at least one SNP markers on chromosome 8p. Of these sixty eight patients studied, twenty two were younger than 50 years old, and forty-six were 50 years or older. Twenty-four percent (16/68) of the tumors were grade III, 29% (20/68) were grade II, and 47% (32/68) were grade I tumors. Forty-three percent (29/68) of tumors were infiltrative types and 57% (39/68) were expansive types. Twenty-one percent (14/68) of tumors had necrosis, and 56% (38/68) of these tumors were less than 2 cm in size. Therefore, the baseline characteristics of these cases are typical of early stage, node-negative breast cancers.

Table 3 Clinical Characteristics of Breast Cancer based on chromosome 8p allelic status

Variable		Total number of patients	Patients without chromosome 8p AI		Patients with chromosome 8p AI		P value*
			Number	Percentage	Number	Percentage	
Overall		68	17	25%	51	75%	
Grade							
	I	32	12	71%	20	39%	p=0.05
	II	20	4	24%	16	31%	
	III	16	1	6%	15	29%	
Histology	expansive	39	8	47%	31	61%	n.s.
	infiltrative	29	9	53%	20	39%	
Necrosis	No	54	15	88%	39	76%	n.s.
	Yes	14	2	12%	12	24%	
Size	< 2 cm	38	10	59%	28	55%	n.s.
	≥ 2 cm	30	7	41%	23	45%	
Age	< 50	22	4	24%	18	35%	n.s.
	> 50	46	13	76%	33	65%	

* Based on Fisher's Exact test

As 75% of all studied tumors contained chromosome 8p allelic imbalance (Figure 3), we concluded that allelic imbalance of chromosome 8p is a common event in early-stage, node negative breast tumors. We have also found that allelic status of chromosome 8p significantly correlated with pathological tumor stage (Table 3). Indeed, 71% (12/17) tumors without AI of 8p were grade I tumors, while only 39% (20/51) with 8p AI were grade I tumors ($p < 0.05$, Fisher's exact test). On the other hand, 6% (1/17) tumors without AI of 8p were grade III tumors, while 29% (15/51) with 8p AI were grade III tumors. Thus, tumors without 8p allelic imbalance were more likely to have a low histological grade. The data in Table 2 show that 8p allelic imbalance, however, did not correlate with infiltrative or expansive histological type, necrosis, tumor size or the age of onset of the disease.

Previous studies indicated that chromosome 8p allelic imbalance occurs in 52-60% of breast tumors (Anbazhagan *et al.*, 1998; Utada *et al.*, 2000). Our study, however, indicated that 8p AI occurred at a much high frequency. It is possible that digital SNP analysis specifically designed for archived clinical specimens, is thus more sensitive in detecting allelic imbalance in such tissues. For more examples of digital SNP analysis, see our earlier published works studying SNPs of human chromosomes 8p and 18q from colorectal cancer tissues (Zhou *et al.*, 2001, 2002).

Discussion

Digital SNP analysis incorporates several features that significantly enhance the accuracy of testing AI in archived clinical specimens, which increases the likelihood that any genuine correlation between genetic aberration and malignant potential will be properly identified. For example, our study with 78 early-stage colorectal cancer patients revealed a strong correlation between 18q loss and vascular invasion in node-negative tumors (Zhou *et al.*, 2001). Subsequent analysis with 198 colorectal cancer patients indicated that early-stage patients could be stratified into three groups with 0%, 27% and 48% chance of recurrence based on allelic status of chromosomes 8p and 18q (Zhou *et al.*, 2002). Note that patients with the early-stage colorectal cancer typically have a good prognosis. However, 30% of these patients develop recurrences and die from the disease. The ability to predict the probability of recurrence based on allelic status of 8p and 18q would greatly benefit the patient and would help the physician in planning for future treatment. It is also important to note that numerous LOH studies have been carried out with traditional methods in colorectal cancer over the past eight years, which did not allow such predictions. On the contrary, our approach allowed us to identify specific associations between allelic loss and certain aspects of tumorigenesis pertinent to prognosis. We believe that the key to our success is the accurate assessment of genetic alterations in clinical specimens through an unbiased and quantitative experimental approach.

Digital SNP analysis can also be used for the analysis of early and small lesions. Because of genetic instability, cancer cell accumulates genetic alterations with each mitotic division. Thus, small and early stage tumors will contain less genetic alterations than large and late stage ones, and the analysis of corresponding mutations should assist in identification of genomic alterations that are relevant to the initiation of tumorigenesis. However, since limited number of tumor cells are present in small lesions, reliable genetic analysis for such specimens often presents a challenge. Since the digital SNP method guarantees on average a smaller amount of testing for a given level of confidence than any other method, it is also suitable for the analysis of small tumors. Shih and coworkers have evaluated allelic imbalance of chromosomes 1p, 8p, 15q and 18q in adenomas less than 2 mm in diameter (Shih *et al.,* 2001). Their results indicate that allelic imbalance occurs early during colorectal tumor development.

Routine analysis of clinical samples with digital SNP analysis is rather simple and straightforward. The entire procedure including the reaction setup, PCR and fluorescent signal measurements can be completed within five hours. Hands-on time for the entire procedure is less than one hour and there is no need for the use of electrophoresis or radioisotopes. With the help of a robotic system, such as BioMek 2000, one research technician can easily perform the analysis of thirty samples per day. Thus, this method can be applied in any properly equipped laboratory.

In summary, we would like to emphasize that a key advantage of digital SNP analysis is that the data output is in digital format, as it measures each parental allele separately. However, this advantage is partly counterbalanced by the requirement of hundreds of PCRs for analysis of each marker for a given sample, thus making it a low-throughput method. Therefore, this method is not suitable for genome-wide SNP analyses because of the cost and labor involved. Yet, digital SNP analysis is amenable to automation and suits well for study of a defined molecular target for disease diagnosis or prognosis.

Acknowledgements

We thank Joyce Fields for editorial assistance. WZ is an assistant professor in the Winship Cancer Institute of Emory University. This work is supported by American Cancer Society Seed Grant IRG-90-016-11, University Research Committee of Emory University, and the Georgia Cancer Coalition fund to WZ.

References

Anbazhagan, R., Fujii, H., and Gabrielson, E. (1998). Allelic loss of chromosomal arm 8p in breast cancer progression. Am. J. Pathol. *152*, 815-819.

Bonnet, G., Tyagi, S., Libchaber, A., and Kramer, F.R. (1999). Thermodynamic basis of the enhanced specificity of structured DNA probes. Proc. Natl. Acad. Sci. USA *96*, 6171-6176.

Brookes, A.J. (1999). The essence of SNPs. Gene *234*, 177-186.

Broude, N.E. (2002). Stem-loop oligonucleotides: a robust tool for molecular biology and biotechnology. Trends Biotechnol. *20*, 249-256.

Chandrasekharappa, S.C., Guru, S.C., Manickam, P., Olufemi, S.E., Collins, F.S., Emmert-Buck, M.R., Debelenko, L.V., Zhuang, Z., Lubensky, I.A., Liotta, L.A. *et al.* (1997). Positional cloning of the gene for multiple endocrine neoplasia-type 1. Science *276*, 404-407.

Colpaert, C., Vermeulen, P., Jeuris, W., van Beest, P., Goovaerts, G., Weyler, J., Van Dam, P., Dirix, L., and Van Marck, E. (2001). Early distant relapse in 'node-negative' breast cancer patients is not predicted by occult axillary lymph node metastases, but by the features of the primary tumour. J. Pathol. *193*, 442-449.

Emmert-Buck, M.R., Lubensky, I.A., Dong, Q., Manickam, P., Guru, S.C., Kester, M.B., Olufemi, S.E., Agarwal, S., Burns, A.L., Spiegel, A.M. *et al.* (1997). Localization of the multiple endocrine neoplasia type I (*MEN1*) gene based on tumor loss of heterozygosity analysis. Cancer Res. *57*, 1855-1858.

Knudson, A.G., Jr. (1978). Retinoblastoma: a prototypic hereditary neoplasm, Semin. Oncol. *5*, 57-60.

Lengauer, C., Kinzler, K.W., and Vogelstein, B. (1997). Genetic instability in colorectal cancers. Nature *386*, 623-627.

Lengauer, C., Kinzler, K.W., and Vogelstein, B. (1998). Genetic instabilities in human cancers. Nature *396*, 643-649.

Li, J., Yen, C., Liaw, D., Podsypanina, K., Bose, S., Wang, S.I., Puc, J., Miliaresis, C., Rodgers, L., McCombie, R. *et al.* (1997). *PTEN*, a putative protein tyrosine phosphatase gene mutated in human brain, breast, and prostate cancer. Science *275*, 1943-1947.

Liu, J., Zabarovska, V.I., Braga, E., Alimov, A., Klien, G., and Zabarovsky, E.R. (1999). Loss of heterozygosity in tumor cells requires re-evaluation: the data are biased by the size-dependent differential sensitivity of allele detection. FEBS Lett. *462*, 121-128.

Lukyanov, K.A., Matz, M.V., Bogdanova, E.A., Gurskaya, N.G., and Lukyanov, S.A. (1996). Molecule by molecule PCR amplification of complex DNA mixtures for direct sequencing: an approach to *in vitro* cloning. Nucleic Acids Res. *24*, 2194-2195.

Marras, S.A., Kramer, F.R., and Tyagi, S. (1999). Multiplex detection of single-nucleotide variations using molecular beacons. Genet. Anal. *14*, 151-156.

Mitra, R.D. and Church, G.M. (1999). *In situ* localized amplification and contact replication of many individual DNA molecules. Nucleic Acids Res. *27*, e34.

Mitra, R.D., Butty, V.L., Shendure, J., Williams, B.R., Housman, D.E., and Church, G.M. (2003). Digital genotyping and haplotyping with polymerase colonies. Proc. Natl. Acad. Sci. U.S.A. *100*, 5926-5931.

Monckton, D.G. and Jeffreys, A.J. (1991). Minisatellite "isoallele" discrimination in pseudohomozygotes by single molecule PCR and variant repeat mapping. Genomics *11*, 465-467.

Royall, R. (1997). *Statistical Evidence: A Likelihood Primer*. Chapman and Hall, London.

Seitz, S., Rohde, K., Bender, E., Nothnagel, A., Kolble, K., Schlag, P.M., and Scherneck, S. (1997). Strong indication for a breast cancer susceptibility gene on chromosome 8p12-p22: linkage analysis in German breast cancer families. Oncogene *14*, 741-743.

Shih, I.M., Zhou, W., Goodman, S.N., Lengauer, C., Kinzler, K.W., and Vogelstein, B. (2001). Evidence that genetic instability occurs at an early stage of colorectal tumorigenesis. Cancer Res. *61*, 818-822.

Steck, P.A., Pershouse, M.A., Jasser, S.A., Yung, W.K., Lin, H., Ligon, A.H., Langford, L.A., Baumgard, M.L., Hattier, T., Davis, T. *et al.* (1997). Identification of a candidate tumour suppressor gene, *MMAC1*, at chromosome 10q23.3 that is mutated in multiple advanced cancers. Nat. Genet. *15*, 356-362.

Thiagalingam, S., Foy, R.L., Cheng, K.H., Lee, H.J., Thiagalingam, A., and Ponte, J.F. (2002). Loss of heterozygosity as a predictor to map tumor suppressor genes in cancer: molecular basis of its occurrence. Curr. Opin. Oncol. *14*, 65-72.

Tyagi, S. and Kramer, F.R. (1996). Molecular beacons: probes that fluoresce upon hybridization. Nat. Biotechnol. *14*: 303-308.

Tyagi, S., Bratu, D.P., and Kramer, F.R. (1998). Multicolor molecular beacons for allele discrimination. Nat. Biotechnol. *16*, 49-53.

Utada, Y., Haga, S., Kajiwara, T., Kasumi, F., Sakamoto, G., Nakamura, Y., and Emi, M. (2000). Allelic loss at the 8p22 region as a prognostic factor in large and estrogen receptor negative breast carcinomas. Cancer *88*, 1410-1416.

Vogelstein, B. and Kinzler, K.W. (1998). *The Genetic Basis of Human Cancer*. McGraw-Hill Health Professions Division, New York.

Vogelstein, B. and Kinzler, K.W. (1999). Digital PCR. Proc. Natl. Acad. Sci. USA *96*, 9236-9241.

Wald, A. (1947). *Sequential Analysis*. Wiley, New York.

Wang, S.S., Esplin, E.D., Li, J.L., Huang, L., Gazdar, A., Minna, J., and Evans, G.A. (1998). Alterations of the *PPP2R1B* gene in human lung and colon cancer.

Science *282*, 284-287.
Zhou, W., Galizia, G., Lieto, E., Goodman, S.N., Romans, K.E., Kinzler, K.W., Vogelstein, B., Choti, M.A., and Montgomery, E.A. (2001). Counting alleles reveals a connection between chromosome 18q loss and vascular invasion. Nat. Biotechnol. *19*, 78-81.
Zhou, W., Goodman, S.N., Galizia, G., Lieto, E., Ferraraccio, F., Pignatelli, C., Purdie, C.A., Piris, J., Morris, R., Harrison, D.J. *et al.* (2002). Counting alleles to predict recurrence of early-stage colorectal cancers. Lancet *359*, 219-225.
Zhou, W. (2003). Mapping genetic alterations in tumors with single nucleotide polymorphisms. Curr. Opin. Oncol. *15*, 50-54.

2.5

Real-Time Quantitative PCR in the Analysis of Gene Expression

Manohar R. Furtado, Olga V. Petrauskene and Kenneth J. Livak

Abstract

Quantitation of changes in transcript levels using real-time PCR, without the need for generating standard curves for each assay, is a useful tool for high throughput analysis of large numbers of genes in cellular regulatory networks. These methods are particularly useful for monitoring changes in expression of low abundance transcripts and messages that are strongly repressed in response to external stimuli. We describe protocols for accurate analyses of real-time PCR and discuss their potential for relative quantitation of gene expression. We also provide examples of practical uses of these methods for studying changes in gene expression caused by interferon induction of human cells.

Background

Differences in gene expression patterns reflect well the functional status of a cell. Gene expression patterns can be altered as a result of changes during the metabolic or developmental stage of a cell, disease pathogenesis, environmental stress or response to therapeutic agents (Levy and Darnell, 2002; Ji *et al*, 2003). Monitoring changes in cellular gene expression is a key to understanding the mechanisms of drug action. Similar studies are also useful in understanding signaling pathway perturbations induced by regulators like siRNA, negative dominant proteins, and extra-cellular stimuli. A major indicator of alterations in gene expression is the fold change (FC) in transcript levels between different cellular states. Currently, the most sensitive and accurate method for detection and measurement of FC in mRNA levels is the fluorescent quantitative real-time RT-PCR (reverse transcription polymerase chain reaction) assay (Holland *et al.*, 1991; Heid *et al*, 1996; Livak, 1997; Liu and Saint, 2002). The PCR assays that measure product in real-time have largely overcome the limitations of classical RT-PCR methods that relied on end-point detection. One of the advantages of these assays is that they enabled quantitation of low abundance mRNAs. In real-time PCR, the amount of target is related to a threshold cycle number (Ct) when the PCR amplification is still in the exponential phase. The Ct values can be used for absolute target quantitation and for relative quantitation to determine FC. Amplification efficiency measurements using a template dilution series (Liu and Saint, 2002; Pfaffl, 2001) can generate highly variable results depending on the conditions employed. At the same time, small differences in measured Ct values translate into significant changes in calculated efficiency. In this article we describe the effect of template dilution ranges on measured amplification efficiencies and provide recommendations for minimizing variations. We also compare the results obtained from calculation of FC using two different methods: (i) pre-determined amplification efficiency (Pfaffl, 2001; Lui and Saint, 2002) and the comparative Ct method also referred as the ΔΔCt method (Livak 1997; Livak and Schmittgen, 2001). Lastly, we describe the use of our TaqMan® Assays-on-Demand® products to measure changes in transcript levels of the genes induced by interferon β (IFN-β).

Protocols

Primer and Probe Design

Primers and probes are designed using bioinformatics tools that choose genomic regions: (i) with optimal annealing parameters; (ii) lacking known single nucleotide polymorphisms (SNPs); and (iii) with minimal local secondary structure to ensure high amplification efficiency. Literature reports have suggested that large amplicons with high GC content and/or strong secondary structure may be amplified with low amplification efficiency (Smith *et al.*, 2002). Therefore our design criteria are also optimized to ensure that such regions are avoided and shortest possible amplicons are synthesized. Usually our targets fall in the 50 to 160 bp length range and have GC content below 65%. Additionally, selectivity criteria are implemented to prevent amplification of paralogous genes. This is achieved by positioning primers in regions that vary among paralogous genes, and in such a way that mismatches are located at the 3' ends of primers to enhance differential amplification (Endrizzi *et al.*, 2002). All designed primers and probes (in total over 20,000) are first tested to ensure that they detect the right transcripts in RNAs isolated from human tissues and cells.

Materials

The human Universal Reference RNA (URR) is a mixture of RNAs isolated from ten human cancer cell lines (Stratagene). It is used for reverse transcription to generate cDNA. Reverse transcription is performed using the High Capacity cDNA Archive kit (Applied Biosystems) that uses random primers. Serial dilutions are made from cDNA derived from total RNA at a starting concentration of 100 ng/μl. The PCR amplification with real-time detection is performed using the Taqman® Universal PCR Master Mix with or without Amperase® (uracil N-glycosylase) as appropriate. PCR amplicons are purified using the QIAquick PCR purification kit (Qiagen) and quantified using the Picogreen® dsDNA quantitation kit (Molecular Probes). The non-human transcript (NHT) is spiked into the URR. It is prepared by transcription with T7 RNA polymerase (Promega) of a 227 bp fragment from the Hepatitis C virus 5' UTR inserted into a TOPO-TA plasmid (Invitrogen). HeLa cells are grown in minimal essential media with 10% fetal calf serum (Invitrogen), split into two halves and one half of the plates is treated with 1000 U/mL interferon-β (RandD Systems) when the cells attain 70% confluency. Cells are harvested 24 hrs post-treatment, RNA is isolated using RNAzol (Invitrogen) and reverse transcribed as described before.

The Ct Slope Method for Determination of Amplification Efficiency

A key attribute of any real-time PCR is the amplification efficiency, that is, how close the reaction is to doubling at every PCR cycle as detected by fluorescence increase (Liu and Saint, 2002; Livak and Schmittgen, 2001). This is described by the equation:

$$Xn = Xo\ (1+E)^n \tag{1}$$

where Xo and Xn are the amounts of the template at zero and n cycles, respectively, and E is the amplification efficiency. Since Xn is proportional to the reporter fluorescence, the above equation can be written as:

$$Rn = R_0\ (1+E)^n \tag{2}$$

where R_0 and R_n are fluorescence readings at 0 and n cycles, correspondingly.

One of the methods for amplification efficiency calculations is the Ct slope method (Liu and Saint, 2002), which includes generating a target dilution series and determining Ct values for each dilution.

cDNA derived from the URR total RNA is used to make 1-log and 3-log dilution series. There are a total of three and four dilutions in the 1-log and 3-log series, respectively. cDNA from each of the dilutions is amplified in a real-time PCR targeting specific gene transcript. Two primers and a TaqMan probe are used in each assay . The Ct values are determined in quadruplicate for each concentration point and averaged.

An outlier analysis is necessary to remove individual Ct values that contribute > 80% of the variance. Efficiency is calculated from the slope of the graph of Ct *vs* log template concentration (Lui and Saint, 2002). For generating larger dynamic ranges, cDNA derived from 100 ng of total RNA is pre-amplified using each of the 700 Assay-on-Demand™ primer/probe mixes. PCR is performed in a 50 μL volume for 40 cycles using our universal cycling conditions (40 cycles of 95°C, 15 sec; 60°C, 1 min) on an ABI PRISM® 7000 Sequence Detection System. Universal Master Mix (Applied Biosystems) with Uracil N-Glycosylase is used in these reactions. This initial amplification is performed to produce sufficient template to allow a broad template dilution range. 5 μL of the product is analyzed on a 4 % agarose gel (NuSieve Seakem) with size standards (25 bp ladder, Invitrogen) to estimate the DNA amount and verify the size of the amplicon. The product is purified using QIAquick PCR purification kit (Qiagen) according to the manufacturer's directions. DNA is eluted in 100 μL of the elution buffer. 10 μL of the eluate is used for Picogreen staining (PicoGreen® dsDNA quantitation kit, Molecular Probes) to estimate amplicon concentration. Figure 1 depicts the format used for measuring assay efficiency, dynamic range and sensitivity.

Sensitivity of Real-Time PCR

10 μL of cDNA is diluted serially 10-fold to obtain a dilution series that spanned over 10-logs. Each of these dilutions is subjected to PCR as before using the TaqMan® Universal PCR Master Mix without Amperase® on an ABI PRISM® 7900® sequence detection system. The Ct values obtained over a 6-log dilution range (Figure 2) are plotted against log of copy number (Figure 3). Ct values in the range of 35 to 37 cycles are used as the upper limit (sensitivity limit) for calculations. The reason for this limit is that Ct reproducibility is low when the target copy number drops below the 5 to 10 copy range because of stochastic effects. The copy number of the targets is estimated from the Picogreen staining as described above, while amplification efficiency, E, is calculated from the slope of the graph using the equation:

$$E = 10^{(-1/\text{slope})} - 1 \tag{3}$$

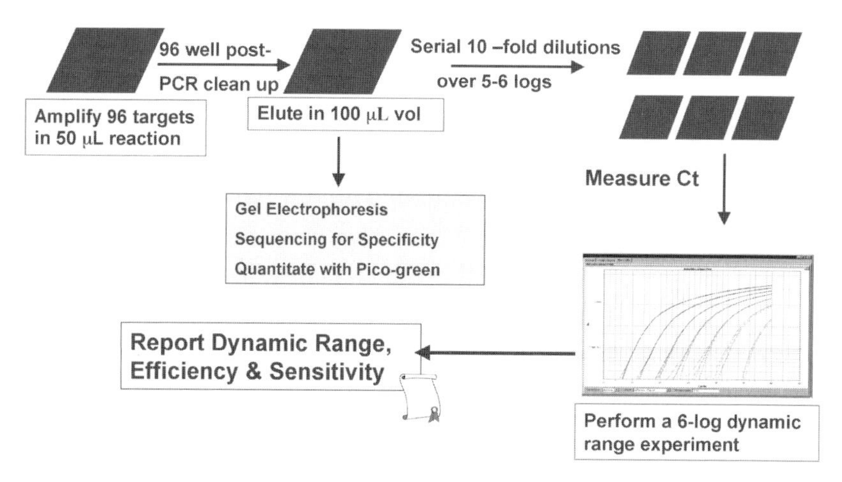

Figure 1. Flow chart depicting experiments on measuring assay efficiency, dynamic range and sensitivity of real-time PCR. Amplification of target amplicon from the URR pool is followed by purification of the PCR product, which is then serially diluted and subjected to real-time PCR in buffers lacking Uracil N-Glycosylase. The Ct values obtained over a 6-log dilution range are used to determine amplification efficiency. Quantitation by Picogreen staining of the PCR amplicon before real-time PCR is used to determine absolute DNA concentration necessary for sensitivity determination and agarose gel electrophoresis is employed to verify amplicon size. **The colour version of this figure is located in the colour plate section at the back of this book.**

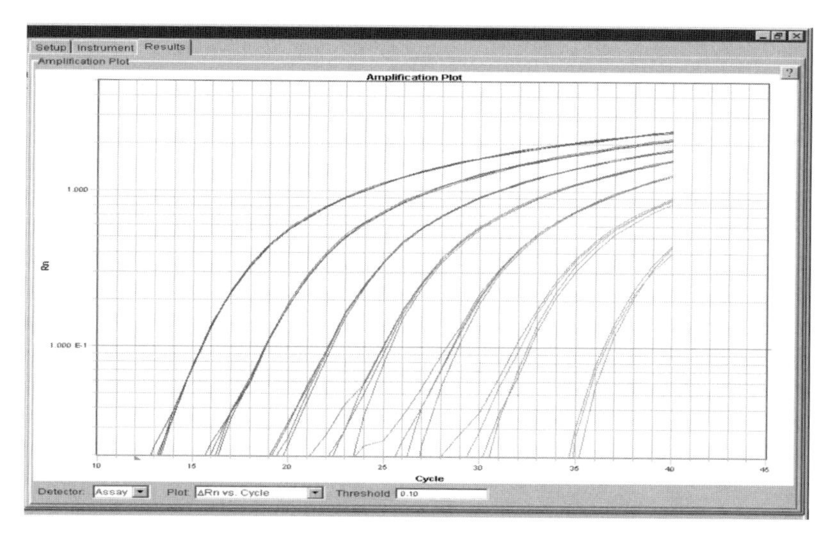

Figure 2. Typical amplification plots obtained in real-time PCR using serially diluted DNA samples. The cycle number (X-axis) corresponding to the point when the amplification plot rises above the threshold set at 0.10 (red line) is the Ct value. **The colour version of this figure is located in the colour plate section at the back of this book.**

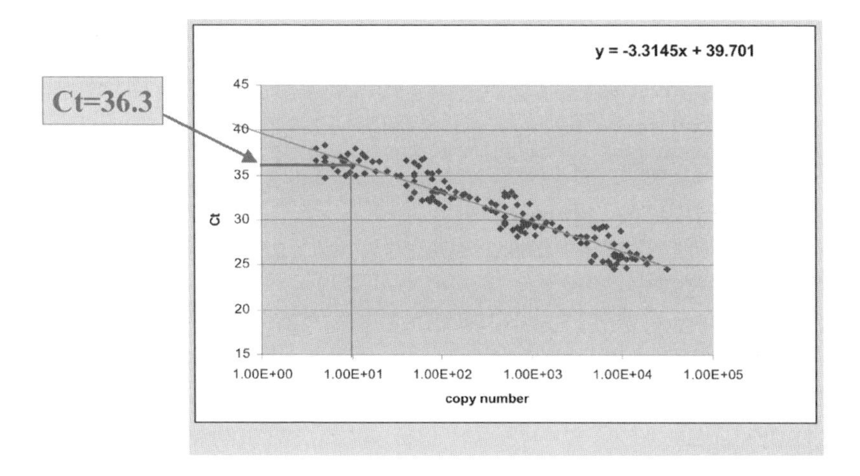

Figure 3. A plot of the amplicon copy number (X-axis) *vs* the Ct value obtained for the corresponding dilution. The results are a compilation of data from 40 different assays. The average Ct obtained for 10 copies is around 36. The slope of the graph is 3.3 indicating an average efficiency value of 100%. **The colour version of this figure is located in the colour plate section at the back of this book.**

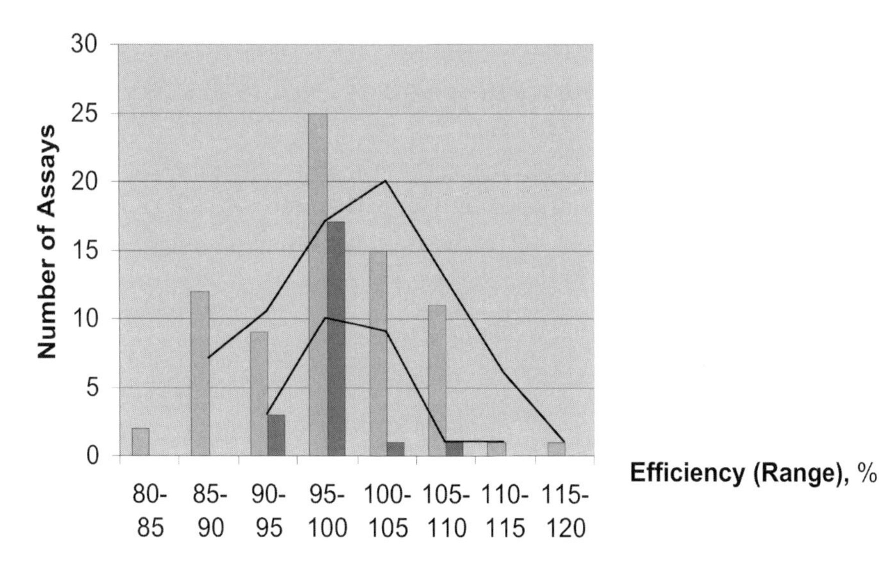

Figure 4. Amplification efficiency distribution for GAPDH target. Measurements correspond to a 2-log (blue) and 5-log (red) dilution series of cDNA. The data show that 2-log range dilution results in amplification efficiencies 82-112%, with average = 97.6% (number of assays 76). At the same time, 5-log range dilution results in amplification efficiencies 92-105%, with average 98.4% (number of assays 22). **The colour version of this figure is located in the colour plate section at the back of this book.**

Factors Affecting Measured Amplification Efficiency

It is very clear that the PCR efficiency variation depends on the width of the dilution dynamic range and on the amount of the data points (Figure 4 and Table 1). Table 1 presents the results of 30 assays, while Figure4 shows the results of the glyceraldehyde-3-phosphate dehydrogenase (GAPDH) gene quantitation. These data show that standard deviation in efficiency measurements is strongly dependent on the dilution range used for testing. Using larger dilution range (5- versus 2-log) minimizes efficiency variations and thus ensures more reliable results. Table 2 illustrates the possible range of efficiency variations if the Ct values differ by 0.5 Ct from the expected ones and a two-concentration point assay is used to measure efficiency. As can be deduced from Table 2, the effect of the fixed 0.5 Ct variation on the range of possible efficiency variations is reduced as the fold dilution range is increased. Therefore, using a broader dynamic range and a larger number of data points is necessary for accurately estimating PCR efficiency.

Example 1: Determination of the Amplification Efficiencies in 700 Real-Time RT-PCR Assays

To verify this concept and to ensure that the pipeline generates assays with high amplification efficiency we performed over 700 different real-time PCRs. In these PCRs we covered a wide range of base

Table 1. Comparison of PCR efficiencies obtained in 30 assays using either a 1-log or a 3-log template dilution range. The distribution of assays in different efficiency bins altered significantly with a much greater number of assays in the > 90% efficiency category when a broad template dilution range was used.

Efficiency	Number of assays	
	1-log	3-log
>90%	2	14
80-90	10	12
70-80	12	3
60-70	6	1
	30	30

Table 2. Effect of dilution range and Ct variations on PCR amplification efficiency determination. Distribution range for measured amplification efficiency was tabulated assuming a ± 0.5 variation in Ct measurement under different dilution ranges. Log fold.

Ct dilution	Variation	Measured Efficiency Range, % (expected 100%)
1.0	1.0	70- 168
2.0	1.0	82-125
3.0	1.0	87-115
4.0	1.0	89-110
5.0	1.0	92-108

composition and sequences, amplicon lengths and amplicon-related secondary structure features. The amplicon sizes ranged from 53 to 186 bp, and GC content varied from 26% to 79% (free energy values ranged from zero to – 19.36 kcal/mole). This number of assays constitutes a statistically significant set for achieving 95% confidence that 99.5% of the assays generated with this design pipeline will yield equivalent results. These assays were all a part of the Applied Biosystems Assays-on-Demand™ product line. The assays displayed amplification values ranging from 90 to 110 % with very few (4) outliers. Repeat testing of these outliers resulted in values in the 90 to 110 % range. The distribution of efficiencies is shown in Figure 5. The distribution approximates a normal distribution with a mean around 100% (98.73%). The distribution is similar to that obtained when a single GAPDH-gene specific assay was performed multiple times (Figure 4).

Example 2: Comparison of Different Methods for Determination of Fold Change (FC)

In a model experiment designed to compare different methods of FC determination, cDNA obtained from Stratagene URR was diluted in a 100-fold dilution range to obtain a dilution series with known amounts of cDNA. A constant amount of cDNA obtained from a non-human transcript was used as a control to measure FC.

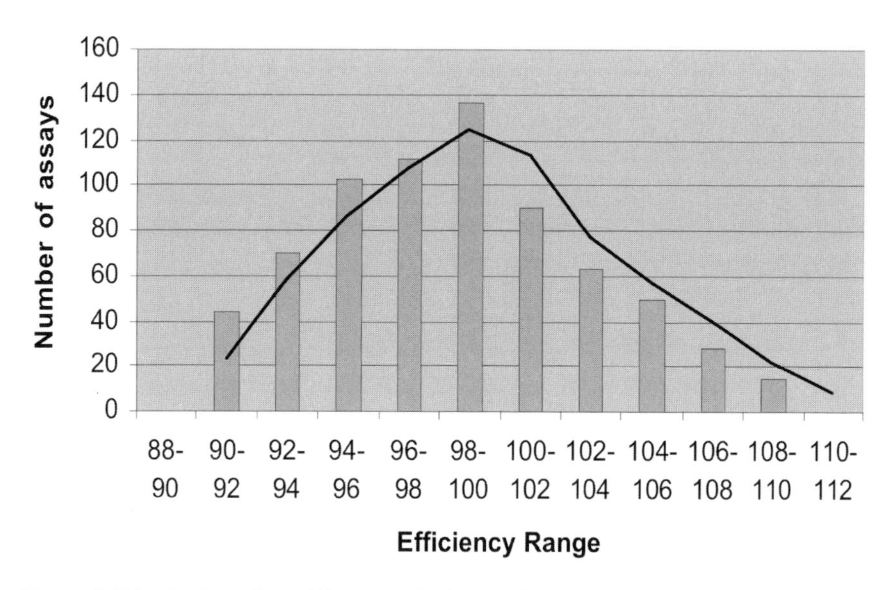

Figure 5. Distribution of amplification efficiencies for about 700 real-time PCR assays. The data display normal distribution with a mean around 100%. **The colour version of this figure is located in the colour plate section at the back of this book.**

FC in transcripts level was determined using the ΔΔCt method (Livak 1997; Livak and Schmittgen, 2001) and the method based on the pre-determined efficiency values (Pfaffl, 2001; Liu and Saint, 2002). The ΔΔCt method assumes that the amplification efficiency for both targets (GAPDH and NHT) are very similar and close to 100%. The FC results obtained for GAPDH target using the ΔΔCt method (Figure 6) showed good correlation with the expected FC with a slope of 0.94 *vs* the expected slope of 1.0. The second method of FC calculations (Pfaffl, 2001; Liu and Saint, 2002) relies on efficiency values pre-determined in prior experiments. Using efficiency for GAPDH target as 97.5% and for the NHT target as 94%, a slope of 0.78 *versus* expected 1.0 has been obtained. This result showed that the second method did not correlate well with the expected FC.

Example 3: Analysis of Changes in Cellular Gene Expression upon β-Interferon Induction

To analyze changes in gene expression in a biologically relevant system we studied changes following β-interferon induction in HeLa cells. In preliminary experiments it was shown that GAPDH does not display changes in transcript levels as a result of interferon treatment. Therefore, we used GAPDH as an endogenous negative control for changes in other transcript levels. Changes in gene expression levels for a set of eight genes involved in the interferon induction pathway (Aaronson and Horvath, 2002) are shown in Table 3. Binding of β–interferon to a cell membrane receptor activates tyrosine kinases that phosphorylate and activate STAT1 and STAT2. These proteins form a complex with IRF9 and bind to the cis-regulatory elements (ISRE) that activate transcription from a set of genes. According to this scheme we analyzed FC for several transcripts involved in the cell response to β-interferon induction (Table 3). These FC values were compared to those reported by the Bryan Williams laboratory (http://www.lerner.ccf.org/labs/williams/oligo.cgi), obtained in the microarray study of gene expression changes in response to β–interferon induction. Table 3 shows that the correlation between the data from microarray analysis and real-time PCR is consistent when the fold changes are relatively low (around 10-fold). Both methods also consistently identified the induced genes. However, the

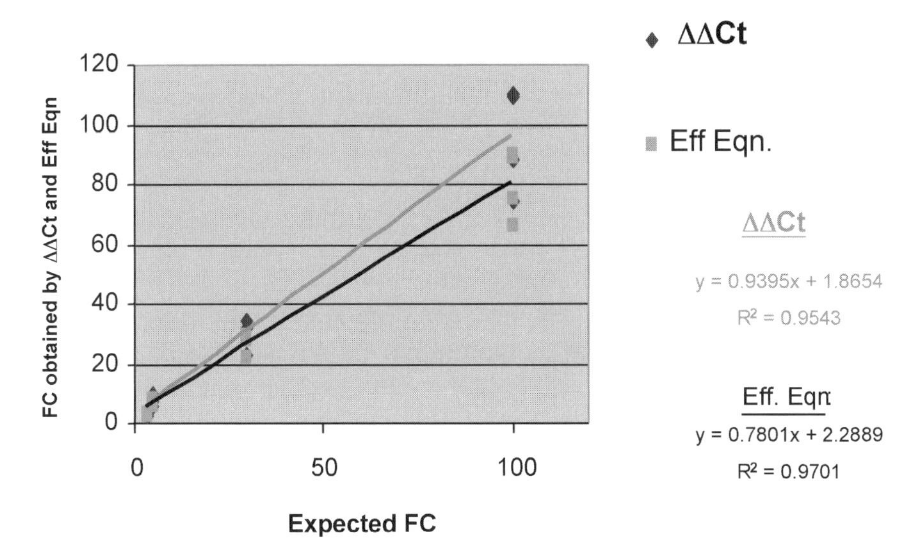

Figure 6. Comparison of the two methods for quantitation of fold change (FC) values with the expected FC. (A) ΔΔCt method, (B) method using pre-determined amplification efficiencies for a target (GAPDH) and a control gene (see text for details). **The colour version of this figure is located in the colour plate section at the back of this book.**

Table 3. Comparison of the fold change (FC) values for eight genes involved in the β-interferon-regulated pathway 24 hr after IFN induction. The data obtained in this study (FC observed) are compared with that reported in the literature using microarray analysis (FC expected).

Gene Name	FC expected	FC observed
PRKR	7	7
L22342	10	12.9
IFI16	12.7	15.1
SSA1	6.6	18.8
ISGF3/IRF9	9.6	12.7
STAT1	11.2	22.3
OAS1	25.2	29.6
IFIT1	125	490

microarray platform seems to underestimate the fold change when the fold changes are higher than 10-fold. For example, for IFIT1 124–fold increase determined by microarray hybridization should be compared to the 490-fold increase obtained by real-time PCR. Such bias in underestimation of fold change by both oligonucleotide and cDNA arrays has been previously reported in other studies (Yuen *et al.*, 2002).

Discussion

Analysis of cellular gene expression requires that transcript (RNA) levels are reliably quantified. Two general methods have been used to analyze data from real-time quantitative PCR experiments: absolute quantitation and relative quantitation. In absolute quantitation, one determines the copy number of a target in a sample of interest with reference to an internal calibrator that co-amplifies with the target (Furtado *et al.*, 1991; 1999; Mulder *et al.*, 1994), or with reference to an external calibrator by relating the threshold cycle (C_t) of the target to a standard curve (Wang and Brown, 1999). Thus, internal calibrators or external calibrator standard curves are necessary for accurate absolute quantitation making the process less amenable to high throughput screening.

For many gene expression experiments, it is not necessary to determine the absolute transcript copy number because relative quantitation (RQ) with measurement of FC is sufficient. For relative quantitation it is necessary to have an endogenous control that is minimally altered in the course of the study/experiment (Hamalainen *et al.*, 2001, Vandesompele *et al.*, 2002). Quantitation standards that can be externally added to the sample and carried through the entire assay are also useful for normalization across large studies. Relative quantitation (RQ) can be performed by using the amplification efficiency values for the target and endogenous control in the PCR equation (Pfaffl, 2001; Liu and Saint, 2002). This requires that efficiency values be determined for each assay. RQ can also be performed using the comparative Ct (also referred to as the ΔΔCt) method using the formula:

$$FC = 2^{-\Delta\Delta Ct} \qquad (4)$$

to calculate normalized fold changes in test samples relative to a calibrator sample (Livak and Schmittgen, 2001; Livak 1997). Mathematical models have been developed to show that the FC values would be altered significantly if the efficiency values deviate from 100% (Pfaffl, 2001).

Our results obtained in several assays indicate that the ΔΔCt method results in FC values that correlate better to expected FC values than other RQ methods. This conclusion was justified in multiple assays indicating that relative gene quantitation using the comparative Ct method expediently generates reliable data.

Concluding Remarks

In our analysis, we have observed that efficiency measurements using the Ct slope method have a high degree of variability. This variability is inversely correlated with the log dilution range. The smaller the dilution range, the more variable and less reliable the measured efficiency values are. In contrast, dilutions of template DNA across 5 to 6 logs of dynamic range provide reliable efficiency values. With good design strategies, it is possible to consistently generate assays that have near

100% efficiency for all targets. Further, our analysis indicates that the ΔΔCt method yields more reliable RQ results compared to procedures which use pre-determined PCR efficiencies.

Acknowledgements

We would like to thank Jennifer McLean and Michael Malicdem for technical assistance and Karl Guegler and Dennis Gilbert for support.

References

Aaronson, D. S., and Horvath, C. M. (2002). A road map for those who don't know JAK-STAT. Science *296*, 1653-1655.

Endrizzi, K., Fischer, J., Klein, K., Schwab, M., Nussler, A., Neuhaus, P., Eichelbaum, M., and Zanger, U. M. (2002). Discriminative quantification of cytochrome P4502D6 and 2D7/8 pseudogene expression by TaqMan real-time reverse transcriptase polymerase chain reaction. Anal. Biochem. *300*, 121-131.

Furtado, M. R., Balachandran, R., Gupta, P., and Wolinsky, S. M. (1991). Analysis of alternatively spliced human immunodeficiency virus type-1 mRNA species, one of which encodes a novel tat-env fusion protein. Virology *185*, 258-270.

Furtado, M. R., Callaway, D. S., Phair, J. P., Kunstman, K. J., Stanton, J. L., Macken, C. A., Perelson, A. S., and Wolinsky, S. M. (1999). Persistence of HIV-1 transcription in peripheral-blood mononuclear cells in patients receiving potent antiretroviral therapy. N. Engl. J. Med. *340*, 1614-1622.

Hamalainen, H. K., Tubman, J. C., Vikman, S., Kyrola, T., Ylikoski, E., Warrington, J. A., and Lahesmaa, R. (2001). Identification and validation of endogenous reference genes for expression profiling of T helper cell differentiation by quantitative real-time RT-PCR. Anal. Biochem. *299*, 63-70.

Heid, C. A., Stevens, J., Livak, K. J., and Williams, P. M. (1996). Real time quantitative PCR. Genome Res *6*, 986-994.

Holland, P. M., Abramson, R. D., Watson, R., and Gelfand, D. H. (1991). Detection of specific polymerase chain reaction product by utilizing the 5'--3' exonuclease activity of Thermus aquaticus DNA polymerase. Proc. Natl. Acad. Sci. USA. *88*, 7276-7280.

Ji, X., Cheung, R., Cooper, S., Li, Q., Greenberg, H. B., and He, X. S. (2003). Interferon alfa regulated gene expression in patients initiating interferon treatment for chronic hepatitis C. Hepatology *37*, 610-621.

Levy, D. E., and Darnell, J. E., Jr. (2002). Stats: transcriptional control and biological impact. Nat Rev Mol Cell Biol *3*, 651-662.

Liu, W., and Saint, D. A. (2002). A new quantitative method of real time reverse transcription polymerase chain reaction assay based on simulation of polymerase chain reaction kinetics. Anal. Biochem. *302*, 52-59.

Livak, K.J. (1997). ABI PRISM 7700 Sequence Detection System, User Bulletin 2, PE Applied Biosystems.

Livak, K. J., and Schmittgen, T. D. (2001). Analysis of relative gene expression data using real-time quantitative PCR and the 2(-Delta Delta C(T)) Method. Methods *25*, 402-408.

Mulder, J., McKinney, N., Christopherson, C., Sninsky, J., Greenfield, L., and Kwok, S. (1994). Rapid and simple PCR assay for quantitation of human immunodeficiency virus type 1 RNA in plasma: application to acute retroviral infection. J. Clin. Microbiol. *32*, 292-300.

Pfaffl, M. W. (2001). A new mathematical model for relative quantification in real-time RT-PCR. Nucleic Acids Res. *29*, e45.

Smith, S., Vigilant, L., and Morin, P. A. (2002). The effects of sequence length and oligonucleotide mismatches on 5' exonuclease assay efficiency. Nucleic Acids Res. *30*, e111.

Vandesompele, J., De Preter, K., Pattyn, F., Poppe, B., Van Roy, N., De Paepe, A., and Speleman, F. (2002). Accurate normalization of real-time quantitative RT-PCR data by geometric averaging of multiple internal control genes. Genome Biol. *3*, RESEARCH0034.

Wang, T., and Brown, M. J. (1999). mRNA quantification by real time TaqMan polymerase chain reaction: validation and comparison with RNase protection. Anal. Biochem. *269*, 198-201.

Yuen, T., Wurmbach, E., Pfeffer, R.L., Ebersole, B.J., and Sealfon, S.C. (2002). Accuracy and calibration of commercial oligonucleotide and custom cDNA microarrays. Nucleic Acids Res. *30*, e48.

2.6

Quantitative Genetic Analysis with Multiplex Ligation-Dependent Probe Amplification (MLPA)

A.O.H Nygren, A. Errami and J.P. Schouten

Abstract

The Multiplex Ligation-dependent Probe Amplification (MLPA) technique provides the means for the quantitative analysis of various changes in gene structure and/or gene copy number of several dozens (40-50) of DNA targets in a single reaction. MLPA requires only 20 ng of human chromosomal DNA and can discriminate sequences differing by a single nucleotide. In MLPA, binary probes containing a sequence-specific part and a universal part are hybridized to their DNA targets and ligated. Each ligated probe is then PCR amplified with a universal primer pair and gives rise to an amplicon of unique size. Relative amount of each amplicon reflects the amount of the corresponding target that is present in the nucleic acid sample. The necessary equipment used in MLPA, a thermocycler and a high resolution electrophoresis apparatus with fluorescent detection, are available in most molecular biology laboratories. Current applications of MLPA include detection of copy number changes of complete chromosomes, specific chromosomal areas and single exons of specific genes. MLPA can also be used for quantitative analysis of small mutations including SNPs and for studies of the methylation status of genomic sequences.

Background

Changes in copy number of certain specific genomic sequences are frequently linked to development of human diseases and syndromes, or to predisposition to certain illnesses. Such changes may include the presence of an extra copy of a complete chromosome as in Down's syndrome, deletions of up to several million base pairs, as in DiGeorge syndrome, and deletions or duplications of smaller chromosomal fragments like complete genes or single exons as in the adenomatosis polyposis coli (APC) gene (Stenson *et al.,* 2003). The need for multiplex detection of disease-causing single nucleotide polymorphisms (SNPs) and small mutations, such as the cystic fibrosis ΔF508 mutation (Harris 1992), becomes more urgent as more mutations are identified. In addition, epigenetic changes (*e.g.* DNA methylation of tumor suppressors like MLH1 or BRCA2) could play an important role in tumorigenesis (Esteller, 2003). Therefore, high throughput analysis of the DNA methylation status of various tumor suppressor genes is also of great value.

At present, different techniques are used for the detection of genomic copy number changes: comparative genomic hybridization (CGH) (Kallioniemi *et al.*, 1992), fluorescent *in situ* hybridization (FISH) (Klinger *et al.*, 1992), BAC arrays (Snijders *et al.*, 2001), Southern blots (Southern 1975) and loss of heterozygosity (LOH) assays (Devilee *et al.*, 2001). The conventional methods for mutation detection that are based on PCR amplification are not able to detect most exon deletions and duplications because the normal allele that is present is also exponentially amplified. Several multiplex PCR amplification techniques such as quantitative multiplex PCR of short fluorescent fragments (QMPSF) (Casilli *et al.,* 2002) have been developed for copy number analysis of genomic sequences. However, the presence of multiple PCR primers in a single reaction tube often reduces robustness of PCR and reliability of quantification. Finally, most methods currently in use require significant amounts of DNA, and it is therefore difficult to perform large-scale analysis of genomic aberrations in tumors due to the limited amounts of sample.

Recently several methods have been developed, which allow multiplex mutation analysis. One example is the Multiplex Amplifiable Probe Hybridization (MAPH) method, in which 40 different specific target sequences are detected and quantified (Armour *et al.,* 2000). MAPH uses oligonucleotide probes with generic ends that hybridize to target nucleic acid sequences. All hybridized probes are then amplified with the use of a single primer pair complementary to generic ends and yield amplicons of unique size specific for each target. The copy number of target sequences is reflected in the relative intensities of amplification products. However MAPH, like Southern blotting, requires immobilization of DNA and tedious removing of unbound probes, making this method difficult to implement in a routine diagnostic setting.

Ligase mediated gene detection (Landegren *et al.,* 1988) and ligase-dependent PCR have been described previously (Carrino *et al.,* 1996; Hsuih *et al.*, 1996; Barany *et al.,* 1997). Our method is based on the same principle, which means that the major sequence-discriminating step in MLPA includes the ligation reaction. However, to make these techniques suitable for high level multiplex analysis, we introduced the following four modifications. 1) MLPA uses much lower concentrations of oligonucleotide probes and longer hybridization times. 2) MLPA uses a hybridization buffer that upon dilution is compatible with enzymatic ligation and PCR amplification. 3) MLPA uses a heat-inactivated thermostable ligase-65 enzyme. 4) An important component of MLPA is the development of a cloning method for preparation of long probe oligonucleotides.

In MLPA, specially designed oligonucleotide probes are hybridized to target DNA sequences, ligated and then amplified in a multiplex PCR. Each gene-specific probe consists of two oligonucleotides and each oligonucleotide has a gene-specific and a universal sequence part. The concentrations of oligonucleotide probes and the extended hybridization

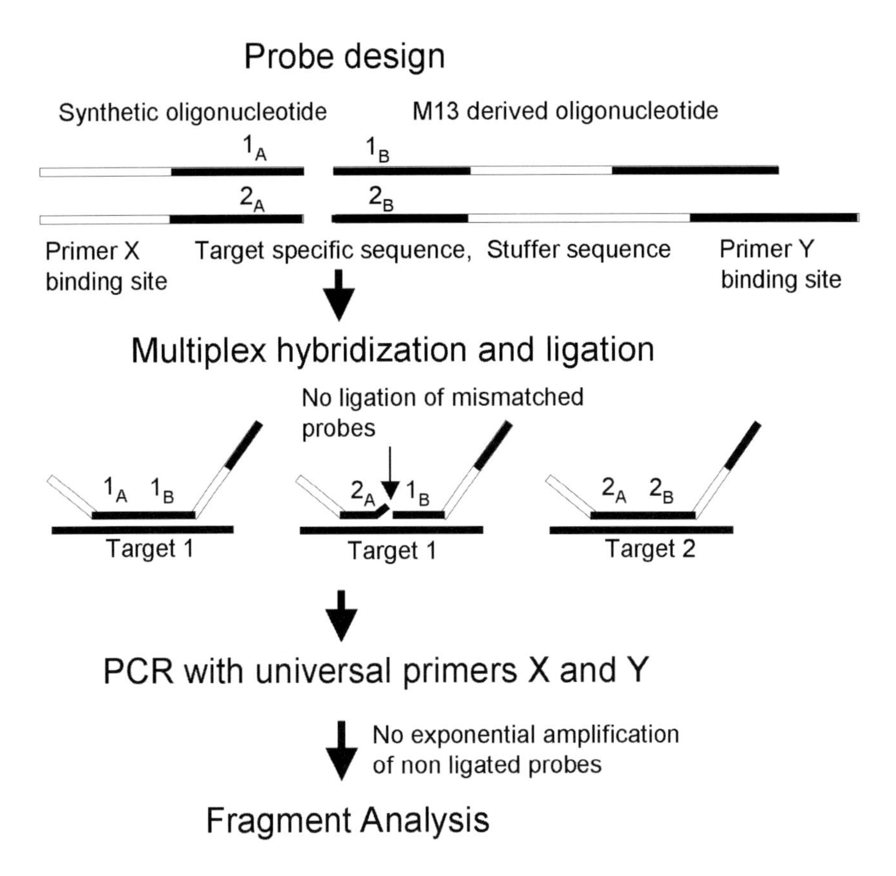

Figure 1. Outline of Multiplex Ligation-dependent Probe Amplification (MLPA). Each MLPA probe consists of two oligonucleotides, one short synthetic and one long M13-derived. Two probe pairs for two targets, A and B, are shown. The short synthetic oligonucleotide consists of a 5' universal primer sequence X and a target-specific 3' sequence. The M13-derived oligonucleotide consists of a 5' target-specific sequence designed to hybridize in juxtaposition to the synthetic probe, a 3' universal primer sequence Y and a stuffer sequence that has a different length for each probe. 1 fmol of each probe oligonucleotide is hybridized to the denatured target DNA. Only perfectly matched probes are ligated by the thermostable Ligase-65 and only ligated probes are exponentially amplified by the universal primer pair X and Y in subsequent PCR. The amplified fragments are separated and analyzed by capillary electrophoresis.

times are sufficient to ensure practically complete hybridization of target sequences and corresponding probe oligonucleotides. Only probes hybridized to a perfectly matched target DNA are ligated and then exponentially amplified in the following PCR (See Figure 1).

MLPA probes that do not find a target sequence cannot be ligated and consequently amplified in the PCR. Also probes hybridized to a target with a mismatch close to the ligation point are not ligated due to the high sensitivity of the thermostable ligase-65. Thus, the relative peak area of each amplification product reflects the relative copy number of the target sequence in the DNA sample and comparison with a control sample allows absolute copy number quantization. The MLPA technique presented here permits relative quantification of up to 45 different target sequences in an easy to perform reaction. In order to obtain reproducible results, MLPA currently needs at least 20 ng human DNA, corresponding to approximately 3000 cells or 6000 single copy target sequences.

MLPA Protocol

Materials and Methods

Male and female control DNAs were obtained from Promega. DNAs from a female with triple chromosome 21 and a female with triple chromosome X were provided by the Department of Clinical Genetics, Free University of Amsterdam. Blood DNAs from individuals diagnosed with familiar adenomatous polyposis (FAP) and healthy controls were provided by the Swedish APC registry (Stockholm, Sweden).

50–500 ng of target DNA in 5 μl of 10 mM Tris (pH 8.5)-0.1 mM EDTA was denatured at 98°C for 5 min in a thermocycler with heated lid (Biometra Uno II). 1.5μl probe mix (1 fmol of each probe) and 1.5 μl MLPA Buffer (1.5 M KCl, 300 mM Tris-HCl, pH 8.5, 1mM EDTA) were mixed and added to the denatured DNA. The mix was incubated at 95°C for 1 min and then overnight at 60°C. After hybridization the sample was diluted to 40 μl with ligase buffer (2.6 mM $MgCl_2$, 5 mM Tris-HCl pH 8.5, 0.013% non-ionic detergents, 0.2 mM NAD) containing 1U Ligase-65 (MRC-Holland) and the ligation was performed at 54°C for 15 minutes. The ligase was then inactivated by incubating the sample at 98°C for 5 min. 10 μl of the ligation mixture was mixed with 30 μl of PCR buffer (20 mM Tris-HCl pH 8.5, 50 mM

KCl, 1.6 mM $MgCl_2$ and 0.01 % non-ionic detergents). While at 60°C, 10 μl containing 10 pmol PCR primer solutions (one unlabeled and one D4 labeled), 2.5 nmol of dNTPs and 2.5 U Taq polymerase (Promega) or SALSA polymerase (MRC-Holland) was added to the mix and 33 cycles of PCR (95°C, 30s; 60°C, 30 s; 72°C, 1 min) was performed followed by a 20 min incubation at 72 °C. The sequence of the labeled primer was 5'-GGGTTCCCTAAGGGTTGGA and that of the unlabeled was 5'-GTGCCAGCAAGATCCAATCTAGA. Samples were analyzed on a Beckman CEQ2000 capillary electrophoresis system. Samples were denatured for 120 s at 90 °C. Electrophoresis conditions were as follows: injection voltage was 2.0 kV and injection time 30 sec. Run time was 60 min at 50 °C and 4.8 kV. The exact number of probes in the kits used in this article and the specific recognition sequences of each probe can be found at www.mrc-holland.com.

To automate the interpretation of the fragment analysis, the relative quantity of the amplified probes in each sample was determined using an Excel template. For this purpose, the relative peak areas for each probe were calculated as fractions of the total sum of peak areas in a sample. Subsequently, the fraction of each peak was divided by the average peak fraction of the corresponding probe in control samples. Finally, the values were normalized using the values obtained for the control probes, which served as a reference for the copy number of 2.0.

Comments on DNA Analysis with MLPA

MLPA Probe Design

MLPA probes are designed in such a way that the length of each amplification product is different. In the currently available MLPA kits, length differences between different amplicons are 6-9 nucleotides and their lengths range in sizes between 130 and 480 nucleotides. These length differences and the size range provide good separation and low background during electrophoresis. Each gene-specific probe consists of two oligonucleotides, one prepared synthetically and one obtained by cloning in the M13 vector. Each oligonucleotide contains a part specific for a certain nucleic acid sequence. The synthetic oligonucleotide has a universal 5' sequence used for PCR amplification and a target-specific 3' end. M13-derived oligonucleotides have a 25-46 nt-long target specific 5' end designed to hybridize in juxtaposition to the 3' end of synthetic oligonucleotide. All M13-derived probes have a universal 3' end used for PCR amplification and a stuffer sequence 19-370 nt-long between the sequence-specific and the universal primer sequence. Since preparation of chemically synthesized oligonucleotides of these sizes is difficult and expensive, cloning in M13 vector is used instead. To this end, the target-specific sequence is cloned into an M13 vector already containing stuffer and primer sequence (Messing *et al.,* 1977). After purification, the single-stranded M13 DNA is made partially double-stranded by hybridization with short complementary oligonucleotides and digested by two different restriction enzymes (*Eco*RV and *Bsm*I) resulting in a probe with a 5' phosphorylated end. Since restriction endonuclease *Bsm*I cuts DNA outside its recognition site, no nucleotides of this recognition sequence are present at the 5' end of the resulting probe oligonucleotide.

DNA Denaturation and Hybridization with the SALSA-Probes

The use of a thermocycler with heated lid is essential and the use of 0.2 ml PCR-tubes is recommended. Make sure that the total amount of human chromosomal DNA is within the recommended limits of 20-500 ng. The volume of the sample should not exceed 5 μl; this is important for hybridization efficiency, which is probe- and salt-concentration dependent. The EDTA concentration in the DNA sample should not exceed 1 mM.

Ligation Reaction

Prepare the ligation mixture immediately before use and store on ice. Reduce temperature of the thermal cycler to 54°C and add 32 μl of the ligation mix to each sample and mix. Final volume is 40 μl. After inactivation of the Ligase-65, the samples can be stored at 4°C for up to one week.

PCR

Polymerase mix should be made less than 1 hr before use and stored on ice. Mix well but do not vortex the polymerase mix. Incomplete mixing of the viscous 50% glycerol enzyme solutions with dilution buffers is a major source of errors. The hot start is not essential. Preparation of the PCR mix on ice will result in almost identical results.

The following issues should be considered for successful MLPA analysis:

1. Never use micropipettes for performing MLPA reactions that have been used for handling MLPA PCR products. Following the PCR, the tubes should not be opened in the vicinity of the thermal cycler.
2. The number of PCR cycles can often be reduced to 30. This will result in a minor improvement of the linearity of relative signal and target copy number. On the contrary, if only approximately 20 ng of human DNA are used, it may be necessary to increase the number of cycles to 36-40. Note that MLPA substantially differs from conventional PCR. At the end of conventional PCR, renaturation of fragments competes with primer binding / elongation. In MLPA, amplification slows down due to primer depletion. As the concentration of each of the 45 amplicons is not extremely high at the end of an MLPA reaction, different renaturation kinetics of the various amplicons does not result in different amplification efficiencies during extra PCR cycles, which is the case in conventional PCR. Thus, the use of 40 PCR cycles results in a small increase in the peak heights/areas but practically has no influence on the relative signals.
3. PCR products can be stored at 4°C for at least one week. As the fluorescent labels used are light-sensitive, the PCR products should be stored in a dark box or wrapped in aluminum foil.
4. MLPA is more prone to PCR contamination than conventional PCR. Since the same primer pair is used in all reactions, a tiny contamination from a previous MLPA PCR may result in a false positive signal.
5. Sometimes PCR products contain a nonspecific broad peak of

unknown nature. Its size depends on the apparatus and fluorescent primer used. On the Beckman CEQ2000 CE apparatus, the peak is usually overlapping with 360 bp-long fragments, but we have seen ABI3100 pictures with the blob being clearly visible at 213 bp. This broad "blob" is also visible when no PCR was performed; however it usually disappears when the signal to noise ratio is increased by using 36 cycles of PCR.

Troubleshooting with Quantitative Analysis of Capillary Electrophoresis Patterns

A. No bands visible.

1. *Is the molecular weight marker pattern OK*? If not, check capillaries and electrophoresis conditions.
2. *Is the PCR primer peak present, and off-scale?* If not, check whether the correct fluorescent label is used.
3. *Are the four control fragments visible*? Each SALSA probe mix will generate 4 fragments of 64, 70, 76 and 82 bp even when ligation is omitted. These bands will be small when more than 50 ng human DNA is used for MLPA. If these bands are stronger than the expected amplicons, either the ligation reaction failed or the amount of DNA was much lower than the recommended minimum of 20 ng. In either case the results obtained may not be reliable. The control mix generates a fifth band of 94 bp, which is ligation-dependent. This is an amplification product of a synthetic MLPA probe specific for the human IL1B gene. This band should be of similar peak area as most of the expected MLPA products.
4. *Primer peak ok, but no control fragments and no probe amplification products*: PCR failed. We recommend repeating the PCR using the same ligation reactions, for instance using a different thermal cycler. Reduce all volumes by 50% to save reagents.
5. *Primer peak ok, control fragments clearly visible, but no probe amplification products*: No DNA was present in your sample, or the ligation reaction failed. The MLPA reagents are quite stable. After storage of a complete MLPA kit including probes, ligase

and polymerase, for more than 12 weeks at room temperature, good results were obtained. However, the Ligase cofactor NAD, present in the ligase buffer A might be inactivated after 15 freeze-thaw cycles. Enzymes may be inactivated by repeated freeze-thaw cycles in freezers operating at –25 / -30°C.

B. Bands visible, but weak signals.
After PCR, the tubes can be placed in the thermal cycler for another 4 cycles, or a new PCR with the same ligation reactions but with 36-40 cycles might help. Reduce all volumes by 50% to save reagents.

C. Signal strength is ok but there are large differences in relative peak areas between different samples.
This might be due to impurities in the DNA samples. The use of lower amounts of DNA often solves the problem. Another likely explanation is overloading of capillaries resulting in saturation of the fluorescence detection device. Broad peaks, especially of longer fragments can be caused by incomplete denaturation during the electrophoresis.

D. Extra peaks.
Incomplete denaturation during the electrophoresis can result in extra peaks and in broadening of the peaks, especially of longer fragments. An increase in the temperature during CE might solve this problem. Note that in some apparatuses such as ABI3700, overloading of capillaries may result in signals in neighboring capillaries.

E. Low peak area of the longer probes.
This can have many causes. Impurities in DNA samples particularly affect long PCR products and products that already have lower than average peak area. Additionally, the electro-kinetic injection procedure of capillary electrophoresis is biased towards injection of smaller fragments. This appears to be more pronounced with ABI sequencers as compared to the Beckman CEQ.

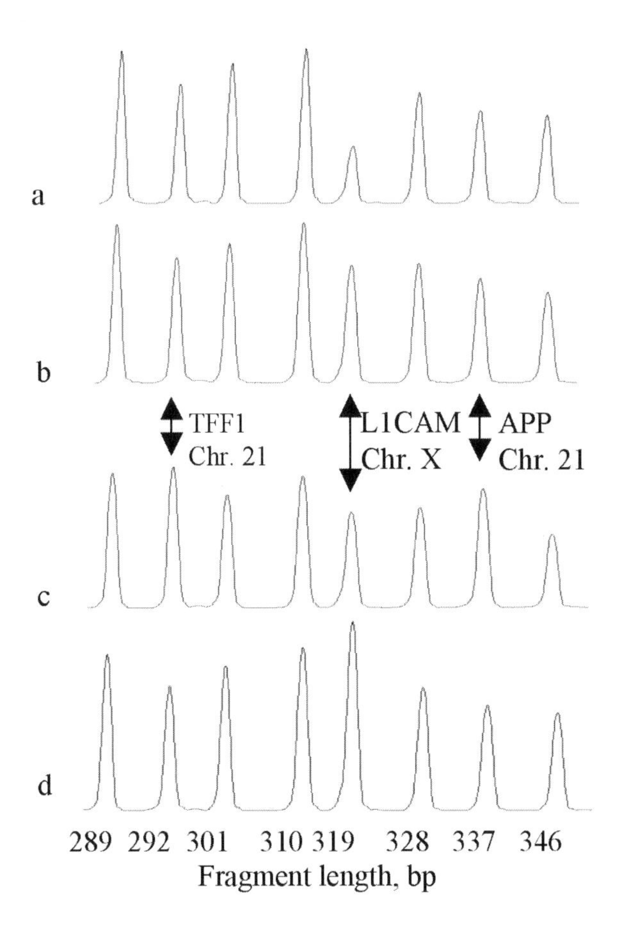

Figure 2. Detection of trisomies by MLPA. 100 ng DNA were analyzed by MLPA using probe mix P001 (MRC-Holland). Mix P001 contains probes for 40 genes which are shown in Table 1. The specific recognition sequences of each probe can be obtained at www.mrc-holland.com. Male (a), female (b) control DNAs, and DNAs from a female with triple chromosome 21 (c) and a female with triple chromosome X (d) were subjected to MLPA. PCR products were analyzed by capillary electrophoresis (Beckman CEQ2000). Only part of the electropherogram is shown. Arrows indicate the positions of a 292 bp amplicon specific for the TFF1 gene on chromosome 21, a 319 bp amplicon specific for the L1CAM gene on the X chromosome and a 337 bp amplicon specific for the APP gene on chromosome 21. Other signals are specific for chromosome 3 (283 bp), chromosome 18 (301 and 346 bp), chromosome 13 (310 bp) and chromosome 1 (328 bp). For quantitative analysis of data in Figure 2 see Table 1.

Examples

Detection of Trisomies

In order to evaluate performance of MLPA in terms of linearity between signal and target copy number we analyzed targets located on human chromosomes X, Y, 13, 18 and 21, for which trisomies are not always lethal. Corresponding probes have been designed and MLPA with male, female or triple chromosome 13 or triple X DNA samples was performed as described above. Capillary electrophoresis patterns obtained are shown in Figure 2. For quantification of each target, corresponding peak area was presented as a fraction of the sum of all peak areas on the display. The ratio of each peak area was then normalized to that obtained with a control sample with known copy number of the corresponding gene. If the presence of two copies of a target gene in diploid genome results in a relative signal of 1.0, the presence of three copies of chromosome 21, for example in a DNA sample obtained from a Down syndrome patient, should result in a signal 1.5 for the chromosome 2-specific probes. The results shown in table 1 show that the relative signals obtained for each probe reflected the relative amount of the target sequences in the DNA sample. The excellent reproducibility of relative signals enabled the detection of a single extra copy of a target sequence per diploid genome.

Detection of Deletions of an Entire Gene or of Single Exons

Adenomatosis polyposis coli (APC) gene located on human chromosome 5q21-q22 is the primary gene implicated in Familiar Adenomatous Polyposis (FAP), an inherited autosomal dominant disease that is characterized by the formation of adenomatous polyps in the small and large intestine and is, due to the large number of polyps, associated with an almost 100% risk of developing colorectal cancer (CRC), (for review see Fearnhead *et al.*, 2001). Most of the APC gene mutations result in expression of proteins with a truncated C-terminus. In order to detect deletions in the APC gene 20 probes were designed, one for each of 15 exons, 3 probes for the promoter

Table 1. Quantitative analysis of amplicons obtained in a 40-plex MLPA using DNAs with trisomies of chromosomes 21 and X respectively. Amplicons specific for 21 and X chromosomes in corresponding DNAs are in bold.

Gene		Relative % area to XX		
	Chr	XX	XXX	Tri-21
BCAR3	1	1	0.91	1.03
F3	1	1	0.96	1.06
PRKCE	2	1	1.03	1.08
MME	3	1	0.98	1.02
XRCC4	5	1	1.1	0.92
IL4	5	1	0.97	1.07
MYC	8	1	0.99	1.03
B2M	15	1	1	0.87
ZNF198	13	1	1.07	1.01
BRCA2	13	1	0.96	0.99
CCNA1	13	1	0.93	1
DLEU1	13	1	1.03	1.02
RB1	13	1	0.95	0.98
ABCC4	13	1	1.07	1
ING1	13	1	1.02	1.07
P85SPR	13	1	1.02	1.07
TYMS	18	1	0.97	1
GATA6	18	1	1	1.03
CDH2	18	1	0.99	0.96
DCC	18	1	1.14	0.97
PMAIP1	18	1	0.96	1.02
BCL2	18	1	1.02	1.07
SERPINB2	18	1	0.96	1
NFATC1	18	1	0.94	0.95
STCH	21	1	0.94	**1.33**
NCAM2	21	1	0.97	**1.4**
APP	21	1	0.96	**1.46**
TIAM1	21	1	1	**1.56**
TMEM1	21	1	0.99	**1.38**
SIM2	21	1	1.13	**1.57**
TFF1	21	1	1.02	**1.54**
S100B	21	1	98	**1.52**
PPEF1	X	1	**1.38**	0.93
AR	X	1	**1.39**	0.96
PDCD8	X	1	**1.44**	0.98
L1CAM	X	1	**1.39**	0.97
SRY	Y	0	0	0
EIF1AY	Y	0	0	0
UTY	Y	0	0	0
CBY	Y	0	0	0

and 5' untranslated region, and two extra probes for exon 15 due to its large size. As a control, 12 probes for other human genes located on different chromosomes were included in the same MLPA assay. Using DNA samples from the Swedish APC registry, a deletion of exons 11-13 was confirmed in one DNA sample and a novel total gene deletion was identified in DNA from a patient suffering from

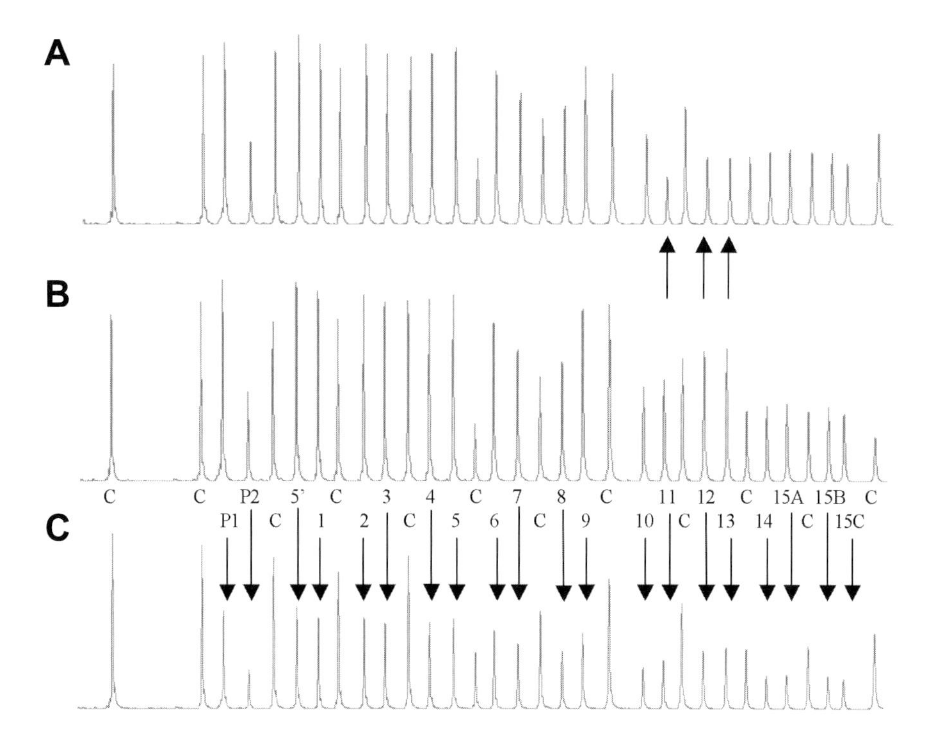

Figure 3. Detection of deletions in the APC gene with MLPA. 100 ng of DNAs from individuals diagnosed with familiar adenomatous polyposis (FAP) were analyzed by MLPA using probe mix P043. Probes were designed for each of the 15 APC exons, (1-15A), two for the promoter, (P1-P2), one specific for the 5' untranslated region (5') and two extra probes for the largest exon 15 (15B and 15C). The 12 controls probes (C) specific for different genes on different chromosomes were also designed and used in a single MLPA assay together with APC probes. The specific recognition sequence of each probe of P043 can be obtained at www.mrc-holland.com. A: A CE pattern obtained from DNA with a heterozygous APC exon 11-13 deletion. B: A CE pattern obtained from a control DNA sample containing two copies of the normal APC gene. C: A CE pattern obtained from DNA with heterozygous deletion of a complete APC gene including promoter and 5'-untranslated region. Arrows point to the decreased signals corresponding to the exon deletions. For quantitative analysis of the data in Figure 3 see Table 2.

Table 2. Quantitative analysis of the data shown in Figure 3 obtained from a DNA of a patient with a heterozygous deletion of a complete APC gene.

Probe	Length (bp)	Ratio (case/ctrl)[a]
C	130	0.94
P1	139	**0.49**
P2	148	**0.55**
C	157	0.83
5'	166	**0.5**
1	175	**0.51**
C	184	0.95
2	193	**0.51**
3	202	**0.5**
C	211	1.05
4	220	**0.53**
5	229	**0.6**
C	238	1.04
6	247	**0.54**
7	256	**0.58**
C	265	0.99
8	274	**0.51**
9	283	**0.56**
C	292	0.94
10	301	**0.55**
11	310	**0.59**
C	319	1.17
12	328	**0.55**
13	337	**0.55**
C	346	1.12
14	355	**0.55**
15A	364	**0.57**
C	373	1.14
15B	382	**0.63**
15C	391	**0.57**
C	400	1.07

[a]Expected ratio for two copies of an exon is 1.00.

FAP (Figure 3) (Kanter-Smoler, unpublished). As seen in table 2, the signal strengths of the probes specific to the APC gene were 2-fold lower in the patient compared to the control, corresponding to only one copy of the APC gene.

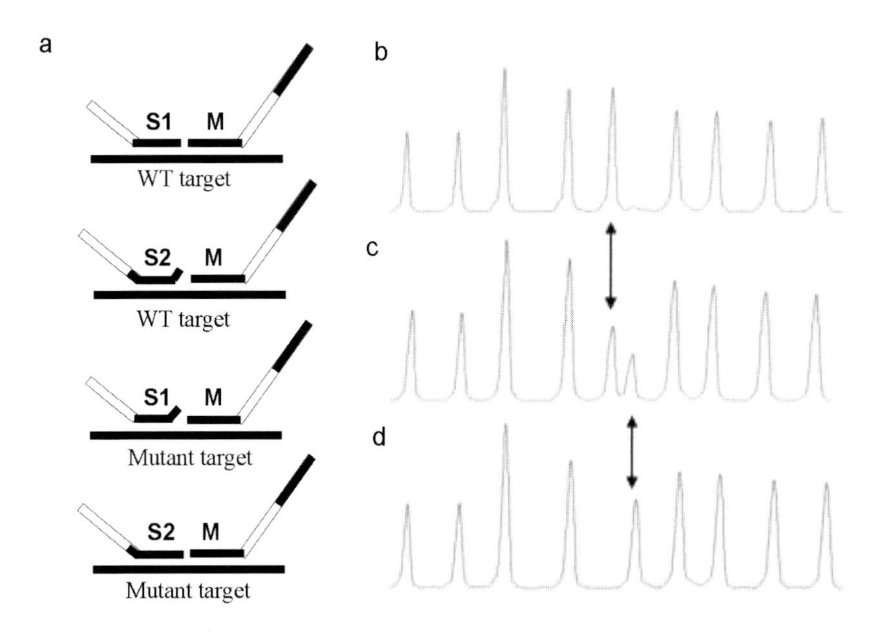

Figure 4. MLPA analysis of SNPs and other small mutations. (a) Outline of the assay. Wild-type (W) and mutant (M) sequences differ in 1 or few nucleotides. The same is true in case of biallelic SNPs. Two short synthetic oligonucleotides S1 and S2 are specific for one of the polymorphic sequences. In addition, S1 and S2 differ in length due to the presence of a 3 nt stuffer sequence between the hybridizing sequence and the PCR primer sequence of oligonucleotide S2. Both oligonucleotides anneal to target sequences W and M and compete with each other for binding. A mismatch adjacent to the ligation site prevents oligonucleotide ligation when S1 is bound to the mutant sequence, or when S2 is bound to the wild type target sequence. (b-d) Detection of the CFTR ΔF508 mutation by MLPA. Normal DNA from a healthy individual (b), heterozygous (c) and homozygous DNAs (d) with the cystic fibrosis transmembrane regulator (CFTR) ΔF508 mutation were tested by MLPA. The probe mix contained 40 different probes including a probe for the wild type CFTR sequence and a probe specific for ΔF508 mutation. Only part of the capillary electrophoresis profile is shown. Arrows indicate wild type (202 bp) and mutation-specific (205 bp) amplification products.

Detection of Small Mutations and Single Nucleotide Polymorphisms

MLPA signal is completely absent if the short probe oligonucleotide has a mismatch at the 3' nucleotide when annealed to the target sequence (Figure 4). This exquisite sensitivity of Ligase-65 for mismatches next to the ligation site can be used to distinguish single base mutations, such as single nucleotide polymorphisms (SNPs) and other small mutations. In order to obtain signals from both alleles, two different synthetic oligonucleotides, each specific for one allele, are added together with a common M13-derived oligonucleotide (Figure 4). These synthetic oligonucleotides have different 3'-end nucleotides, corresponding to the respective nucleotides in wild-type and mutant alleles or SNPs and differ also by 3 nucleotides in length. These two oligonucleotides compete with each other for binding to the target sequences, as a single mismatch at the end of the probe oligonucleotide is usually not sufficient to completely destabilize the hybrids. However, ligation of hybrids with mismatches next to the ligation site occurs with very low efficiency. Therefore, these probes are not ligated and can not be amplified by PCR. As a result, corresponding products are missing from the profile. Figure 4 shows the results of MLPA with three DNA samples: control normal DNA, DNA heterozygous for the ΔF508 mutation, and DNA from a cystic fibrosis patient homozygous for ΔF508 mutation. Note that the probe specific for the ΔF508 mutant allele resulted in a 3 bp longer amplification product than the product from the wild-type DNA.

DNA Methylation Studies Using MLPA

MLPA can be used as a tool to analyze in a semi-quantitative manner methylation status of several CpG regions in one tube. The probe design is similar to that previously described except that a site for a methylation sensitive restriction enzyme (*e.g. Hpa*II or *Hha*I) should be located within or near the ligation site. Upon digestion of DNA with the restriction enzyme, the signal will be generated only if the site is methylated, and will be absent if the CpG site is unmethylated (Figure 5).

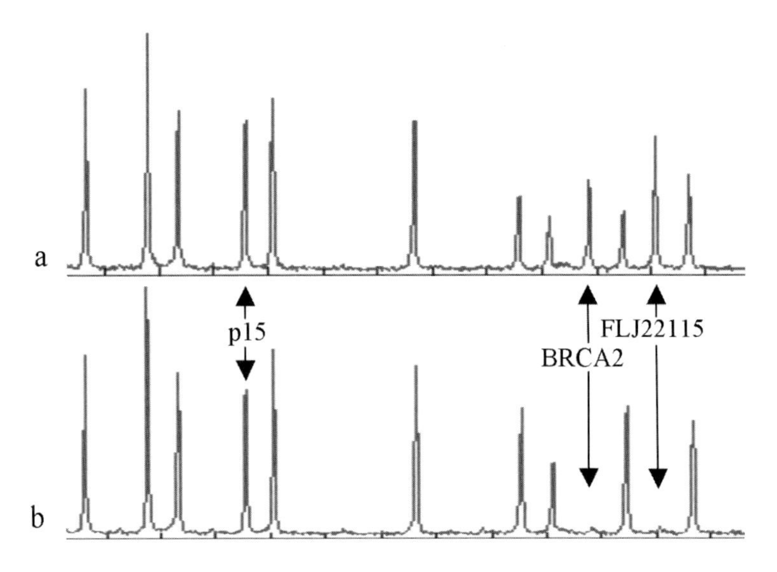

Figure 5. Detection of the methylation status of promoters with MLPA. The probe is designed in such a way that the ligation site is located within or in close proximity to the recognition site for a methylation-sensitive restriction enzyme, such as *Hpa*II or *Hha*I. Upon digestion of DNA with the restriction enzyme, the signal is generated only if the CpG site is methylated, because the corresponding site will not be digested. Only part of the electropherogram is shown. Arrows indicate the position of a 211 bp amplicon specific for the p15 promoter region on chromosome 9, a 301 bp amplicon specific for the BRCA2 promoter region on chromosome 13 and a 319 bp amplicon specific for the FLJ22115 ORF on the chromosome 20. It is not methylated and thus serves as a positive control for the digestion (negative control for MLPA). The other signals do not contain an *Hpa* II restriction site and serve as positive controls in MLPA. The gene names and specific probe recognition sites can be obtained at www.mrc-holland.com. a, A CE pattern obtained from a DNA sample without digestion with the methylation sensitive restriction enzyme *Hpa* II. b, A CE pattern obtained from the same DNA sample after digestion with *Hpa* II. The results show that the p15 promoter is methylated in both alleles while the promoter of BRCA2 is not methylated.

Discussion

Probe Specificity

When 40 probes, each consisting of two oligonucleotides, are present in one tube with genomic DNA, not 40 but 1600 different target sequences are theoretically able to hybridize with two oligonucleotides and make them a substrate for ligation. However, the results show that specificity of MLPA is very high and no non-specific products are generated by PCR (Figures 2-5). In control experiments we also showed that no product was detected when the short synthetic oligonucleotide was omitted or replaced by an identical oligonucleotide with a mismatch at the 3' end (data not shown). Several factors contribute to exquisite specificity of MLPA: (i) the need for the two oligonucleotides to anneal in juxtaposition on a target nucleic acid; (ii) hybridization and ligation performed at 54–60°C which reduces unspecific binding and increases specificity; (iii) the use of thermophilic ligases requiring NAD (*e.g.* Ligase-65) that are very sensitive to mismatches next to the ligation site.

Probe Signal and Reproducibility

Twenty eight probes specific for targets on chromosomes X, 21, 18 and 13 gave on average a 1.44-fold higher signal on the respective trisomy DNA samples as compared with control DNA (Table 1) (Schouten *et al.*, 2002). This is close to the theoretical 1.5-fold increase, and compares well with the 1.31-fold higher relative signal obtained with cDNA microarrays upon analysis of human trisomy DNA (Pollack *et al.,* 1999). Reproducibility of results in MLPA depends on several factors: the purity of DNA samples, the method of PCR product analysis (capillary electrophoresis being superior to fluorescent detection on slab gels), apparatus and software used for peak detection, and on the method used for normalization of data. We found that DNA degradation barely influenced results, but the presence of PCR inhibitors, for example traces of phenol, did. In general, the standard deviation of results obtained was between 4 and 10% for each probe.

We have shown that signal strength depends on the amount of the target sequences (Table 1). In model experiments we showed that an almost linear increase of the signal can be obtained with up to 5-fold increase of synthetic targets added to a DNA sample (data not shown). Addition of higher amounts of synthetic target resulted in an additional non-linear increase in signal. We also observed that the nature of the first nucleotide incorporated by DNA polymerase (first nucleotide following the primer) has a large effect on the amount of amplification products. When the first nucleotide to be incorporated is adenine, a smaller peak is usually obtained (Schouten *et al.,* 2002), while incorporation of cytosine results in 2–3-fold higher signals. This effect may be due to the higher stabilization of annealed primers by elongation with a C or G during the first seconds of PCR as compared with the incorporation of an A.

Applications

Our results show that MLPA was successfully used for detection of aberrant number of copies of both sex and autosomal chromosomes. The possibility to analyze quantitatively several chromosomal markers in one assay starting with a very small amount of DNA makes the MLPA analysis unique among other approaches. As we showed, this method is also capable of detecting deletions and amplifications of complete genes, single exons and small mutations including SNPs. Finally, our results show that methylation status of different genes can be studied using MLPA approach.

It should be emphasized that MLPA is an ideal method for screening of poor quality DNA, such as DNA isolated from frozen or paraffin-embedded formaldehyde-fixed cancer tissues. Indeed, the recognition sequence for each probe is only 60 to 70 nucleotides long and up to 45 independent markers are tested in parallel. Therefore even the degraded DNA can be analyzed using this approach. For example, MLPA is already routinely used for detection of amplifications of the Her2/neu gene from paraffin-embedded formaldehyde-fixed breast cancer tissues. (Nederlof *et al.*, unpublished).

MLPA can also be used to generate DNA fingerprints to test clonality of tumors and to identify and compare cell lines. It is well known that chromosomal rearrangements and copy number changes may result in large differences in behavior between supposedly 'identical' cell lines obtained from different laboratories.

Despite its versatility, there are some limitations to MLPA applications. First, development of MLPA probes for new assays is difficult and time-consuming. Each probe requires design, preparation of M13 clone, and purification of single-stranded DNA and digestion of this DNA with enzymes. However, once clones have been made, there is practically unlimited source of DNA probes. MLPA is more sensitive to contaminants than conventional PCR, therefore to obtain reliable data, all DNA samples should be of equal quality and isolated in a similar way.

MLPA is not suitable to detect new mutations; additionally, unknown polymorphism close to the ligation site can give a false positive signal. MLPA, as it is performed today, cannot be used to study single cells. On heterogeneous cell populations, MLPA can be used to detect deletions or amplifications of a certain gene if more than 60% cells are aberrant.

In conclusion, MLPA method has proved to be very sensitive, reproducible and capable of simultaneous quantitative analysis of up to 45 genes. Its exquisite specificity overweighs difficulties in probe preparations and holds promise to be a widely used technique in quantitative studies of multiple genes.

References

Armour, J.A., Sismani, C., Patsalis, P. C. and Cross, G. (2000). Measurement of locus copy number by hybridization with amplifiable probes. Nucleic Acids Res. *28*, 605–609.

Barany, F. and Lubin, M. (1997). Patent application WO9745559A1. Detection of nucleic acid sequence differences using coupled ligase detection and polymerase chain reactions.

Bodmer, W.F, Bailey, C.J, Bodmer, J., Bussey H. J., Ellis, A., Gorman, P., Lucibello, F. C., Murday, V.A., Rider, S.H., Scambler, P. (1987). Localization of the gene for familial adenomatous polyposis on chromosome 5. Nature. *328*, 614-616.

BRCA1 Exon 13 Duplication Screening Group. (2000). The exon 13 duplication in the BRCA1 gene is a founder mutation present in geographically diverse populations. Am. J. Hum. Genet. *67*, 207–212.

Carrino, J .J. (1996). Patent application WO9615271A1. Multiplex ligations-dependent amplification.

Casilli, F., Di Rocco, ZC., Gad, S., Tournier, I., Stoppa-Lyonnet, D., Frebourg, T., and Tosi M. (2002). Rapid detection of novel BRCA1 rearrangements in high-risk breast-ovarian cancer families using multiplex PCR of short fluorescent fragments. Hum Mutat. *20*, 218-26.

Devilee, P., Cleton-Jansen, A.-M. and Cornelisse,C. J. (2001). Ever since Knudson. Trends Genet. *17*, 569–573.

Esteller M. (2002). CpG island hypermethylation and tumor suppressor genes: a booming present, a brighter future. Oncogene. *21*, 5427-40.

Esteller M. (2003). Relevance of DNA methylation in the management of cancer. Lancet Oncol. 4, 351-358.

Fearnhead, N.S., Britton, M. P., and Bodmer, W. F. (2001) The ABC of APC. Hum Mol Genet. 10, 721-33.

Feo, S., Di Liegro, C., Jones, T., Read, M. and Fried, M. (1994). The DNA region around the c-myc gene and its amplification in human tumor cell lines. Oncogene. *9*, 955–961.

Gille, J.J., Hogervorst, F.B., Pals, G., Wijnen, J.T., van Schooten, R.J., Dommering, C.J., Meijer, G.A., Craanen, M.E., Nederlof, P.M., de Jong, D., McElgunn, C.J., Schouten, J.P., and Menko, F.H,. (2002). Genomic deletions of MSH2 and MLH1 in colorectal cancer families detected by a novel mutation detection approach. Br J Cancer. *87*, 892-897.

Harris, A. (1992). Cystic fibrosis gene. Br Med Bull. *48*, 738-753.

Housby, J. H. and Southern, E. M. (1998). Fidelity of DNA ligation. Nucleic Acids Res. *26*, 4259–4266.

Hsuih, T. C. H., Park, Y. N., Zaretski, C., Wu, F., Tyagi, S., Kramer, F. R., Sperling, R. and Zhang, D. Y. (1996). Novel, ligation-dependent PCR assay for detection of hepatitis C virus in serum. J. Clin. Microbiol. *34*, 501–507.

Jones, C., Payne, J., Wells, D., Delhanty, J. D., Lakhany, S. R. and Kortenkamp, A. (2000). Comparative genomic hybridization reveals extensive variation among different MCF-7 cell stocks. Cancer Genet. Cytogenet. *117*, 153–158.

Kallioniemi, A., Kallioniemi, O. P., Rutovitz, D., Gray, J. W., Waldman, F. and Pinkel, D. (1992). Comparative genomic hybridization for molecular cytogenetic analysis of solid tumors. Science *258*, 818–821.

Kauraniemi, P., Barlund, M., Monni, O. and Kallioniemi, A. (2001). New amplified and highly expressed genes discovered in the ERBB2 amplicon in breast cancer by cDNA microarrays. Cancer Res. *61*, 8235–8240.

Klinger, K., Landes, G., Shook, D., Harvey, R., Lopez, L., Locke, P., Lerner, T., Osathanondh, R., Leverone, B., and Houseal, T. (1992). Rapid detection of chromosome aneuploidies in uncultured amniocytes by using fluorescence *in situ* hybridization (FISH). Am. J. Hum. Genet. *51*, 55–65.

Knudson, A.G. Jr. 1984. Genetic predisposition to cancer. Cancer Detect Prev. 7, 1-8.

Landegren U., Kaiser R., Sanders J., and Hood L. (1988). A ligase-mediated gene detection technique. Science. *241*, 1077-1080.

Leyland-Jones, B. and Smith, I. (2001). Role of herceptin in primary breast cancer. Oncology, *61*(Suppl. 2), 83–91.

Liang, P. and Pardee, A. B. (1992). Differential display of eukaryotic messenger RNA by means of the polymerase chain reaction. Science. *257*, 967–971.

Messing, J., Gronenborn, B., Muller-Hill, B. and Hans Hopschneider, P. (1977). Filamentous coliphage M13 as a cloning vehicle: insertion of a HindII fragment of the lac regulatory region in M13 replicative form *in vitro*. Proc. Natl Acad. Sci. USA. *74*, 3642–3646.

Norman, A. M., Thomas, N. S. T., Kingston, H. M. and Harper, P. S. (1990). Becker muscular dystrophy: correlation of deletion type with clinical severity. J. Med. Genet. *27*, 236–239.

Pauletti, G., Godolphin, W., Press, M. F. and Slamon, D. J. (1996). Detection and quantitation of HER-2/neu gene amplification in human breast cancer archival material using fluorescence *in situ* hybridization. Oncogene, *13*, 63–72.

Petrij-Bosch, A., Peelen, T., van Vliet, M., van Eijk, R., Olmer, R., Drusedau, M., Hogervorst, F. B., Hageman, S., Arts, P. J., and Ligtenberg, M. .J. (1997). BRCAI genomic deletions are major founder mutations in Dutch breast cancer patients. Nature Genet. *17*, 341–345.

Pollack, J. R., Perou,C.M., Alizadeh, A. A., Eisen, M. B., Pergamenschikov, A., Williams, C. F., Jeffrey, S. S., Botstein, D. and Brown, P. O. (1999). Genome-wide analysis of DNA copy-number changes using cDNA microarrays. Nature Genet. *23*, 41–46.

Schouten, J. P. (1985) Hybridization selection of covalent nucleic acid-protein complexes. 2. Cross-linking of Escherichia coli mRNAs and DNA sequences by formaldehyde treatment of intact cells. J. Biol. Chem., *260*, 9929–9935.

Schouten, J.P., McElgunn, C.J., Waaijer, R., Zwijnenburg, D., Diepvens, F., and Pals, G. (2002). Relative quantification of 40 nucleic acid sequences by multiplex ligation-dependent probe amplification. Nucleic Acids Res. *30*, e57.

Snijders, A. M., Nowak, N., Segraves, R., Blackwood, S., Brown, N., Conroy, J., Hamilton, G., Hindle, A. K., Huey, B., Kimura, K. *et al.*, (2001). Assembly of microarrays for genome-wide measurement of DNA copy number. Nature Genet. *29*, 263–264.

Southern, E. M. (1975) Detection of specific sequences among DNA fragments separated by gel electrophoresis. J. Mol Biol. *98*, 503-17.

Stenson, P. D., Ball, E. V., Mort, M., Phillips, A. D., Shiel, J. A., Thomas, N. S., Abeysinghe, S., Krawczak, M., and Cooper, D. N. (2003) Human Gene Mutation Database [HGMD(R)]: 2003 update. Hum Mutat. *21*, 577-581.

Spek, C.A., Verbon, A., Aberson, H., Pribble, J.P., McElgunn, C.J., Turner, T., Axtelle, T., Schouten, J., Van Der Poll, T., and Reitsma, P.H. (2003). Treatment with an anti-CD14 monoclonal antibody delays and inhibits lipopolysaccharide-induced gene expression in humans *in vivo*. J Clin Immunol. *23*, 132-140.

Tong,J., Cao,W. and Barany, F. (1999). Biochemical properties of a high fidelity DNA ligase from Thermus species AK16D. Nucleic Acids Res. *27*, 788–794.

Vos, P., Hogers, R., Bleeker, M., Reijans, M., van de Lee, T., Hornes, M., Frijters, A., Pot, J., Peleman, J. and Kuiper, M. (1995). AFLP: a new technique for DNA fingerprinting. Nucleic Acids Res. *23*, 4407–4414.

Wijnen, J., van der Klift, H., Vasen, H., Khan, P. M., Menko, F., Tops, C., Meijers Heijboer, H., Lindhout, D., Moller, P. and Fodde, R. (1998). MSH2 genomic deletions are a frequent cause of HNPCC. Nature Genet. *20*, 326–328.

Section 3

Isothermal Methods of DNA Amplification

3.1

Homogenous Real-Time Strand Displacement Amplification

David M. Wolfe, Sha-Sha Wang, Keith Thornton, Andrew M. Kuhn, James G. Nadeau and Tobin J. Hellyer

Abstract

Strand Displacement Amplification (SDA) is an isothermal DNA amplification technique that may be coupled with real-time homogenous fluorescent detection for a variety of different applications. Here we describe the principles of SDA and provide protocols for the qualitative detection of pathogens in clinical specimens as well as the analysis of Single Nucleotide Polymorphisms (SNPs). Both procedures employ a novel universal detection format that permits the use of the same Detector Probe(s) across multiple assays, thereby reducing the cost and complexity of assay development. Coupled with the BD ProbeTec™ ET System, SDA offers a simple, streamlined workflow that is amenable to high-throughput clinical or research applications.

Background

Isothermal SDA-based detection of specific nucleic acid sequences involves the coordinated activity of two enzymes, a DNA polymerase and restriction endonuclease, to bring about exponential amplification of a target DNA fragment (Figure 1; Walker *et al.*, 1992a,b). This process relies upon the ability of: 1) the restriction enzyme to nick the unmodified strand of a hemiphosphorothioate form of its recognition sequence; and 2) the DNA polymerase to initiate replication at the site of the nick and to displace the downstream, non-template strand of DNA into solution. SDA offers the potential for extremely rapid doubling times with the possibility of 10^{10}-fold amplification in as little as 15 min (Spargo *et al.*, 1996). The isothermal nature of the reaction process reduces the complexity of instrumentation required relative to PCR and the technology is well suited to applications requiring a robust amplification platform. In addition to its use in clinical diagnostics (Little *et al.*, 1999), SDA has also proven amenable to miniaturization for industrial/research chip-based multiplexed applications (Westin *et al.*, 2000).

Figure 1. Schematic diagrams of SDA. S1 and S2: SDA Primers; B1 and B2: Bumper Primers; *Bso*BI restriction enzyme recognition sites within SDA Primers are designated with thick lines; nicking by *Bso*BI is indicated by small arrows; nicked/partial *Bso*BI restriction sites are shown with dotted double lines.
A: Generation of target amplicon: 1) Bumper Primer B1 hybridizes to single-stranded DNA target (obtained by prior heat denaturation) upstream of SDA Primer S1. 2) DNA polymerase extension from the 3' ends of B1 and S1 results in the displacement of the S1 extension product into solution. 3) Capture of the S1 extension product by SDA Primer S2 and upstream Bumper Primer B2. 4) Extension from the 3' end of B2 displaces the downstream S2 extension product into solution. 5) Hybridization of an S1 primer to the displaced S2 extension product. 6) Extension from the 3' end of the hybridized S1 primer results in formation of a double-stranded molecule with nickable restriction sites at either end. 7) Nicking and DNA polymerase extension from the nick sites displaces single-stranded molecules into solution that possess partial *Bso*BI sites at either end and which feed into the exponential phase of SDA depicted in Figure 1B.
B: Exponential amplification: 1) Displaced single-stranded molecules generated by the sequence of events depicted in Figure 1A hybridize to SDA Primers S1 and S2. 2) DNA polymerase extension, nicking, displacement and re-generation of the nick site. 3) Displaced single-stranded molecules with partial *Bso*BI sites at either end circulate back into step (1) to bring about exponential amplification.

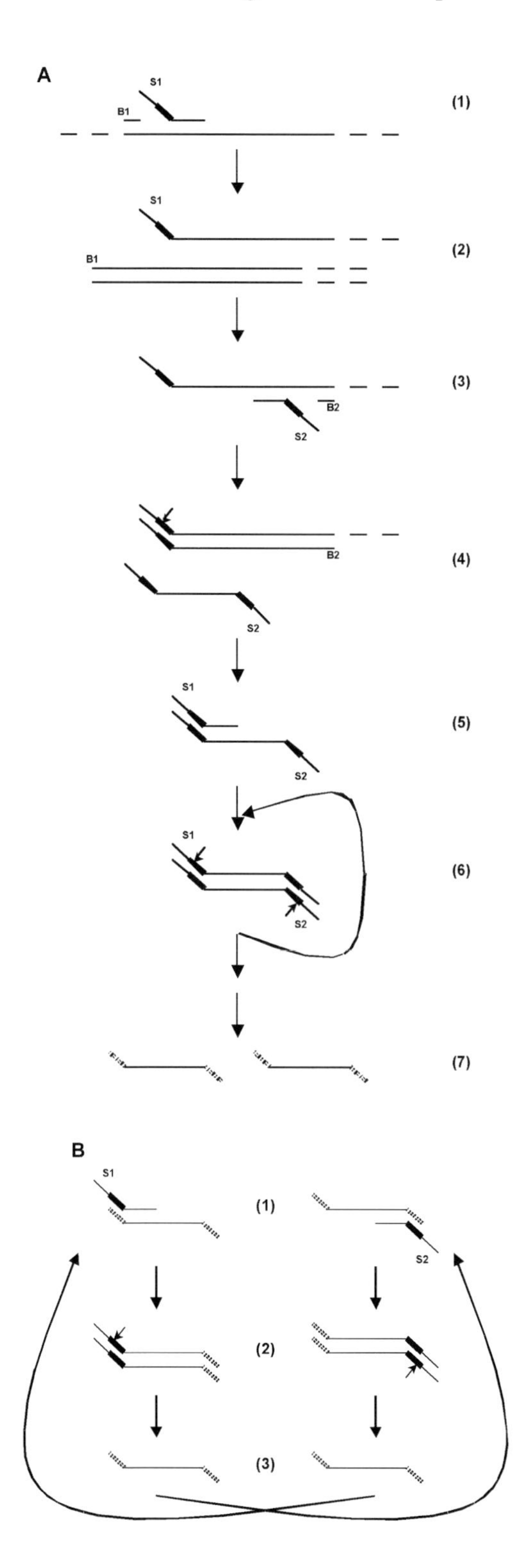
A
S1
B1
(1)
S1
B1
(2)
B2
S2
(3)
B2
S2
(4)
S1
S2
(5)
S1
S2
(6)
(7)
B
S1
(1)
S2
(2)
(3)

The original mesophilic amplification process described in 1992 in two pioneering papers by Walker *et al.* was subsequently modified to employ thermophilic enzymes that permit use of higher temperatures, with a corresponding improvement in the specificity of primer hybridization, faster reaction kinetics and reduced incubation times (Spargo *et al.*, 1996; Milla *et al.*, 1998). In 1999, Nadeau *et al.* reported the development of a real-time homogeneous fluorescent reporter system for the detection of SDA products. This technology now forms the basis of the BD ProbeTec™ ET System (BD Diagnostics, Sparks, MD) that is specifically designed for high-throughput applications in the clinical laboratory (Little *et al.*, 1999). The instrument comprises a temperature-controlled kinetic fluorescence reader that monitors the target-dependent increase in fluorescence associated with conformational changes in energy transfer (ET) reporter probes. The probes are labeled with both a fluorophore (donor) and a fluorescence-quenching moiety (acceptor) such that, in the absence of target, fluorescence of the donor is quenched. When target is amplified, the probes undergo a conformational change, resulting in separation of the donor and acceptor and an increase in fluorescence.

Using this technology, qualitative assays have been developed for the detection of *Chlamydia trachomatis* and *Neisseria gonorrhoeae* as well as for other clinically important analytes (Van Der Pol *et al.*, 2001; Akduman *et al.*, 2002; Johansen *et al.*, 2002; Maugein *et al.*, 2002; Iinuma *et al.*, 2003; Mazzarelli *et al.*, 2003; Verkooyen *et al.*, 2003). Through the use of two optical channels, the BD ProbeTec ET instrument is designed to facilitate simultaneous amplification and detection of two different DNA sequences. For diagnostic applications, this permits the incorporation of an Internal Amplification Control (IAC) sequence that co-amplifies with the target nucleic acid and which may be used to monitor for inhibitory specimens and verify negative results (Johansen *et al.*, 2002; Maugein *et al.*, 2002; Iinuma *et al.*, 2003; Mazzerelli, *et al.*, 2003). Competition for rate-limiting reagents between an IAC and a native target sequence also forms the basis for quantitative SDA (Nycz *et al.*, 1998; Nadeau *et al.*, 1999). In this chapter, we have limited our discussion to the qualitative detection of DNA analytes and the analysis of SNPs, although SDA may also be coupled with a reverse transcriptase enzyme for the detection of RNA sequences (Mehrpouyan *et al.*, 1997; Nycz *et al.*, 1998; Hellyer *et al.*, 1999; Nadeau *et al.*, 1999; Nuovo, 2000; Westin *et al.*, 2000).

Recently, we have refined the real-time SDA detection format originally described by Nadeau *et al.* (1999) to permit use of the same pair of fluorescent Detector Probes across multiple assays. This Universal Detection system is based upon the target-dependent extension of an unlabeled Adapter Primer that comprises a target-specific sequence and 5' generic tail (Figure 2). During the course of amplification, the complement of the generic tail is synthesized and captured by a fluorescently labeled Detector Probe that then undergoes a conformational change to generate a detectable signal. Such a Universal Detection format has been applied to the qualitative detection of DNA targets as well as the analysis of SNPs (Wang *et al.*, 2003). The principles behind SDA-based SNP analysis are illustrated in Figure 3. The system is based upon the relative efficiency of DNA polymerase extension from the 3' ends of Adapter Primers that differ by a single nucleotide. Each Adapter Primer is specific for one of two allelic variants (*i.e.* either allele A or allele B at a given locus) and the relative signal intensity generated by each primer is used to determine the genotype of the sample. In order to reduce the efficiency of extension of a mismatched sequence and thereby maximize discriminatory power, the SNP site is located one base from the 3' end of the Adapter Primers. Use of the same 5' tail sequences for Adapter Primers across different assays permits discrimination of multiple SNPs using a single pair of labeled Detector Probes.

Here we outline protocols for the qualitative detection of *Chlamydophila pneumoniae* DNA in clinical specimens as well as the analysis of SNPs within the human β_2-adrenergic receptor (β_2AR) gene. The SDA protocols described below are specifically adapted to work on the BD ProbeTec ET System utilizing the same hardware as the existing commercial assays (Figure 4; Little *et al.*, 1999). Modification of these procedures may be necessary for use of alternative fluorescent readers.

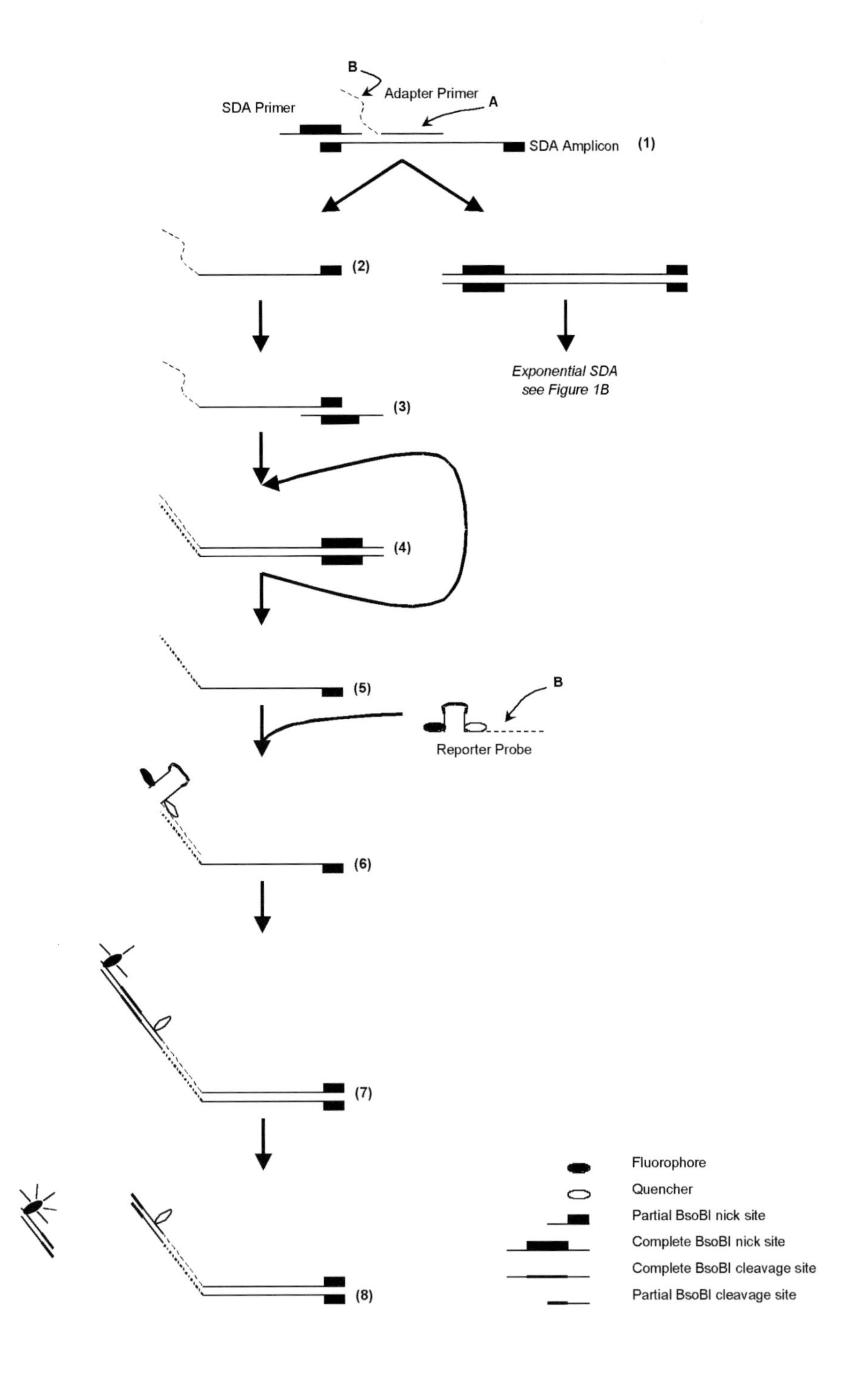
B
Adapter Primer
A
SDA Primer
SDA Amplicon
(1)
(2)
Exponential SDA
see Figure 1B
(3)
(4)
(5)
B
Reporter Probe
(6)
(7)
(8)
Fluorophore
Quencher
Partial BsoBI nick site
Complete BsoBI nick site
Complete BsoBI cleavage site
Partial BsoBI cleavage site

Protocol for Qualitative Detection of *C. pneumoniae* DNA in Sputum

Materials

We recommend use of Nunc™ Lockwell™ C8 Polysorp® microwells (Nalge Nunc International) that have the necessary optical characteristics required for use with the BD ProbeTec ET System and which are formatted conveniently in breakable strips. Other accessories for this instrument are available from BD (http://catalog.bd.com). For homebrew applications using the BD ProbeTec ET System, data must be downloaded via floppy disc to an off-line computer for analysis.

Sample Processing - N-acetyl-L-cysteine (NALC) 1% (v/v) in water (prepared fresh daily); Wash Buffer (4 M urea, 200 mM CAPS (3-[cyclohexylamino]-1-propanesulfonic acid); phosphate-buffered saline (PBS), pH 8.0; QIAamp® DNA Blood Mini Kit (QIAGEN, GmbH); ethanol (95-100% v/v).

Amplification/Detection - SDA Buffer (274 mM bicine; 96 mM KOH; 19.2% dimethyl sulfoxide (DMSO); 13.7 mM magnesium acetate);

Figure 2. SDA with Universal Detection: An Adapter Primer (comprising an allele-specific 3' sequence (A) and 5' generic tail (B)) hybridizes to the amplified target downstream of an SDA primer (1). DNA polymerase extension from the 3' ends of both the Adapter Primer and upstream SDA Primer results in displacement of the Adapter Primer extension product into solution (2), where it is captured by a complementary SDA Primer (3), and generation of a double-stranded molecule that feeds into the exponential phase of SDA as depicted in Figure 1B. Extension from the 3' ends of the SDA Primer and captured Adapter Primer extension product generates the complement of the 5' Adapter Primer tail sequence and a double-stranded restriction recognition site (4). Nicking of the restriction site and extension from the nick displaces a single-stranded copy of the Adapter Primer complement into solution (5). The displaced sequence is captured by a complementary fluorescent Detector Probe that possesses the generic sequence (B) at its 3' end. Extension from the 3' ends of the Detector Probe and its target sequence (6) results in opening of the detector hairpin, separation of the fluorophore and quencher and the generation of target-specific fluorescence (7). Additional increase in fluorescent signal is obtained by complete separation of quencher and fluorophore via cleavage of the double-stranded Detector Probe restriction site (8).

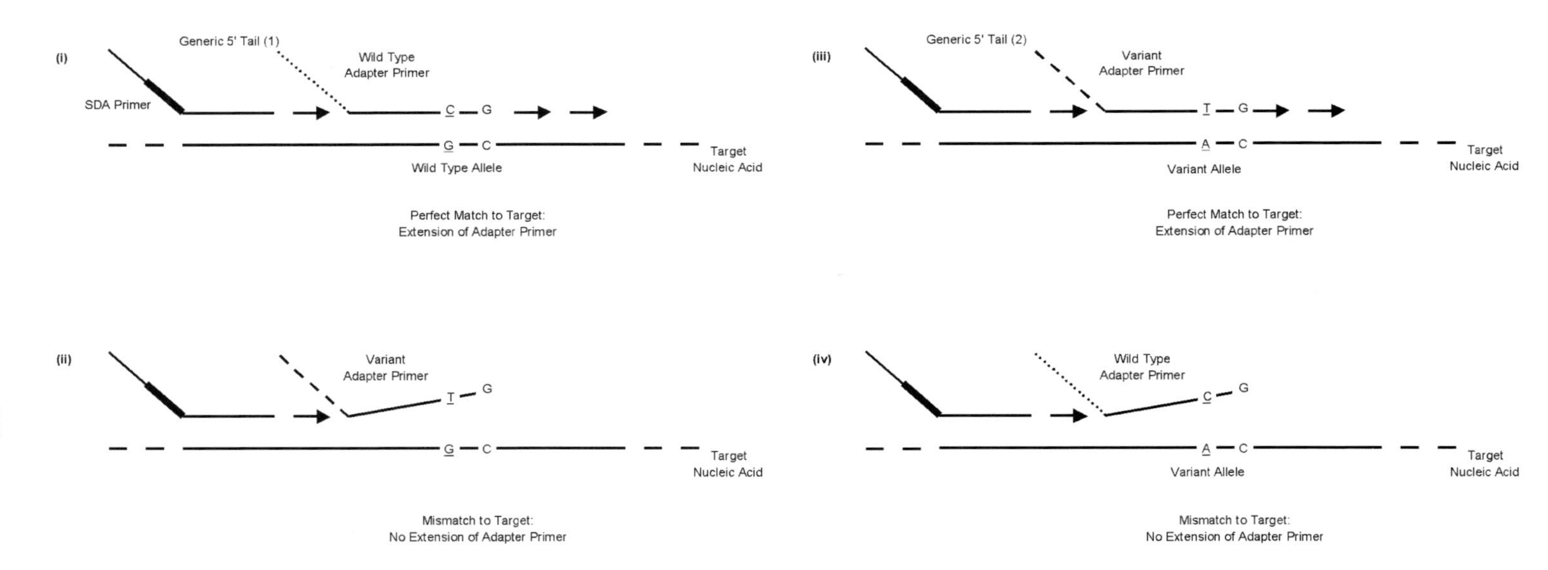

Figure 3. Allele-specific signal generation: Each SDA reaction contains two Adapter Primers with different 5' tail sequences [(1), (2)] and two fluorescent Detector Probes labeled with different dyes. Each Adapter Primer is specific for one of two allelic variants (A or B) at the targeted SNP locus. Hybridization and extension of a correctly matched Adapter Primer [(i), (iii)] results in generation of allele-specific fluorescent signal (Figure 2). The diagnostic nucleotide (underlined) is located one base (N-1) from the 3' end of the Adapter Primer to maximize discriminatory power when it is mismatched with the target sequence. A mismatch at the N-1 position substantially reduces the efficiency of extension of the Adapter Primer relative to that of the perfectly matched Adapter [(ii), (iv)]. A ratio of fluorescent signals obtained in each optical channel is used to determine the identity of the nucleotide at the targeted locus.

Priming Mix (172 mM potassium phosphate, pH 7.6; 2450 nM SDA Primer 1; 975 mM SDA Primer 2; 1200 nM Adapter Primer 1; 738 nM Adapter Primer 2; 245 nM Bumper Primers; 2450 nM Detector Probe-1 (Dabcyl/Rhodamine (ROX)); 738 nM Detector Probe 2 (Fluorescein (FAM)/Dabcyl); 2.45 mM 2'-Deoxycytidine 5'-O-(1-Thiotriphosphate), S-isomer (dC_STP); 0.49 mM dATP; 0.49 mM dGTP; 0.49 mM dTTP; 245 ng/μL bovine serum albumin (BSA); Amplification Mix (~110 U *Bst* polymerase; ~370 U *Bso*BI restriction enzyme; 150 ng/μL BSA; 43 mM potassium phosphate, pH 7.6; 25% v/v glycerol).

The SDA Buffer may be prepared in bulk and stored at ambient temperature or 4°C. The sequences of the primers, probes and internal amplification control are listed in Table 1. The primers described here are designed to detect all clinically relevant species within the *Chlamydiaceae* Family (*Chlamydia trachomatis*, *Chlamydophila pneumoniae* and *Chlamydophila psittaci*). As the most important respiratory pathogen among this group of organisms, only *C. pneumoniae* was analyzed in the example provided here. To prevent unwarranted primer interactions that may inhibit amplification of the specific target sequence or internal control, do not mix the primers together at high concentrations. Prepare the Priming Mix from individual stocks of each primer/probe on the day of assay. The Amplification Mix should also be prepared on the day of experimentation and stored on ice until use.

Step 1: Sample Preparation

Lower respiratory samples are frequently the specimen of choice for detection for *C. pneumoniae*. However, the viscosity of sputum makes it a difficult sample type with which to work and recovery of nucleic acid that is free of inhibitors of nucleic acid amplification requires preprocessing of the sample to liquefy the material and facilitate concentration of target microorganisms. Here we describe use of a NALC-based liquefaction procedure in conjunction with the QIAamp DNA Blood Mini Kit to isolate purified DNA for the detection of *C. pneumoniae*. This procedure permits the concentration of microorganisms from a large volume of sputum, thereby enhancing

Table 1. Oligonucleotide and IAC sequences for detection of *C. pneumoniae* and other members of the *Chlamydiaceae* Family

	Sequence Name	5'-3' Sequence
Bumper Primers	rnpLB	ATA AgA AAA gAT ACT ggA
	rnpRB	TTT TCT CTT gCT TCA gAT
SDA Primers	rnpLP	ACC gCA TCg AAT gAC TgT *CTC ggg* AAg gCT ACg gAA AgT
	rnpRP	CgA TTC AgC TgC AgA CgT *CTC ggg* TTC AgC CTg TCT ATA A
Adapter Primers	rnpAD	ACg TTA gCC ACC ATA Cgg ATA CAT AgC ggA gTg TTT TCT gTT g
	rnpIAC AD2	AgC TAT CCg CCA TAA gCC ATg Cgg AgT gTT gAC TTA gT
Detector Probes	TBD10.2 D/R	(Dabcyl)-TAg Cg*C CCg Ag*C gCT-(Rox)-ACg TTA gCC ACC ATA Cgg AT
	AltD6.9 F/D	(Fam)-AgT TgC *CCC gAg* gCA ACT-(Dabcyl)-AgC TAT CCg CCA TAA gCC AT
Internal Amplification Control[a]	rnpIAC2	AAg gCT ACg gAA AgT gAC TAA gTC AAC ACT CCg CTA TAA AAT TTA TTT TAT AgA CAg gCT gAA C

[a] pUC19-based plasmid clone containing mutated *C. pneumoniae* target sequence. The region shown spans the SDA Primer binding sites.

Target hybridization regions in SDA Primers and Adapter Primers are underlined; *Bso*BI sites are italicized.

D/R = dabcyl-rhodamine; F/D = fluorescein-dabcyl.

analytical sensitivity and offering the potential for improved clinical performance.

1. Mix 0.75 mL sputum with 1.0 mL fresh NALC solution in water.
2. Vortex the tubes to mix and liquefy the specimen.
3. Centrifuge at maximum speed (≥10,000 x *g*) in a microcentrifuge for 3 min to recover the organisms.
4. Carefully decant the supernatant and add 1 mL Wash Buffer.
5. Re-centrifuge as above and carefully decant the supernatant.
6. Resuspend the pellet in 200 μL PBS and transfer 200 μL of the resulting suspension to a clean 2 mL centrifuge tube.
7. Process the specimen according to the Blood and Body Fluid Spin Protocol provided in the QIAGEN product handbook (http://www1.qiagen.com/).
8. Elute the purified DNA in 200 μL QIAGEN AE Buffer.

Step 2: Amplification and Detection

Arrange two sets of empty microwells in the special metal trays designed for the BD ProbeTec ET instrument. Designate one tray as the Priming Plate, the other as the Amplification Plate (the microwells should be arranged in the same orientation on both). Use a plate map to record the location of each sample. Include wells for positive and negative amplification controls.

Following the workflow depicted in Figure 4, add 40 μL purified target DNA to 100μL SDA Assay Buffer. Denature the target by heating the mixture in a boiling water bath for 5 min. During the denaturation step, aliquot 40 μL Priming Mix to each Priming Microwell. Cool the denatured samples to room temperature and transfer 100 μL of the buffer-target mixture to the corresponding well. Cover the Priming Microwells and incubate at ambient temperature for 20 min. During this incubation, aliquot 40 μL Amplification Mix to each empty Amplification Microwell. At the end of the 20 min incubation, place the Priming Microwells on the 72°C BD ProbeTec ET Priming Heater and the Amplification Microwells on the adjacent 54°C Warming Heater. Incubate for 10 min, then transfer 100 μL liquid from each Priming

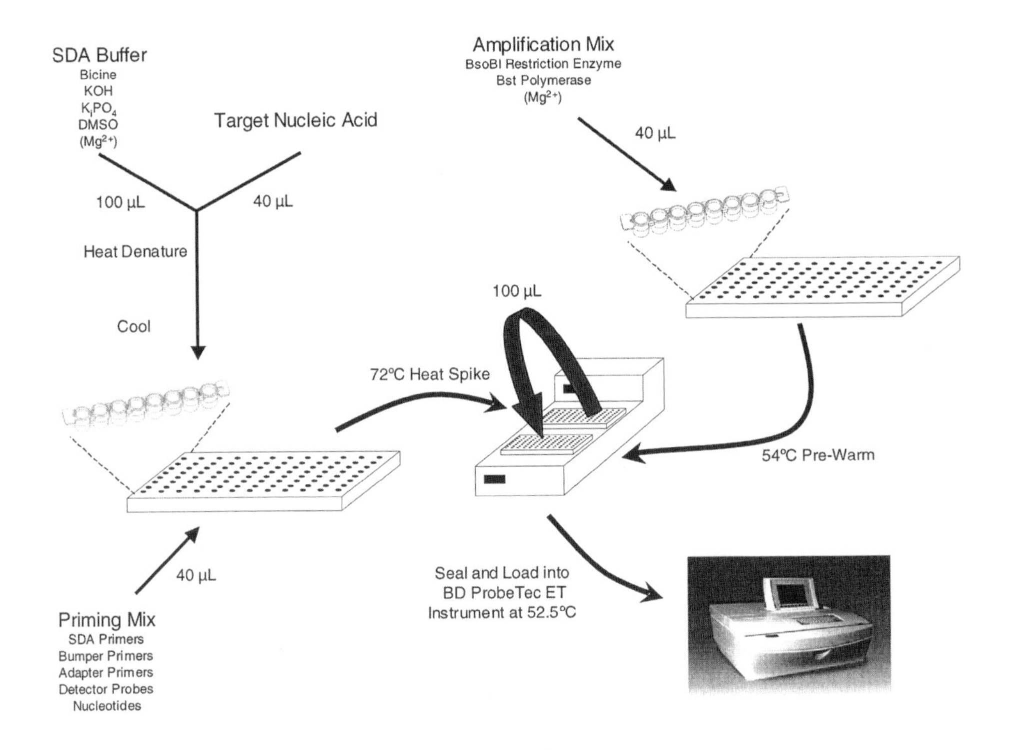

Figure 4. Workflow diagram for homebrew applications on the BD ProbeTec ET System. Target DNA is mixed with SDA Buffer and then heated to denature. After cooling, denatured target is added to microwells containing Priming Mix. At the same time, Amplification Mix is dispensed into a second set of microwells. The Priming Microwells are heated at 72°C for 10 min, while the Amplification Microwells are warmed at 54°C. Liquid is transferred using a multichannel pipette from the Priming to the Amplification Microwells. These are sealed and loaded into the instrument that is set at 52.5°C. Data are collected over 60 passes in 1 hour and downloaded to a computer for analysis.

Microwell to the corresponding Amplification Microwell and mix the contents by pipetting up and down three times. Seal the Amplification Microwells and load into the BD ProbeTec ET instrument that is set at the standard operating temperature of 52.5°C. Seal and discard the Priming Microwells. Collect fluorescence readings over a 1 hour incubation period and download the data to an off-line computer for analysis. Discard the sealed Amplification Microwells containing the SDA products.

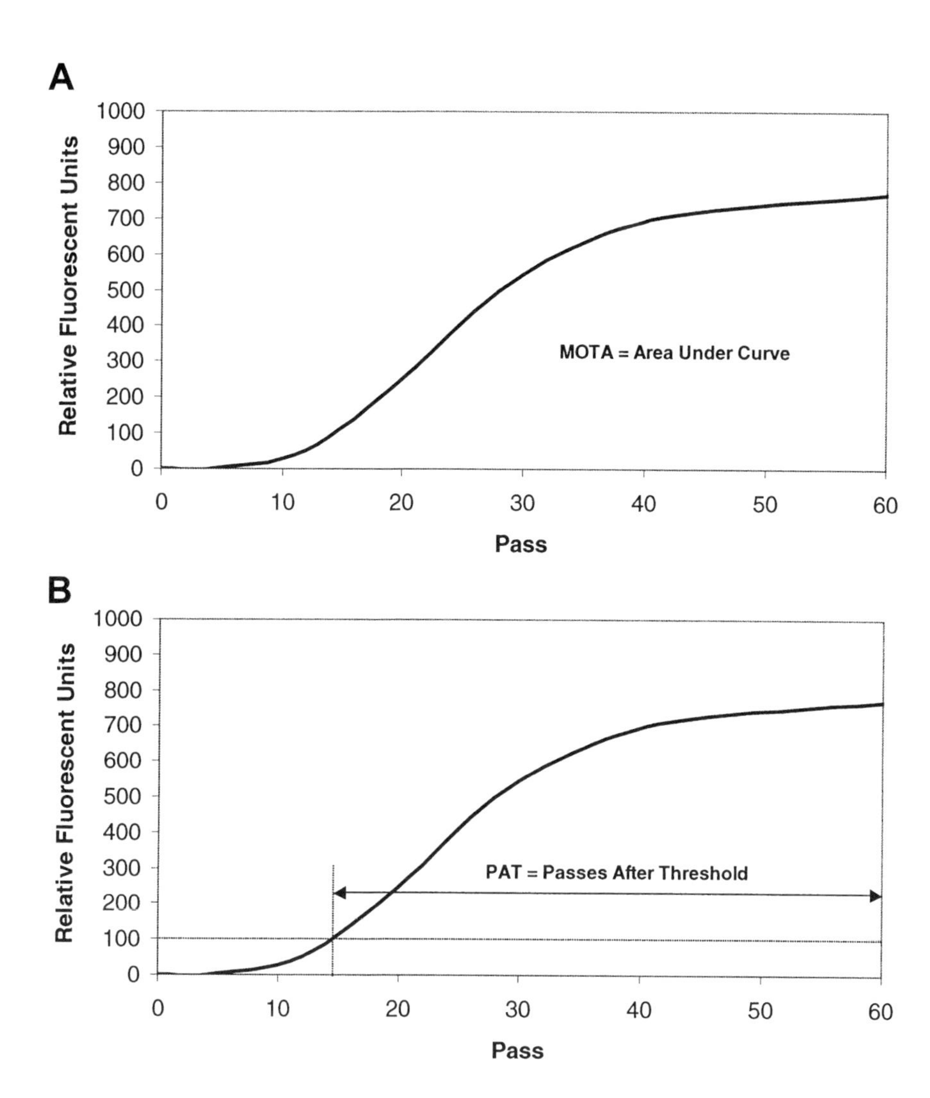

Figure 5. Qualitative algorithms used with the BD ProbeTec ET System. MOTA and PAT scores are calculated based on background-corrected fluorescent signals which are derived by determining the average fluorescent signal in each optical channel during the first few passes of the instrument and subtracting theses values from all subsequent readings.

A: MOTA scores represent the area under the amplification curve. Positive and negative results are determined by applying a predetermined MOTA cut-off. Reactions in which the MOTA threshold is not achieved in either optical channel (target sequence or IAC) are recorded as indeterminate.

B: PAT scores are calculated from the point at which the signal curve crosses a predetermined RFU threshold. All PAT values >0 are considered positive. Reactions in which the PAT threshold is not achieved in either optical channel are recorded as indeterminate.

Table 2. Detection of *C. pneumoniae* EBs in seeded sputum.

EBs			Mean Score (%CV)[a]	
Per mL Sputum	Per SDA Reaction	Number (%) Positive (n = 10)	MOTA	PAT
980	75.0	10 (100)	56718 (63)	47.0 (30)
490	37.5	10 (100)	70565 (62)	48.2 (29)
245	18.8	8 (80)	17080 (114)	31.2 (61)
122	9.4	6 (60)	7689 (80)	22.3 (52)
61	4.7	1 (10)	1477 (153)	3.0 (17)
0	0	0	327 (33)	0 (0)

[a] Coefficient of variation = (standard deviation / mean) x 100

Step 3: Data Analysis

During the 1 hour amplification, the BD ProbeTec ET reader collects 60 readings from both optical channels (corresponding to FAM and ROX). These data may be analyzed using different metrics according to the specific application. Qualitative results are typically expressed in terms of MOTA (Metric Other Than Acceleration) or PAT (Passes After Threshold) scores (Figure 5). MOTA values represent the total area under the signal curve and positive and negative results are determined by applying a MOTA cut-off. PAT values represent the number of readings after the signal curve has crossed a predetermined threshold. Although neither of these metrics is quantitative, in both cases, higher values indicate more efficient amplification. The same metrics, though with separately derived cut-offs, are applied to amplification of both the target sequence and IAC. Reactions that fail to yield an amplification signal for either sequence are considered indeterminate and should be repeated.

Example

Figure 6 shows the results of a titration of *C. pneumoniae* elementary bodies (EBs) in sputum. Residual sputum specimens from patients suspected of tuberculosis were pooled and seeded with different levels of *C. pneumoniae*. Samples were processed and amplified according to the method described above. Results were expressed in terms of

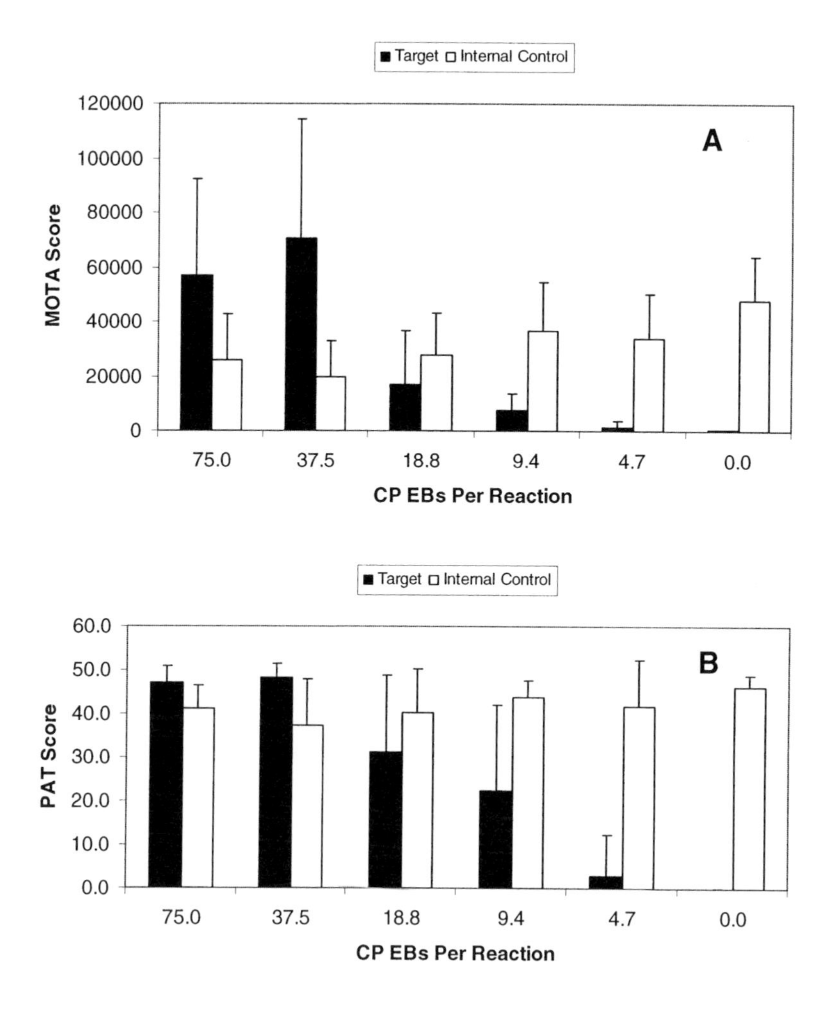

Figure 6. Detection of *C. pneumoniae* elementary bodies (CP EBs) in pooled seeded sputum (10 replicate SDA reactions per spike level). Bars represent mean MOTA (A) or PAT (B) scores. Error bars indicate standard deviations.

MOTA scores and PAT values. While both algorithms yielded the same qualitative results (Table 2), lower coefficients of variation (CV) were observed using the PAT algorithm that is more robust than MOTA to well-well variations in amplification efficiency. PAT scores are therefore a more preferable metric than MOTA values for qualitative diagnostic applications. No indeterminate results were observed in this experiment. These data demonstrate that the protocol described here permits reliable detection of <500 EBs of *C. pneumoniae*/mL sputum.

Protocol for SNP Analysis of the Human β_2AR Gene

The following protocol describes a method for SDA-based genotyping of the human β_2AR gene from blood specimens. This gene is involved in the response of patients to treatment for asthma and a total of six SNPs have been described that permit discrimination of all the major β_2AR haplotypes (Drysdale *et al.*, 2000). The target-specific regions of the SDA primers employed in detecting these SNPs are based upon the published sequence of the β_2AR gene (GenBank accession number M15169) and have been described elsewhere (Wang *et al.*, 2003).

Materials

Sample processing - PPT™ plasma preparation tubes containing K_2EDTA as the anticoagulant (BD Vacutainer Systems).

Amplification/Detection – Universal SDA Buffer (248 mM bicine; 149 mM KOH; 112 mM K_iPO_4, pH 7.6; 19.2% DMSO); Priming Mix (490-2450 nM SDA Primers, 245 nM Bumper Primers, 490-1960 nM Adapter Primers, 735-2450 nM fluorescent Detector Probes, 0.49 mM dATP, 0.49 mM dGTP, 0.49 mM dTTP, 2.45 mM dC_sTP; 245 ng/μL BSA); Amplification Mix (~110 U *Bst* polymerase; ~360 U *Bso*BI restriction enzyme; 17.5 mM magnesium acetate; 175 ng/μL BSA; 23% v/v glycerol). The same Universal SDA Buffer and Amplification Mix formulations are used for each of the six β_2AR assays. For two of the systems that target particularly GC-rich regions of the β_2AR gene (-47 and +46 loci), additional glycerol and DMSO are added to the Priming Mix to increase the stringency of reaction conditions and bring the final concentrations in the Amplification Microwell to 14% and 12.5% v/v respectively. The additional co-solvent alleviates primer-primer interactions and helps reduce non-specific amplification.

Note that the formulations used for the β_2AR assays differ from those employed in the above example for detection of *C. pneumoniae*. Most significantly, for the β_2AR assays, the magnesium required as a cofactor for the activity of the SDA enzymes is included in the

Amplification Mix, rather than the SDA Buffer. Importantly, however, in both cases the magnesium and K_iPO_4 are kept separate until being combined in the final reaction mixture. This precludes the possibility of precipitation of magnesium phosphate that might be detrimental to assay performance. The Universal SDA Buffer may be prepared in bulk and stored at ambient temperature or 4°C. To prevent unwarranted primer interactions that may inhibit amplification of the specific target sequence, do not mix the primers together at high concentrations. Prepare the Priming and Amplification Mixes from individual components on the day of assay and store on ice until use.

Step 1: Sample Preparation

Collect blood samples using PPT plasma preparation tubes. Specimens may be tested immediately or stored at -20°C. Mix the desired volume (up to 40 µL) of anticoagulated blood directly with 100 µL Universal SDA Buffer. If less than 40 µL of blood is used, bring the total volume to 140 µL with the addition of molecular biology grade water. Heat the blood-buffer mixture for 5 min in a boiling water bath to lyse the cells and denature the DNA. Cool the samples and centrifuge at ≥10,000 x *g* for 1 min to pellet cellular debris.

Step 2: Amplification and Detection

The SNP assay procedure is similar to that described for the qualitative detection of *C. pneumoniae* DNA. Prepare two sets of empty microwells as outlined above and designate one as the Priming Plate, the other as the Amplification Plate. Dispense 40 µL Priming Mix into each Priming Microwell. Directly add 100 µL of the blood-buffer supernatant to the Priming Microwells and incubate at ambient temperature for 5 min. During this incubation, dispense 40 µL of Amplification Mix into the corresponding Amplification Microwells. Place the Priming Microwells on the 72°C BD ProbeTec ET Priming Heater and the Amplication Microwells on the adjacent 54°C Warming Heater. Incubate for 10 min, then transfer 100 µL liquid from each Priming Microwell to the corresponding Amplification Microwell and mix the contents by pipetting up and down three times. Seal the Amplification Microwells and load into the BD ProbeTec ET instrument set at the standard operating temperature of 52.5°C. Seal and discard the Priming Microwells. Collect fluorescence readings over 1 hour and download the data to an off-line computer for analysis. Discard the sealed Amplification Microwells containing the SDA products.

Step 3: Data Analysis

We have developed special SNP analysis software (Wang, *et al.* 2003; BD Diagnostics) to analyze diplex (*i.e.* dual dye) BD ProbeTec ET data and generate genotyping results. Raw optical data are processed by dark correction and dynamic normalization using internal fluorescence calibrators, followed by step detection/repair and smoothing with running average and median filters to remove periodic noise. The background fluorescence is then calculated over a predetermined number of readings from early in the course of amplification and subtracted from all subsequent data points. An algorithm is applied to the background-corrected data to determine the ratio (r) of fluorescent signals from the two Detector Probes at each pass during the reaction according to the equation: $r = ln(\text{ROX/FAM})$, where ROX and FAM correspond to the signals in Relative Fluorescence Units (RFU) obtained from the rhodamine- and fluorescein-labeled probes, respectively. A maximum density function is then used to determine the most likely value of r for the sample and this number is output as the Maximum Density (ρ-Max) score. For the β_2AR assays we have developed, ρ-Max values >+1 correspond to high ROX signals and are considered indicative of a homozygous Allele A genotype. Similarly, ρ-Max values <-1 correspond to high FAM fluorescence and a homozygous Allele B genotype. ρ-Max values between -1 and +1 indicate the detection of similar fluorescence intensities in the two optical channels and a heterozygous Allele A/B genotype.

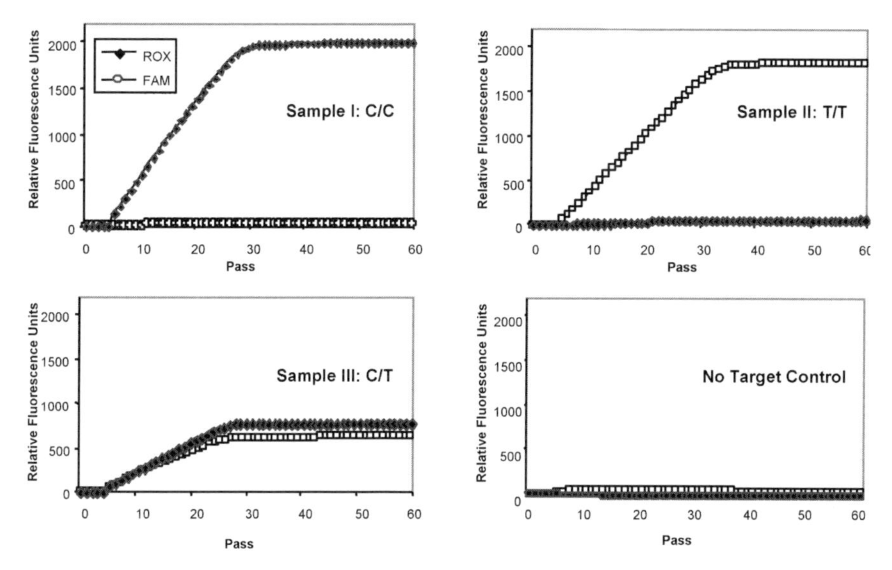

Figure 7. Genotyping of human blood specimens using the BD ProbeTec ET System. Genotyping of the -367 β_2AR SNP locus (C↔T polymorphism) was performed on three blood samples using 1 μL whole blood per reaction. Graphs show the relative increase in fluorescence over time in each of the two optical channels (corresponding to fluorescein, FAM and rhodamine, ROX).

Example

Whole blood was collected from healthy volunteers and genotyped using the β_2AR assay procedure described above. SDA reactions were performed using 1 μL unprocessed blood spiked directly into the reaction mixture. Figure 7 shows representative amplification curves generated for the –367 β_2AR SNP locus, the amplicon for which has a GC content of 79%. Sample I exhibited a strong ROX signal and weak FAM fluorescence, with a ρ-Max score of +4.4, indicative of a homozygous C/C allele pair. In contrast, Sample II, showed high levels of FAM fluorescence but a weak signal in the ROX channel. The ρ-Max score for this sample was –3.6, indicating the presence of a homozygous T/T allele pair. With blood Sample III, there were approximately equivalent levels of fluorescence in both optical channels, yielding a ρ-Max score of +0.2. This value lies between –1 and +1 and corresponds to a heterozygous genotype (C/T allele pair). In the absence of target, no fluorescent signal was generated and the sample was therefore recorded as indeterminate.

Figure 8 depicts the ρ-Max scores obtained with all six β_2AR assays using blood samples from three different individuals. SDA-based genotyping results were in complete concordance with those obtained by DNA sequence analysis of PCR products. All homozygous alleles exhibited ρ-Max scores >+1 or <-1 while heterozygous alleles had values that were close to zero, between –1 and +1.

Discussion

The data presented here demonstrate the capabilities of SDA for the real-time qualitative detection of nucleic acids from infectious disease analytes and for genotypic analysis of clinical specimens. Analytical studies demonstrated a lower limit of detection of <500 EBs of *C. pneumoniae* per milliliter of seeded sputum. While other homogenous assay systems for the detection and/or diagnosis of *C. pneumoniae* have been reported (Kuoppa *et al.*, 2002; Apfalter *et al.*, 2003; Reischl *et al.*, 2003; Welti *et al.*, 2003), a particular advantage of the SDA-based microwell format is the large volume of sample that may be analyzed. Use of higher input sample volumes has particularly important implications for infectious disease diagnosis because larger volumes increase the likelihood of delivering target DNA to the reaction, thereby improving clinical sensitivity. Typical PCR protocols utilize only 2-10μL of DNA extract whereas the SDA protocol described here employs >20μL purified nucleic acid and can easily be modified to increase or decrease the sample volume as required.

A further advantage of the BD ProbeTec ET SDA-based assay format we describe here is the incorporation of an IAC that co-amplifies with the target sequence and which serves to verify negative results and identify inhibitory specimens. This is particularly important when testing specimens for clinical diagnosis. Failure to detect amplification of the internal control alerts the user of the need to repeat the assay and/or to reprocess the specimen to remove the assay inhibitors, thus substantially reducing the likelihood of reporting false-negative results.

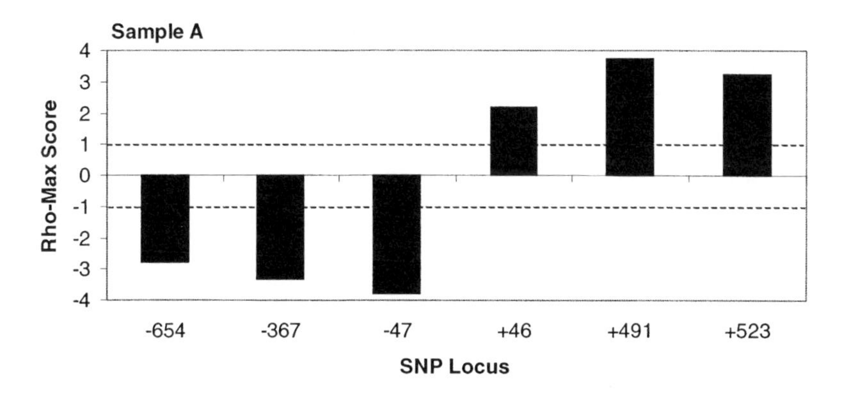

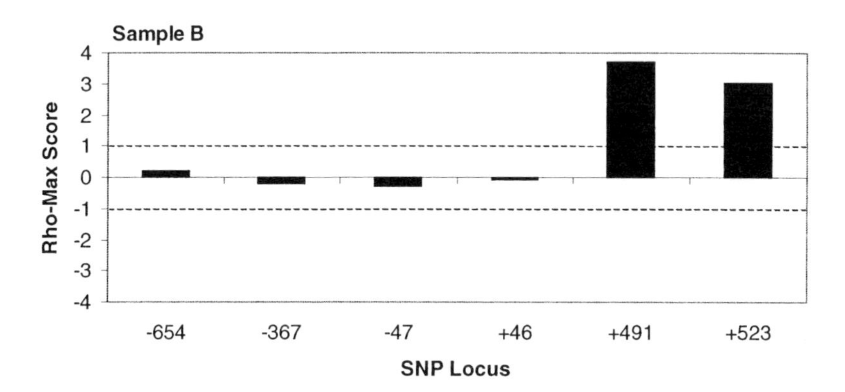

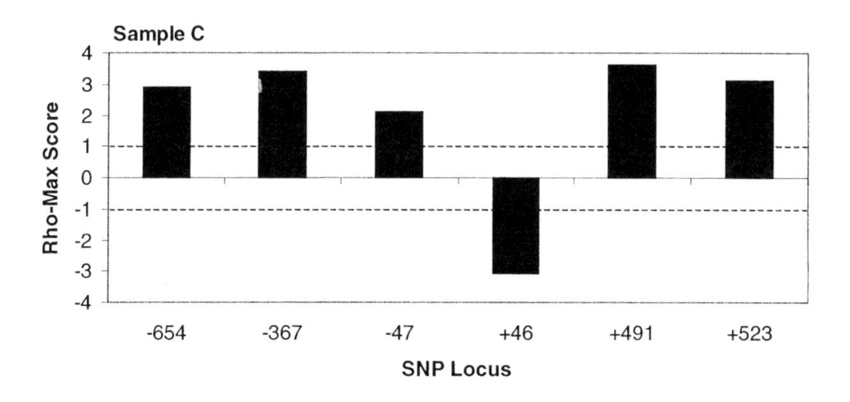

Figure 8. SDA-based SNP analysis was conducted on six polymorphic loci within the human β_2AR gene using unprocessed blood from three individuals. Dashed lines represent the –1 and +1 ρ-Max cutoff values segregating homozygous and heterozygous genotypes. ρ-Max > +1: Homozygous Allele pair A/A; ρ-Max < -1: Homozygous Allele pair B/B; ρ-Max –1 to +1: Heterozygous Allele pair A/B.

SDA-based assays using the BD ProbeTec ET System have been widely used for the detection of *N. gonorrhoeae*, *C. trachomatis* and *Mycobacterium tuberculosis* in a variety of sample types (Little *et al.*, 1999; Van Der Pol *et al.*, 2001; Akduman *et al.*, 2002; Johansen *et al.*, 2002; Maugein *et al.*, 2002; Iinuma *et al.*, 2003; Mazzarelli *et al.*, 2003; Verkooyen *et al.*, 2003). For these commercialized applications, the reaction components are dried in Priming and Amplification Microwells that are rehydrated with the test sample. This reagent configuration, combined with the convenient microwell strip format, provides a simple workflow and rapid time-to-results that are particularly amenable to high-throughput clinical laboratories.

We have recently adapted the Universal Detection format that was originally developed for qualitative detection of nucleic acids to permit discrimination of SNPs and other genetic variations. To demonstrate the utility of SDA technology for genotyping of clinical specimens we targeted six polymorphisms within the human β_2AR gene that in combination determine all the major haplotypes of the β_2AR gene (Drysdale *et al.*, 2000). Important to this demonstration of feasibility was the development of universal buffer conditions and use of the same pair of universal fluorescent Detector Probes across all six assays. This generic approach to assay development has significant potential to reduce the amount of time and cost associated with implementation of tests for novel genetic markers.

In the example provided here, correct genotypes were established with all six SDA-based β_2AR systems using 1 μL whole blood in each reaction. An important benefit of SDA technology is the ability to genotype samples directly while circumventing the time-consuming and costly process of sample preparation. This has significant benefit to busy laboratories in which nucleic acid purification is often the rate-limiting step in determining assay throughput. Additional experiments have shown that the β_2AR assays are all tolerant to $\geq$20 μL unprocessed whole blood, equivalent to $\geq$14% of total reaction volume (Wang *et al.*, 2003). In the present example, blood samples were collected using K_2EDTA as the anticoagulant although we have also demonstrated the compatibility of the β_2AR assays with unprocessed blood containing heparin or citrate. In other experiments, we have also successfully genotyped individuals using throat swabs, buccal washes and urine as well as dried blood specimens.

The sequence of the β_2AR gene is highly GC-rich and represents a significant challenge for genetic analysis. The target regions for the six SNP assays ranged in GC content from 53 to 86% whereas published reports indicate a relatively low success rate in genotyping such highly GC-rich DNA sequences using conventional PCR-based techniques (Vieux *et al.*, 2002). While we were successful in discriminating all the polymorphisms in the example, we did observe differences in analytical sensitivity between the six assays that correlated broadly with GC content. Nevertheless, SDA offers the potential of a simple, standardized protocol for genotypic analysis that is applicable to the vast majority of SNPs and across a broad range of specimen types.

In summary, SDA is a versatile nucleic acid amplification technology that may be coupled with real-time homogeneous fluorescent detection for a wide variety of applications. For example, we have developed SDA-based assays for infectious disease diagnosis and genotyping of clinical specimens using a novel Universal Detection format that helps to reduce cost and speeds assay development.

Acknowledgements

We thank Gordon Franklin and Mark Hall for technical assistance.

References

Akduman, D., Ehret, J.M., Messina, K., Ragsdale, S., and Judson, F.N. (2002). Evaluation of a Strand Displacement Amplification assay (BD ProbeTec-SDA) for detection of *Neisseria gonorrhoeae* in urine specimens. J. Clin. Microbiol. *40*, 281-283.

Apfalter, P. Barousch, W., Nehr, M., Makristathis, A., Willinger, B., Rotter, M., and Hirschl, A. (2003). Comparison of a new quantitative *ompA*-based real-time PCR TaqMan assay for detection of *Chlamydia pneumoniae* DNA in respiratory specimens with four conventional PCR assays. J. Clin. Microbiol. *41*, 592-600.

Drysdale C.M., McGraw D.W., Stack C.B., Stephens J.C., Judson R.S., Nandabalan K., Arnold, K., Ruano, G., and Liggett, S.B. (2000). Complex promoter and coding region β_2-adrenergic receptor haplotypes alter receptor expression and predict *in vivo* responsiveness. Proc. Natl. Acad. Sci. USA *97*, 10483-10488.

Hellyer, T.J., DesJardin, L.E., Teixeira, L., Perkins, M.D., Cave, M.D., and Eisenach, K.D. (1999). Detection of viable *Mycobacterium tuberculosis* by reverse transcriptase-Strand Displacement Amplification of mRNA. J. Clin. Microbiol. *37*, 518-523.

Iinuma, Y., Senda, K., Fujihara, N., Saito, T., Takakura, S., Shimojima, M., Kudo, T., and Ichiyama, S. (2003). Comparison of the BDProbeTec ET System with the Cobas Amplicor PCR for direct detection of *Mycobacterium tuberculosis* in respiratory samples. Eur. J. Clin. Microbiol. Infect. Dis. *22*, 338-371.

Johansen, I.S., Thomsen, V.Ø., Johansen, A., Andersen, P., and Lundgren, B. (2002). Evaluation of a new commercial assay for pulmonary and nonpulmonary tuberculosis. Eur. J. Clin. Microbiol. Infect. Dis. 21: 455-460.

Kuoppa, Y., Boman, J., Scott, L., Kumlin, U., Eriksson, I., and Allard, A. (2002). Quantitative detection of respiratory *Chlamydia pneumoniae* infection by real-time PCR. J. Clin. Microbiol. *40*, 2273-2274.

Little, M.C., Andrews, J., Moore, R., Bustos, S., Jones, L., Embres, C., Durmowicz, G., Harris, J., Berger, D., Yanson, K., Rostkowski, C., Yursis, D., Price, J., Fort, T., Walters, A., Collis, M., Llorin, O., Wood, J., Failing, F., O'Keefe, C., Scrivens, B., Pope, B., Hansen, T., Marino, K., Williams, K., and Boenisch, M. (1999). Strand Displacement Amplification and homogeneous real-time detection incorporated in a second-generation DNA probe system, BDProbeTecET. Clin. Chem. *45*, 777-784.

Maugein, J., Fourche, J., Vacher, S., Grimond, C., and Bebear, C. (2002). Evaluation of the BDProbeTec™ ET DTB assay for direct detection of *Mycobacterium tuberculosis* complex from clinical samples. Diagn. Microbiol. Infect. Dis. *44*, 151-155.

Mazzarelli, G., Rindi, L., Piccoli, P., Scarparo, C., Garzelli, C., and Tortoli, E. (2003). Evaluation of the BDProbeTec ET System for direct detection of *Mycobacterium tuberculosis* in pulmonary and extrapulmonary samples: a multicenter study. J. Clin. Microbiol. *41*, 1779-1782.

Mehrpouyan, M., Bishop, J.E., Ostrerova, N., Van Cleve, M., and Lohmann, K.L. (1997). A rapid and sensitive method for non-isotopic quantitation of HIV-1 RNA using thermophilic SDA and flow cytometry. Mol. Cell. Probes *11*, 337-347.

Milla, M.A., Spears, P.A., Pearson, R.E., and Walker G.T. (1998). Use of restriction enzyme *Ava*I and exo$^-$ Bst polymerase in Strand Displacement Amplification. Biotechniques *24*, 392-396.

Nadeau, J.G., Pitner, J.B., Linn, C.P., Schram, J.L., Dean, C.H., and Nycz, C.M. (1999). Real-time, sequence-specific detection of nucleic acids during Strand Displacement Amplification. Anal. Biochem. *276*, 177-187.

Nuovo, G.J. (2000). *In situ* strand displacement amplification: an improved technique for the detection of low copy nucleic acids. Diagn. Mol. Pathol. *9*, 195-202.

Nycz, C.M., Dean, C.H., Haaland, P.D., Spargo, C.A,. and Walker, G.T. (1998). Quantitative reverse transcription Strand Displacement Amplification: quantitation of nucleic acids using an isothermal amplification technique. Anal. Biochem. *259*, 226-234.

Spargo, C.A., Fraiser, M.S., Van Cleve, M., Wright D.J., Nycz, C.M., Spears, P.A., and Walker, G.T. (1996). Detection of *M. tuberculosis* DNA using thermophilic Strand Displacement Amplification. Mol. Cell. Probes *10*, 247-256.

Reischl, U., Lehn N., Simnacher, U., Marre, R., and Essig, A. (2003). Rapid and standardized detection of *Chlamydia pneumoniae* using LightCycler real-time Fluorescence PCR. Eur. J. Clin. Microbiol. Infect. Dis. *22*, 54-57.

Van Der Pol, B., Ferrero D.V., Buck-Barrington, L., Hook, E., Lenderman, C., Quinn, T., Gaydos, C.A., Lovchik, J., Schachter, J., Moncada, J., Hall, G., Tuohy, M.J., and Jones, R.B. (2001). Multicenter evaluation of the BDProbeTec ET System for detection of *Chlamydia trachomatis* and *Neisseria gonorrhoeae* in urine specimens, female endocervical swabs, and male urethral swabs. J. Clin. Microbiol. *39*, 1008-1016.

Verkooyen, R.P., Noordhoek, G.T., Klapper, P.E., Reid, J., Schirm, J., Cleator, G.M., Ieven, M., and Hoddevik, G. (2003). Reliability of nucleic acid amplification methods for detection of *Chlamydia trachomatis* in urine: results of the First International Collaborative Quality Control Study among 96 laboratories. J. Clin. Microbiol. *41*, 3013-3016.

Vieux, E.F., Kwok P.Y., and Miller R.D. (2002). Primer design for PCR and sequencing in high-throughput analysis of SNPs. Biotechniques *32* (*Suppl.*), S28-S32.

Walker, G.T., Fraiser M.S., Schram, J.L., Little, M.C., Nadeau, J.G., and Malinowski, D.P. (1992a). Strand Displacement Amplification – an isothermal, *in vitro*, DNA amplification technique. Nucleic Acids Res. *20*, 1691-1696.

Walker, G.T., Little, M.C., Nadeau, J.G., and Shank, D.D. (1992b). Isothermal *in vitro* amplification of DNA by a restriction enzyme/DNA polymerase system. Proc. Natl. Acad. Sci. USA *89*, 392-396.

Wang, S.-S., Thornton, K., Kuhn, A.M., Nadeau, J.G., and Hellyer, T.J. (2003). Homogeneous Real-time Detection of Single Nucleotide Polymorphisms by Strand Displacement Amplification on the BD ProbeTec ET System. Clin. Chem. *49*, 1599-1607.

Welti, M., Jaton, K., Altwegg, M., Sahli, R., Wenger, A., and Bille, J. (2003). Development of a multiplex real-time quantitative PCR assay to detect *Chlamydia pneumoniae*, *Legionella pneumophila*, and *Mycoplasma pneumoniae* in respiratory tract secretions. Diagn. Microbiol. Infect. Dis. *45*, 85-95.

Westin, L., Xu, X., Miller, C., Wang, L., Edman, C.F., and Nerenberg, M. (2000). Anchored multiplex amplification on a microelectronic chip array. Nature Biotechnol. *18*, 199-204.

3.2

Loop-Mediated Isothermal Amplification (LAMP) of DNA Analytes

Tsugunori Notomi, Kentaro Nagamine, Yasuyoshi Mori and Hidetoshi Kanda

Abstract

A novel DNA amplification method, which is termed the loop-mediated isothermal amplification (LAMP), employs the self-recurring strand-displacement DNA synthesis primed by a specially designed set of the target-specific primers and amplifies DNA more than 10^9-fold in less than an hour. The basic LAMP method has the following favorable features: (i) All reactions are conducted under isothermal conditions using only one enzyme, large fragment of *Bst* DNA polymerase; (ii) an extremely high specificity is ensured by using four primers, a pair of ordinary, single-domain primers plus a pair of composite, double-domain ones, that recognize six distinct sites on target DNA; (iii) rapidity of generation of a large amount of amplified products; (iv) simplicity of protocol employing customary reagents, easy optimization and low cost. Consequently, LAMP is highly selective for the target DNA and highly efficient at amplification, enabling detection of just a few copies of the designated DNA analyte on excessive unrelated background in less than an hour at a single temperature. Thus, in some aspects, this method is advantageous over other amplification techniques and it should be useful in DNA diagnostics.

Background

In recent years, genetic testing has begun to be widely employed for the diagnosis of hereditary diseases, pathogenic agents and an inherited predisposition towards lifestyle disorders. Such testing involves cell lysis, nucleic acids extraction, their selective amplification and, finally, detection of nucleic acid sequences related to diseases. With current methods, however, genetic testing is a time-consuming, costly and complicated procedure.

Since nucleic acid analytes that form the basis of genetic tests are frequently available in tiny amounts, the first step in their identification involves target amplification. For this, the polymerase chain reaction (PCR) (Saiki *et al.*, 1985, 1988) is widely used, but it requires a high-precision thermal-cycling equipment. Isothermal amplification methods, which do not require changes of temperature, have also been developed (Guatelli *et al.*, 1990; Walker *et al.*, 1992a,b; Liu *et al.*, 1996; Lizardi *et al.*, 1998; Zhang *et al.*, 1998; Dean *et al.*, 2002), yet none of them have so far succeeded in replacing the PCR (for reviews, see: Compton, 1991; Isaksson and Landegren, 1999; Andras *et al.*, 2001; Schweitzer and Kingsmore, 2001).

We have recently developed a new highly-effective isothermal method of DNA amplification (Notomi *et al.*, 2000; Nagamine *et al.*, 2001) that requires only one enzyme, DNA polymerase. Given that major reaction steps proceed via the DNA loops formation, this method was named as the loop-mediated isothermal amplification (LAMP).

Basic Principles of the LAMP Method

General schematics of the LAMP method are shown in Figure 1. This method is based on the use of a specially designed set of four target-

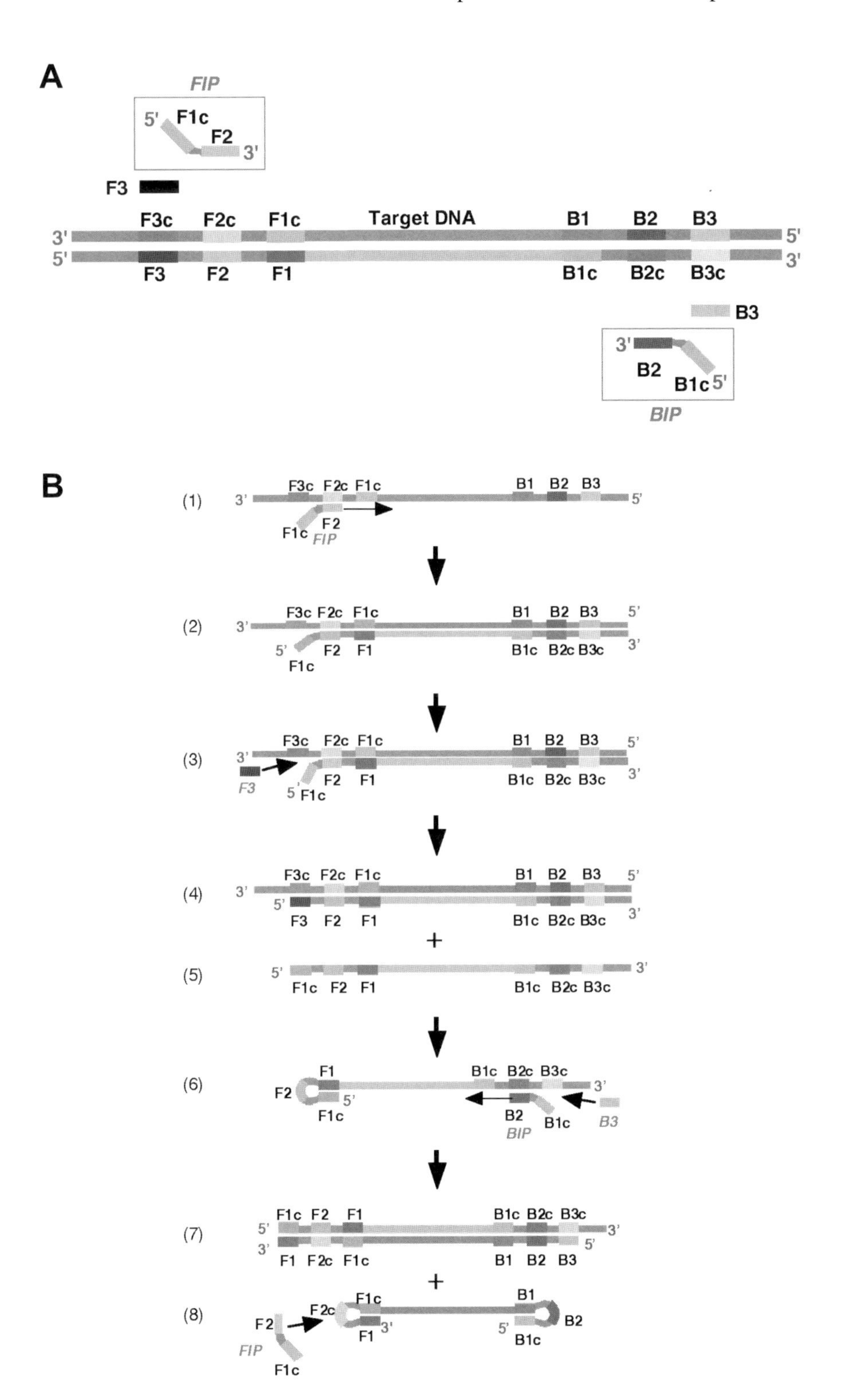
A
FIP
5'
F1c
F2
3'
F3
F3c
F2c
F1c
Target DNA
B1
B2
B3
3'
5'
5'
3'
F3
F2
F1
B1c
B2c
B3c
B3
3'
B2
B1c
5'
BIP
B
(1)
(2)
(3)
(4)
(5)
(6)
(7)
(8)

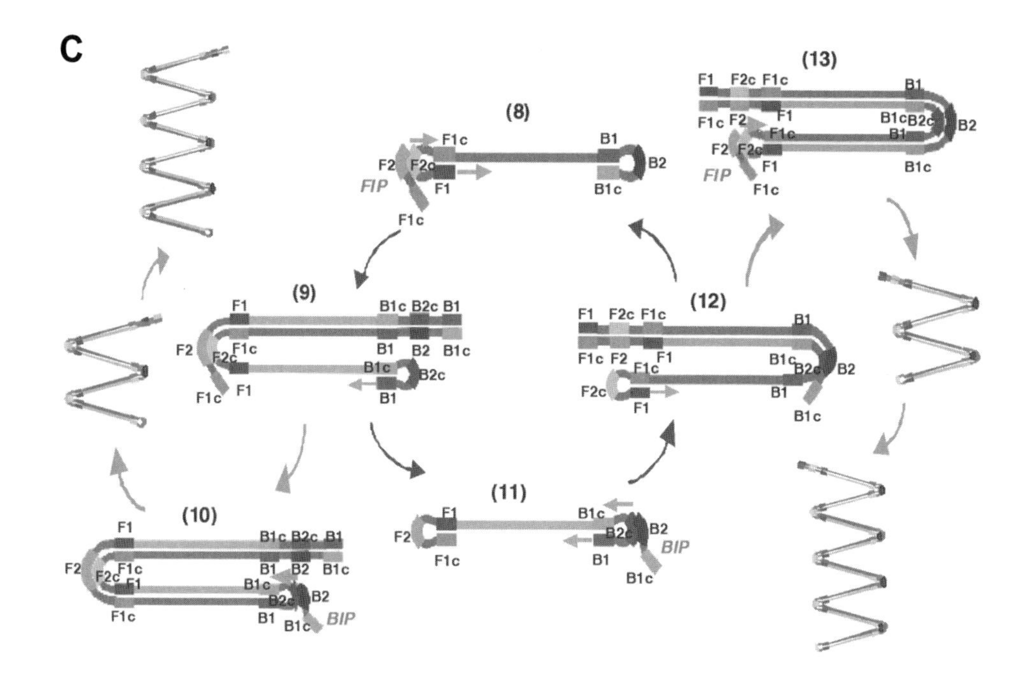

Figure 1. Schematic representation of the LAMP mechanism.
A: Construction of primers. The forward inner primer (FIP) consists of the F2 sequence that is complementary to the F2c region on the target gene at the 3' end plus the identical sequence as the F1c region on the target gene at the 5' end. The backward inner primer (BIP) consists of the B2 sequence that is complementary to the B2c region on the target gene at the 3' end plus the identical sequence as the B1c region on the target gene at the 5' end. The sequence of forward outer primer F3 is complementary to the F3c region on the target gene. The sequence of backward outer primer B3 is complementary to the B3c region on the target gene.
B: LAMP pre-amplification stage (prior denatured target is shown). Through the action of DNA polymerase with strand-displacement activity, a DNA strand complementary to the template DNA is synthesized, starting from the 3' end of the F2 region of the FIP primer. Then, the F3 primer anneals to the F3c region in the target DNA and initiates strand-displacement DNA synthesis, releasing the FIP-linked complementary strand. Then, this released DNA single strand forms a stem-loop structure at the 5' end because of the complementary F1c and F1 regions. This single-stranded DNA serves as template for BIP-initiated DNA synthesis and subsequent B3-primed strand-displacement DNA synthesis. The displaced BIP-linked complementary DNA strand forms a dumbbell-like construct featuring stem-loop structures at each end. This construct initiates the multi-round LAMP amplification stage.
C: LAMP amplification stage. The single-stranded DNA dumbbell (8) is quickly converted to the double-stranded DNA form with single terminal loop by self-primed DNA synthesis. The FIP primer anneals to this loop and primes the strand-displacement DNA synthesis releasing the previously synthesized DNA strand (9). The released DNA single strand forms

specific primers, which are able to recognize six distinct sites flanking the amplified DNA sequence (Figure 1A): F1c, F2c, F3c sites on the 3' side and B1, B2, B3 sites on the 5' side (F is for forward and B is for backward directions; the index letter c is used to distinguish between complementary DNA sequences). One pair of primers, F3 and B3, represents the ordinary, single-domain primers supplemented by an additional pair of composite, double-domain primers, FIP and BIP, which play the major role in the LAMP reactions (FIP and BIP primers are taken in excess, compared to F3 and B3 primers).

LAMP starts with the pre-amplification stage (Figure 1B) requiring the participation of all four primers and resulting in the single-stranded dumbbell-like key intermediate (8) with loops at both ends. This DNA construct is a starting material for the amplification stage of LAMP (Figure 1C), which proceeds with the participation of two composite primers only. As Figure 1B shows, at the pre-amplification stage, one of the inner primers (*i.e.* FIP) works first, thus initiating the replicative DNA synthesis by *Bst* DNA polymerase. Next, the outer primer, F3, takes its place displacing the newly synthesized DNA strand and releasing the original target. With similar performance of BIP and B3, this stage is completed by yielding the single-stranded DNA dumbbell.

Self-primed DNA synthesis then rapidly occurs at the 3' terminus of DNA dumbbell and the inner primer attaches later to the loop, initiating the strand-displacement DNA synthesis (Figure 1C). Finally, this cycle produces the enlarged DNA product featuring an inverted double-stranded DNA mirror repeat with a terminal loop as well as a replica of the starting DNA dumbbell. This replica will initiate similar round of transformations mirroring the previous ones whereas the enlarged product passes through repeated elongation reactions, which

a stem-loop structure at the 3' end owing to complementary B1c and B1 regions. Next, self-primed DNA synthesis occurs at the 3' end of the B1 region, which results in the two times longer double-stranded DNA form with single terminal loop (10) and a replica of the initial DNA dumbbell (11). The 'turnover' chain of events may similarly proceeds with this replica starting from the 3' end of the B1 region. Numerous reiterations of all these steps and alike will finally yield large LAMP amplicons consisting of repeats of the target DNA sequence.

ultimately generate long concatemeric amplicons of various lengths and cauliflower-like structures with multiple loops (not shown in Figure 1C for simplicity; see Notomi *et al.*, 2000 for more detailed schematics).

If necessary, only uniform single-stranded products of the LAMP amplification, which might be preferable for certain hybridization/detection techniques, can be isolated via extra steps (Nagamine *et al.*, 2002a). Thus, the basic LAMP procedure is methodologically rather simple: it utilizes just one enzyme and a single set of self-sustained loop-mediated elongation reactions carrying out at a constant temperature (for more details, see http://loopamp.eiken.co.jp).

Key Features of the LAMP Method

As a potent method of target DNA amplification, the LAMP method has several advantages that are outlined next. (1) Amplification proceeds under isothermal conditions using a single enzyme. (2) Absence of special reagents makes this procedure easily optimized and inexpensive. (3) High sequence specificity is established by the multiple-recognition scheme. (4) Reliable detection of LAMP amplicons becomes possible within 1 hr starting from a few copies of DNA analytes. Even more rapid amplification is achievable by employing extra primers (Nagamine *et al.*, 2002b).

Protocols

Step 1: Rational Design of LAMP Primers

In order to ensure successful DNA amplification by the LAMP method, prior optimization of all primers is crucial. Each domain of composite primers is carefully designed with respect to its base composition, GC-content and secondary structure using commercial primer-design software, as it is done for PCR primers. The following key points should also be kept in mind for the optimal primer design:

Table 1. Reaction mixture 1 (2× composition).

Tris-HCl (pH 8.8)	40 mM
KCl	20 mM
$MgSO_4$	16 mM
$(NH_4)_2SO_4$	20 mM
Tween 20[a]	0.2 %
Betaine[b]	1.6 M
dNTPs	2.8 mM each

[a] Tween 20 is used to suppress a pseudoamplification caused by the primer-primer dimer formation, when primers anneal each other nonspecifically, thus serving as incorrect templates for the amplification reaction, similar to that observed in PCR.
[b] Betaine was found to improve the amplification efficiency and additionally suppress non-specific amplification.

1. Both ends of the inner primer should not be AT-rich.
2. The T_m value for each domain (primer plus complement; calculated by the nearest-neighbor method by Breslauer *et al.*, 1986) should be about 55-65°C.
3. Length of the loop should be 30 to 70 bp.
4. Length of the amplified DNA region (from F2 to B2 sites inclusively) should not be more than 200 bp.
5. Primers purity could be crucial for rapidity and reproducibility of amplification and use of HPLC-purified FIP and BIP primers is recommended.

Step 2: Preparation of Reaction Mixtures

The reaction mixture 1 (stock solution) containing all necessary components except primers and DNA polymerase is prepared according to Table 1. The reaction mixture 2 is described in Table 2 and it should be prepared on ice, being used directly for amplification.

Table 2. Reaction mixture 2[a]	
Reaction mixture 1 (2× composition)	12.5 μl
Primers	
FIP	40 pmol
BIP	40 pmol
F3	5 pmol
B3	5 pmol
Bst DNA polymerase	1 μl/8 U
Add distilled water	to 25 μl

[a] Per one reaction tube.

Step 3: Amplification Procedure

1. Pour 23 μl of the reaction mixture 2 into a reaction tube.
2. Add 2 μl of the target DNA sample. For higher sensitivity, use of prior denatured target (*e.g.* by heating at 95°C for 5 min in 10 mM Tris-HCl or TE buffer, pH 7 to 8) is recommended, although LAMP is also possible with nondenatured DNA analytes (Nagamine *et al.*, 2001).
3. Carefully mix up the solution by gentle tapping or pipetting and then spin it down.
4. Incubate the reaction tubes at 60-65°C for 30-60 min. Note that optimal experimental conditions may vary depending on the primers used.
5. Terminate the LAMP reaction by incubation at 80°C for 2 min, which inactivates the enzyme.
6. Optional: isolation of single-stranded DNA fragments from LAMP amplicons. In some cases (*e.g.* for better detection with DNA microarrays), it is desirable to finally obtain homogeneous single-stranded products of LAMP amplification. To this end, extra steps should be preformed with the use of *Tsp*RI restriction enzyme to digest amplification products and a supplementary primer hybridized to the 3'-overhanging 9-nt-long sequence at *Tsp*RI cleavage site to displace single-stranded DNA via primer extension (Nagamine *et al.*, 2002a).

Step 4: Detection

Detection of LAMP amplicons can be performed by agarose gel electrophoresis using 0.5-2.0 μl of the amplified products per lane and gel staining by ethidium bromide or SYBR Green I. Usually, a ladder-like gel-electrophoretic patterns are observed, which represent the concatemers of target DNA as major products. Note that the fluorescence detection of LAMP products can be performed without gel-electrophoretic separation using common DNA-binding fluorescent dyes. Accordingly, DNA targets can readily be visualized without the need for expensive diagnostic reagents or equipment. Since the reaction byproduct of DNA synthesis, pyrophosphate, is also produced in large amounts and it precipitates with magnesium in reaction buffer, the generated turbidity can easily be detected with an inexpensive photometer or even with the naked eye (Mori *et al.*, 2001). For other possible detection methods, see http://www.eiken.co.jp/genome.

Precautions

As the LAMP reaction is extremely sensitive (amplification can be as high as 10^{12}-fold), any contamination by amplified products can lead to false positives. To avoid this, it is strongly recommended to prepare and keep all reagents and samples on a clean bench in a separate room quite far from that used for amplification and detection of LAMP products.

Examples

For the proof-of-principle demonstration, the ~200-bp-long target fragment of hepatitis B virus (HBV) DNA (HBs region coding for surface antigen) that carries the EarI recognition site in the middle was cloned into pBR322 vector at BamHI site and the EcoRV-linearized recombinant plasmid was used in LAMP reactions. Figure 2A shows that after 45 min of LAMP large amount of amplified products is produced in this model experiment starting from as little as 6 HBV DNA copies to yield the characteristic ladder-like pattern. It was

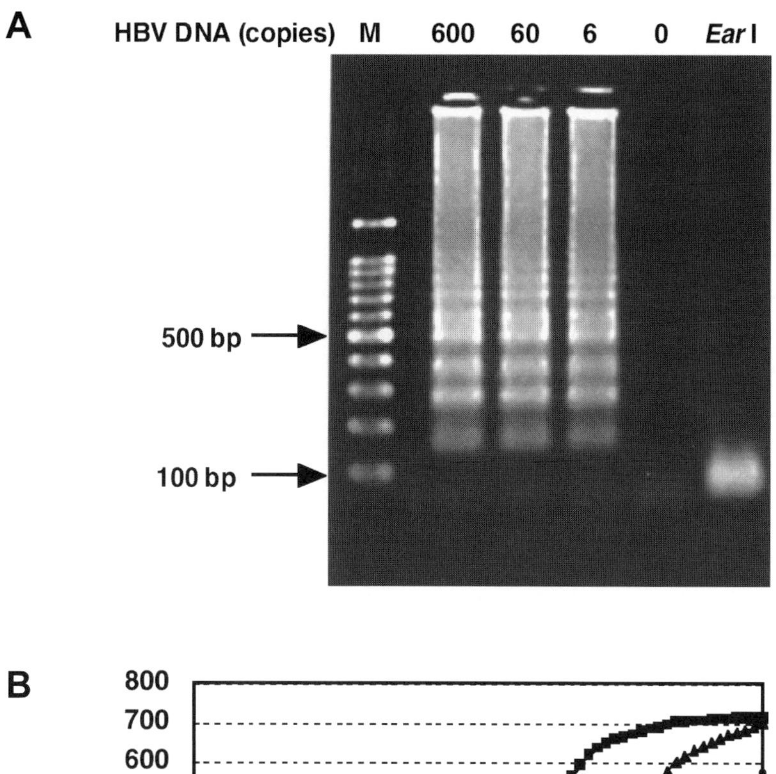

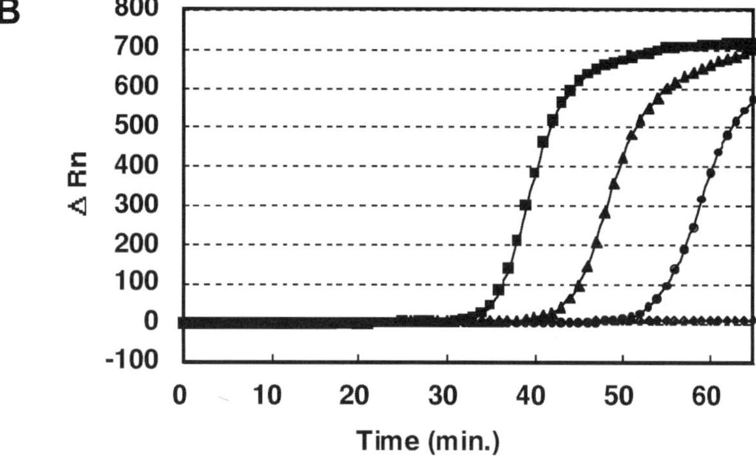

Figure 2. Fluorescence-based LAMP detection of HBV DNA (type adr; HBs region; prior denatured; obtained from the infected patients).
A: Gel-electrophoretic separation of LAMP products generated from different input amounts of cloned target DNA, as indicated. The very last lane on the gel represents the EarI digest of LAMP amplicons; M is the 100-bp DNA ladder (size marker). LAMP was carried out at 60°C for 45 min in a total 25 μl reaction mixture containing 1.6 μM each of FIP and BIP, 0.2 μM each of F3 and B3. 5 μl of LAMP products and 1 μl of them digested with restriction enzyme were run in 2% agarose gel (0.5x TBE

estimated that viral DNA was maximally amplified ~10^{12} times. Cleavage of LAMP products with EarI restriction enzyme resulted in two overlapping bands with the expected size. Southern blot hybridization and sequencing confirmed the HBV origin of LAMP products thus obtained (data not shown).

One more, analytically relevant experiment shown in Figure 2B proves the ability of LAMP to detect the HBV DNA in real time using the fluorescent staining. Figure 3 demonstrates the turbidity-based real-time LAMP detection of lambda DNA. Both these express-analysis approaches promptly provide with the signal/response within 35-60 min, feature the detection sensitivity of ~100 DNA marker molecules as a minimum and suggest a possibility of being readily applicable to the purpose of a quantitative analysis.

Concluding Remarks

Examples presented here demonstrate the simplicity, rapidity and high sensitivity of LAMP: this isothermal method of DNA amplification allows to readily detect a very limited amount of clinically important DNA analytes in less than an hour at a single temperature and on the excessive background of human genomic DNA. As a robust alternative to PCR, rapid and non-complicated LAMP-based DNA diagnostics suit well for the bedside or clinical genetic analyses (Iwamoto *et al.*, 2003) as well as for testing/detection of plant viruses (Fukuta *et al.*, 2003a,b), animal infectious disease (Enosawa *et al.*, 2003) and

buffer) followed by staining with SYBR Green I (Molecular Probes Inc.). The primers used:5'-GATAAAACGCCGCAGACACATCCTTCCAACCTCTTGTCCTCCAA-3' (HBV BIP), 5'-CCTGCTGCTATGCCTCATCTTCTTTGACAAACGGGCAACAT ACCTT-3' (HBV FIP), 5'-CAAAATTCGCAGTCCCCAAC-3' (HBV B3) and 5'-GGTGGTTGATGTTCCTGGA-3' (HBV F3).
B: Real-time detection of LAMP amplicons generated at 60°C from HBV DNA targets in the presense of 100 ng human genomic DNA (the LAMP conditions were as in (A) except 0.25 μg/ml of ethidium bromide was added to reactions). Amplification was carried out with 60 (circle), 600 (triangle), and 6000 (square) copies of cloned HBV DNA. Detection was done with the ABI PRISM 7700 sequencer using the ROX fluorescence channel. ΔRn is the increase in fluorescence (arbitrary units).

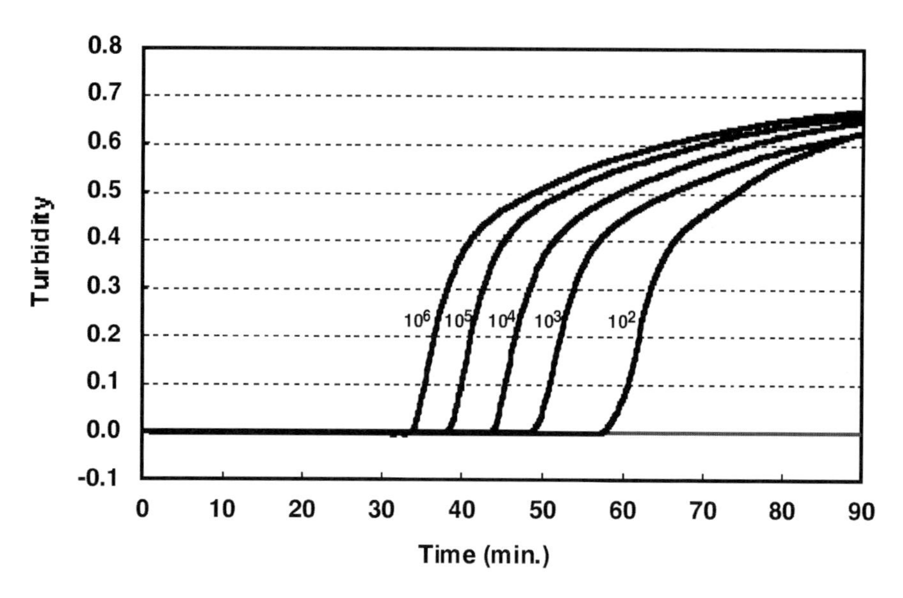

Figure 3. Turbidity-based LAMP detection of lambda DNA (Takara Shuzo, Japan; prior denatured) with the real-time monitoring. Amplification was carried out in a total 25 μl reaction mixture containing 1.6 μM each of FIP and BIP, 0.2 μM each of F3 and B3 at 65°C for 90 min. The primers used: 5'- AGGCCAAGCTGCTTGCGGTAGCCG GACGCTACCAGCTTCT-3' (lambda-FIP), 5'-CAGGACGCTGTGGCATTGCAGA TCATAGGTAAAGCGCCACGC-3' (lambda-BIP), 5'-AAAACTCAAATCAACAG GCG-3'(lambda-F3), 5'-GACGGATATCACCACGATCA-3' (lambda-B3). Various numbers of copies of lambda DNA were amplified, as indicated. Detection was done with the Loopamp turbidimeter LA-200 (Teramecs, Japan).

environmental or food contaminations (Maruyama *et al.*, 2003). Since reverse transcriptase is capable of strand-displacement synthesis too, we were also able to demonstrate the LAMP process with the RNA analytes, if this enzyme is added to the reaction system (Notomi *et al.*, 2000; Fukuta *et al.*, 2003b).

By adding extra primers that bind to amplification products, we have recently developed an advanced version of LAMP, which is twice more rapid than the original procedure (Nagamine *et al.*, 2002b). It has also recently been found that LAMP is capable of accurately and rapidly detecting the single nucleotide polymorphisms (SNPs) (Iwasaki *et al.*, 2003). Considering that SNPs are associated with constitutional factors, such as disease susceptibility, drug responsiveness and adverse drug reactions, the LAMP method holds promise for tailoring the medications for individual patients.

References

Andras, S.C., Power, J.B., Cocking, E.C., and Davey, M.R. (2001). Strategies for signal amplification in nucleic acid detection. Mol. Biotechnol. *19*, 29-44.

Breslauer, K.J., Frank, R., Blocker, H., and Marky, L.A. (1986). Predicting DNA duplex stability from the base sequence. Proc. Natl. Acad. Sci. USA *83*, 3746-3750.

Compton, J. (1991). Nucleic acid sequence-based amplification. Nature *350*, 91-92.

Dean, F.B., Hosono, S., Fang, L., Wu, X., Faruqi, A.F., Bray-Ward, P., Sun, Z., Zong, Q., Du, Y., Du, J., Driscoll, M., Song, W., Kingsmore, S.F., Egholm, M., and Lasken, R.S. (2002). Comprehensive human genome amplification using multiple displacement amplification. Proc. Natl. Acad. Sci. USA *99*, 5261-5266.

Enosawa, M., Kageyama, S., Sawai, K., Watanabe, K., Notomi, T., Onoe, S., Mori, Y., and Yokomizo, Y. (2003). Use of loop-mediated isothermal amplification of the IS900 sequence for rapid detection of cultured Mycobacterium avium subsp. paratuberculosis. J. Clin. Microbiol. *41*, 4359-4365.

Fukuta, S., Kato, S., Yoshida, K., Mizukami, Y., Ishida, A., Ueda, J., Kanbe, M., and Ishimoto, Y. (2003a). Detection of tomato yellow leaf curl virus by loop-mediated isothermal amplification reaction. J. Virol. Methods *112*, 35-40.

Fukuta, S., Iida, T., Mizukami, Y., Ishida, A., Ueda, J., Kanbe, M., and Ishimoto, Y. (2003b). Detection of Japanese yam mosaic virus by RT-LAMP. Arch. Virol. *148*, 1713-1720.

Guatelli, J.C., Whitfield, K.M., Kwoh, D.Y., Barringer, K.J., Richman, D.D., and Gingeras, T.R. (1990). Isothermal, *in vitro* amplification of nucleic acids by a multienzyme reaction modeled after retroviral replication. Proc. Natl. Acad. Sci. USA *87*, 1874-1878 [see erratum in Proc. Natl. Acad. Sci. USA *87*, 7797].

Isaksson, A. and Landegren, U. (1999). Accessing genomic information: alternatives to PCR. Curr. Opin. Biotechnol. *10*, 11-15.

Iwamoto, T., Sonobe, T., and Hayashi, K. (2003). Loop-mediated isothermal amplification for direct detection of Mycobacterium tuberculosis complex, M. avium, and M. intracellulare in sputum samples. J. Clin. Microbiol. *41*, 2616-2622.

Iwasaki, M., Yonekawa, T., Otsuka, K., Suzuki, W., Nagamine, K., Hase, T., Tatsumi, K., Horigome, T., Notomi, T., and Kanda, H. (2003). Validation of the LAMP method for SNP genotyping using whole blood. Genome Lett. *2*, 108-115.

Liu, D., Daubendiek, S.L., Zillman, M.A., Ryan, K., and Kool, E.T. (1996). Rolling circle DNA synthesis: small circular oligonucleotides as efficient templates for DNA polymerases. J. Am. Chem. Soc. *118*, 1587-1594.

Lizardi, P.M., Huang, X., Zhu, Z., Bray-Ward, P., Thomas, D.C., and Ward, D.C. (1998). Mutation detection and single-molecule counting using isothermal rolling-circle amplification. Nature Genet. *19*, 225-232.

Maruyama, F., Kenzaka, T., Yamaguchi, N., Tani, K., and Nasu, M. (2003). Detection of bacteria carrying the stx2 gene by *in situ* loop-mediated isothermal amplification. Appl. Environ. Microbiol. *69*, 5023-5028.

Mori, Y., Nagamine, K., Tomita, N., and Notomi, T. (2001). Detection of loop-mediated isothermal amplification by turbidity derived from magnesium pyrophosphate formation. Biochem. Biophys. Res. Commun. *289*, 150-154.

Nagamine, K., Watanabe, K., Ohtsuka, K., Hasa, T., and Notomi, T. (2001). Loop-mediated isothermal amplification reaction using a nondenatured template. Clin. Chem. *47*, 1742-1743.

Nagamine, K., Kuzuhara, Y., and Notomi, T. (2002a). Isolation of single-stranded DNA from loop-mediated isothermal amplification products. Biochem. Biophys. Res. Commun. *290*, 1195-1198.

Nagamine, K., Hase, T., and Notomi, T. (2002b). Accelerated reaction by loop-mediated isothermal amplification using loop primers. Mol. Cell. Probes *16*, 223-229.

Notomi, T., Okayama, H., Masubuchi, H., Yonekawa, T., Watanabe, K., Amino, N., and Hase, T. (2000). Loop-mediated isothermal amplification of DNA. Nucleic Acids Res. *28*, e63.

Saiki, R.K., Scharf, S., Faloona, F., Mullis, K.B., Horn, G.T., Erlich, H.A., and Arnheim, N. (1985). Enzymatic amplification of beta-globin genomic sequences and restriction site analysis for diagnosis of sickle cell anemia. Science *230*, 1350-1354.

Saiki, R.K., Gelfand, D.H., Stoffel, S., Scharf, S.J., Higuchi, R., Horn, G.T., Mullis, K.B., and Erlich, H.A. (1988). Primer-directed enzymatic amplification of DNA with a thermostable DNA polymerase. Science *239*, 487-491.

Schweitzer, B. and Kingsmore, S. (2001). Combining nucleic acid amplification and detection. Curr. Opin. Biotechnol. *12*, 21-27.

Walker, G.T., Fraiser, M.S., Schram, J.L., Little, M.C., Nadeau, J.G., and Malinowski, D.P. (1992a). Isothermal *in vitro* amplification of DNA by a restriction enzyme/DNA polymerase system. Proc. Natl. Acad. Sci. USA *89*, 392-396.

Walker, G.T., Little, M.C., Nadeau, J.G., and Shank, D.D. (1992b). Strand displacement amplification--an isothermal, *in vitro* DNA amplification technique. Nucleic Acids Res. *20*, 1691-1696.

Zhang, D.Y., Brandwein, M., Hsuih, T.C.H., and Li, H. (1998). Amplification of target-specific, ligation-dependent circular probe. Gene *211*, 277-285.

3.3

Ligation-Mediated Rolling Circle DNA Amplification for Non-Gel Detection of Single Nucleotide Polymorphisms (SNPs)

Xiaoquan Qi

Abstract

This chapter describes a flexible, non-gel based method of SNP (single nucleotide polymorphism) detection. The method is grounded on L-RCA (ligation-rolling circle amplification)–thermostable ligation of circularisable padlock-like DNA probes for allelic SNP discrimination with subsequent RCA procedure for signal enhancement. First, the target DNA fragment is pre-amplified by PCR from a small amount of genomic DNA (*i.e.* 50 ng). Then, the SNPs are targeted by use of allele-specific short padlock probes. Circularised by enzymatic ligation padlock probe is amplified by DNA polymerase via linear or branched RCA reaction. The final reaction mixtures are stained with SYBR Gold to be immediately visualised by UV illumination. A compatible buffer system for all enzymes involved is used, allowing the successive enzymatic reactions to be initiated and detected in the same tube or microplate well, so that the experiment can be scaled up easily for high-throughput detection.

Background

Single-nucleotide polymorphisms (SNPs) are most common DNA variation in the genomes (Wang *et al.*, 1998; Cargill *et al.*, 1999; Bryan *et al.*, 1999). They are usually biallelic although there are four possible nucleotide substitutions in principle (Brookes, 1999). SNPs are mainly used in construction of highly saturated genetic maps, human disease diagnostics and SNP-assisted plant breeding. These applications require to genotype thousands of SNPs from large numbers of individual DNA samples in an accurate, rapid and cost-effective manner (Griffin and Smith, 2000).

Circularisation of padlock probes (Nilsson *et al.*, 1994) with common or thermostable DNA ligases (Barany, 1991; Luo *et al.*, 1996) is applied to specifically discriminate point mutations in target DNA sequences. Circularised padlock probes are efficiently amplified by the single-primer linear RCA (rolling circle amplification) with ϕ29 DNA polymerase (Blanco *et al.*, 1989; Banér *et al.*, 1998; Lizardi *et al.*, 1998) or, even more robustly, by the two-primer hyperbranched RCA (HRCA) with Vent® (exo-) DNA polymerase (Lizardi *et al.*, 1998). By incorporation of energy transfer probes, as the second RCA primers, into the reactions (Faruqi *et al.*, 2001; Pickering *et al.*, 2002; Alsmadi *et al.*, 2003), signals are specifically enlarged and then are detected by a corresponding fluorescence detection system, enabling genotyping of SNPs without preamplification of target sequences. Based on the gap-filling strategy (Lizardi *et al.*, 1998), a method for multiplexed genotyping of SNPs directly from genomic DNA was developed (Hardenbol *et al.*, 2003). However, all these methods are required expensive labeled primers and/or rather complicated diagnostic/ recording equipment. The following chapter describes an alternative ligation-mediated RCA-based method for SNP genotyping of complex genomes without requirement of any elaborate detector device. This easily adaptable method is particularly suitable for genotyping of a few SNPs in a very large population (*e.g.* >500 individuals).

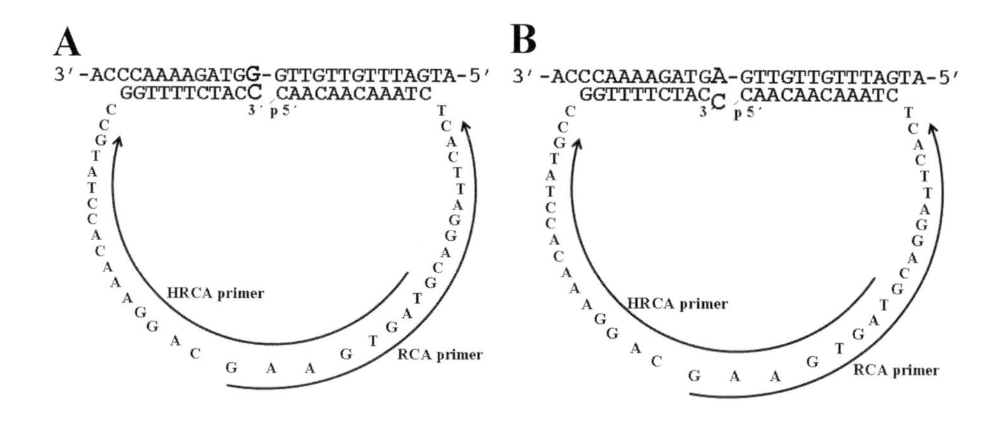

Figure 1. Detection of SNPs with short padlock probes (arrows indicate the 5' to 3' direction of the oligonucleotides). The padlock probe I with 11 and 12 nt target-complementary terminal sequences matches the normal G-type allele (at 3,417 bp) in the S75 wild-type *AsbAS1* gene, thus leading to its successful ligation (A), and mismatches with the A-type allele in mutant #109 causing ligation failure (B).

In brief, the single-copy gene was preamplified from a small amount of genomic DNA (ca. 50 ng) by PCR. Short padlock probes matching the target DNA sequences at both 3' and 5' ends were designed. A perfect match at the target region between the target DNA and the padlock probe allows multi-round ligation/circularisation of the padlock probe in the presence of thermostable ligase, while a single base mismatch at the 3' end of the probe causes failure of ligation (Figure 1). The circularised padlock probe is then amplified with ϕ29 DNA polymerase using primers derived from the central region (backbone) of the padlock probe. The RCA-generated DNA is detected by UV illumination right after staining of reaction samples with SYBR Gold dye. Compatible buffers are used for these reactions and the whole process is carried out in a single tube or microplate well. This method therefore has potential for high-throughput automated analysis at low cost.

Protocol

Materials

The diploid oat samples, wild-type *Avena strigosa* accession S75 and the *sad1* mutant #109, described by Papadopoulou *et al.* (1999) and Haralampidis *et al.* (2001) were used. Point mutation in the *AsbAS1* gene of mutant #109 is shown in Table 1. Scoring of saponin-deficient seedlings was conducted under UV illumination according

Table 1. Targets, primers and probes.

Name	Sequence (5' → 3')
Genotypes	3,417 bp ↓
S75	...atttgttgttg**G**gtagaaaaccca...
#109	...atttgttgttg**A**gtagaaaaccca...
PCR primers	
F	catatccattatgacgacgaatcaacc
R	tctataccaacctgtgccttcattcc
Padlock probes	
I	p-**caacaacaaatc**tcacttaggac*gtagtgaagcaggaaacacctatg*cc**ggttttctacc**
II	p-**caacaacaaatc**ttctgagttcc*aacgcagtaagatgttgcacgatg*cc**ggttttctact**
RCA primers	
I	cttcactacgtcctaagtg
II	tactgcgttggaactcaga
HRCA primers	
I	gtagtgaagcaggaaacacctatg
II	aacgcagtaagatgttgcacgatg

"p": 5' phosphorylation; The underlined nucleotides are complematory to RCA primer, and those in *italic* correspond to sequences of the second primer for HRCA. Nine nucleotides that are both in *italic* and underlined lie in the overlapped region.

to Papadopoulou *et al.* (1999). Genomic DNA was isolated from the wild type S75, *sad1* mutant #109, and the 83 F_2 progeny derived from a cross between S75 and #109 by using the DNeasy 96 Plant Kit (Qiagen). Taq DNA polymerase was provided by Roche Molecular Biochemicals; Vent (exo^-) DNA polymerase was purchased from New England Biolabs; ϕ29 DNA polymerase was gifted by M. Salas; thermostable DNA ligase, Ampligase®, was obtained from Epicentre Technologies. For UV-illumination and data quantitation, Gel Doc 1000 system (Bio-Rad) with maximum emission intensity at 302 nm was used. SYBR® Gold stain, which is highly sensitive and rather universal dye for fluorescent detection of both single- and double-stranded nucleic acids, was obtained from Molecular Probes.

Various oligonucleotide probes and primers (Table 1) were synthesised by Life Technologies or MWG-Biotech. PCR primers F and R were designed for amplification of the SNP flanking region in *AsbAS1*. The 60-nt-long padlock probes I and II were designed to target the G allele in S75 or A allele in mutant #109 at position 3,417 bp in the *AsbAS1* gene, respectively. The corresponding RCA primers I and II were designed for RCA with ϕ29 DNA polymerase and HRCA primers I and II for HRCA with Vent (exo^-) DNA polymerase.

Standardised reaction conditions should be used throughout, and both positive and negative controls have to be included in the experiment in order to achieve reliable allelic discrimination. A compatible set of reaction buffers was used. This enables us to sequentially add ligation and RCA/HRCA mixtures into the initial PCR tubes. The PCR buffer is 10 mM Tris-HCl (pH 8.3), 50 mM KCl, 1.5 mM $MgCl_2$. In ligation, this buffer is readily adjusted by addition of Mg-containing ligation mixture to 6.7 mM Tris-HCl (pH 8.3), 33.3 mM KCl, 6.0 mM $MgCl_2$. For RCA, buffer system is changed to 7.5 mM Tris-HCl (pH 8.3), 37.5 mM KCl, 4.9 mM $MgCl_2$. In the HRCA buffer, 6 mM Tris-HCl (pH 8.3), 21 mM KCl and 3.8 mM Mg^{+2} are used.

Step 1: Amplification of DNA Targets by Asymmetric PCR

The SNP region was pre-amplified by asymmetric PCR using the forward/reverse primers, F and R, in the 9:1 ratio. The 20 µl PCR sample contained 50 ng genomic DNA, 0.45 µM forward primer, 0.05 µM reverse primer, 200 µM dNTPs, 10 mM Tris-HCl (pH 8.3), 50 mM KCl, 1.5 mM $MgCl_2$ and 0.5 U of Taq DNA polymerase. After initial denaturation at 94°C for 3 min, amplification was carried out with 35 cycles of 1) denaturation at 94°C for 30 s, 2) annealing at 58°C for 30 s, and 3) extension at 72°C for 45 s. PCR products were split into two 10 µl aliquots for subsequent ligation with two different, allele-specific padlock probes.

Although it is technically possible to operate directly with genomic DNA, we prefer to pre-amplify the designated DNA target by PCR as this step has a number of advantages. Firstly, it narrows down the unrelated non-target sequences so that shorter padlock probes can be used. Secondly, the concentration of target DNA is substantially increased after PCR, thus only a small amount of genomic DNA (*e.g.* ≤50 ng of oat genomic DNA) is required per assay. Thirdly, the use of asymmetric PCR provides with the probe-accessible and mass-normalised targets, which is significant when a high-throughput DNA isolation method is used and measurement/adjustment of DNA concentrations is impracticable. The normalised PCR products are also crucial for achieving uniformity of signals and reliable allelic discrimination.

Step 2: Ligation of Padlock Probes

5 µl aliquots of ligation mixture containing 15 mM $MgCl_2$, 1.5 mM nicotinamide adenine dinucleotide (NAD), 2 U of Ampligase® (thermostable DNA ligase) and 0.6 µM allele-specific padlock probes (Table 1) were added to each 10 µl post PCR sample containing the amplified target DNA. Multi-round ligation was conducted with 10 cycles of 1) denaturation at 94°C for 30 s and 2) ligation at 37°C for 5 min. For effective circularisation with this procedure, 50-60-nt-long padlock probes with 11-13 nt target-complementary sequences at their

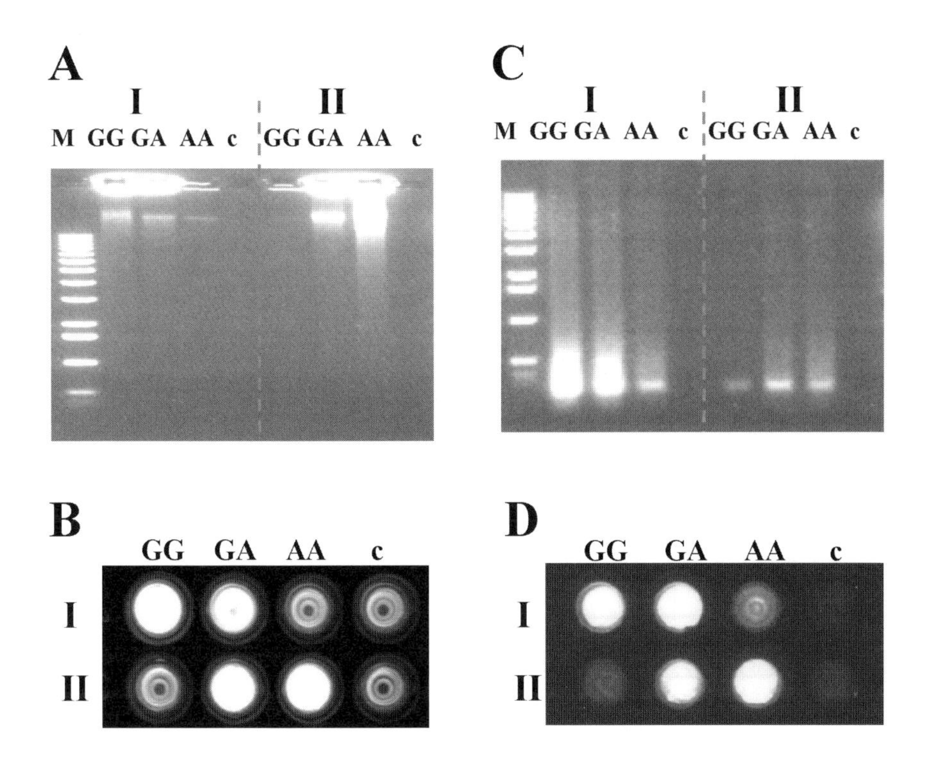

Figure 2. Model L-RCA genotyping. Padlock probes I and II are the allele-specific padlock probes (Table 1). "GG" and "AA" are the genotypes of S75 and #109, respectively. "GA" is a simulated heterozygous genotype generated by mixing equal amounts of DNA from S75 and #109. "c" is the negative control lacking target DNA. "M" is the 1-kb DNA Ladder (Life Technologies).
A: RCA with ϕ29 DNA polymerase. The allele-specific RCA primers I and II were used to amplify the circularised padlock probes, respectively. Products were separated in a 0.6% agarose gels and stained with ethidium bromide.
B: Detection of the ϕ29-generated RCA products in a microplate, as described in Protocol.
C: HRCA with Vent (exo-) DNA polymerase. The allele-specific primer pairs, RCA primer I/HRCA primer I and RCA primer II/HRCA primer II, were used to amplify the circularised padlock probes I and II, respectively. Products were separated in a 0.8% agarose gel and stained with ethidium bromide.
D: Detection of the Vent-generated HRCA products in a microplate, as described in Protocol.

3' and 5' termini are recommended (ligation could be more efficient with longer probe; Qi *et al.*, 2001), which suit well for the SNP detection in ≤10 kb PCR fragments.

Step 3a: RCA Reaction

A 5 μl aliquot of 0.6 μM allele-specific RCA primer I or II was added to the 15 μl ligation mixture to initiate the RCA reaction. The samples were heated to 70°C and cooled to room temperature prior to the addition of 5 μl of RCA mix (800 μM dNTPs, 50 mM Tris-HCl/pH 8.3, 250 mM KCl, 7.5 mM $MgCl_2$, 4 μg BSA, 0.7 μg phage T4 gene-32 protein and 60 ng ϕ29 DNA polymerase), followed by incubation at 37°C for 3 hr. The RCA reactions were terminated by heating of samples at 70°C for 10 min. Linear RCA amplification of the allele-specific circularised padlock probes using ϕ29 DNA polymerase produced long single-stranded DNA amplicons that tended to remain in the loading wells of a 0.6% agarose gel after electrophoresis (Figure 2A).

Step 3b: HRCA Reaction

Alternatively, the HRCA process can be used for detection of circularised padlock probes by generating double-stranded DNA amplicons (Figure 2C). A 10 μl aliquot containing 400 μM dNTPs, 50 mM Tris-HCl (pH 8.3/25°C), 25 mM KCl, 5 mM $MgSO_4$, 25 mM $(NH_4)_2SO_4$, 0.05% Triton X-100, 0.7 μg phage T4 gene-32 protein, 2.5 U Vent (exo^-) DNA polymerase and 0.3 μM allele specific RCA primer/HRCA primer pair was added to the 15 μl ligation mixture for amplification of circularised padlock probes. The reaction was first incubated at 92°C for 3 min and then at 63°C for 3 hr. Note that in our own experience the RCA reaction with ϕ29 DNA polymerase shows more allelic specificity in signal amplification than the HRCA reaction with Vent (exo^-) DNA polymerase. Importantly, signals were efficiently amplified in both RCA and HRCA reactions. Topological constraints (Banér *et al.*, 1998; Kuhn *et al.*, 2002) did not significantly affect the probe' rolling replication in our experiments: multi-round ligation was applied in the protocol, which may help to release the circularised padlock probe for efficient RCA/HRCA (Banér *et al.*, 1998).

Step 4: Signal Detection under UV Illumination

The products of both RCA and HRCA reactions are well detectable under UV illumination in a tube or microplate wells after their staining with SYBR Gold dye (Figure 2 B,D). Prior to detection, 100 µl of SYBR Gold solution ($10^4\times$ stock concentrate in DMSO; 10^3-fold diluted by TAE/pH 7.6 buffer) was added to 20-25 µl reaction samples. After brief mixing, stained DNA was visualised under UV-illumination (302 nm) and the fluorescence was quantified using the Gel Doc system. The whole genotyping procedure (from PCR step to signal detection) can be completed in ca. 7 hr.

Example

Oat mutant #109 (one of the two *sad1* mutants) and the S75 wild type were used as a test system for development of the non-gel L-RCA based SNP-genotyping protocol. Mutant #109 is one of the sodium azide-generated saponin-deficient (*sad*) mutants of diploid oat (*Avena strigosa* accession S57) that have been produced to analyze the specific secondary metabolite biosynthesis of this plant (Papadopoulou *et al.*, 1999; Haralampidis *et al.*, 2001). The *sad1* mutations involve the *A. strigosa* gene *AsbAS1*, which is known to encode β-amyrin synthase, a key enzyme in biosynthesis of triterpenoid saponins (Haralampidis *et al.*, 2001). Sequence analysis of the wild type (S57) and *sad1* mutant DNA revealed a single-base change of G to A in the *AsbAS1* gene (Table 1).

We used our protocol to expediently genotype the G/A SNP at the 3,417-bp position of *sad1* in the F_2 population derived from a cross between the wild-type *A. strigosa* accession S75 and *sad1* mutant #109 (83 samples total; Figure 3). In this figure, one can clearly see that 44 analytes were GA heterozygotes, while 17 and 22 of these crossbreeds were GG and AA homozygotes, respectively. Therefore, it is expected that 61 F_2 individuals–17 GG homozygotes plus 44 GA heterozygotes containing the normal *AsbAS1* gene–should be capable of saponin production, whereas 22 AA homozygotes could not do this.

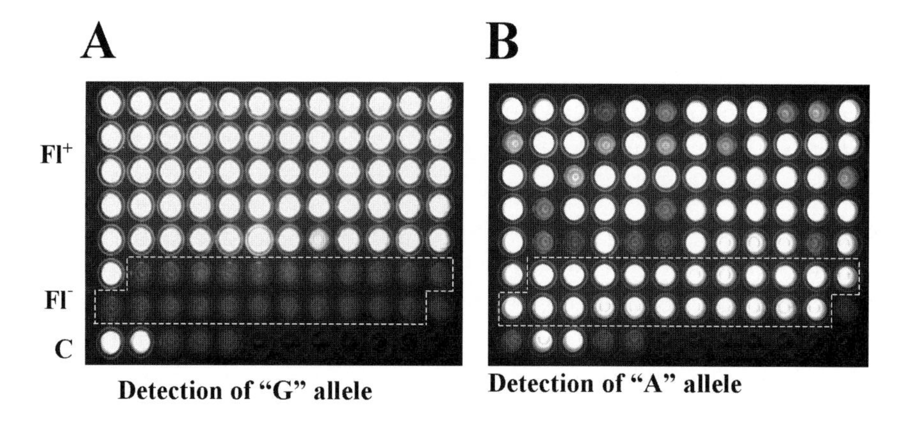

Figure 3. L-RCA based non-gel SNP genotyping in a F_2 population derived from a cross between S75 and mutant #109. Seedlings were scored and grouped for the presence and absence of saponin by fluorescent test. "Fl⁺" indicates the saponin-competent seedlings with fluorescent root tips and "Fl⁻" indicates the saponin-deficient seedlings. "C" represents the controls: S75, S75/#109, #109 and two negative controls lacking genomic DNA (from left to right). Padlock probe I and RCA primer I were used in detection of the "G" allele (A), and padlock probe II and RCA primer II were used for detection of the "A" allele (B). RCA was conducted with ϕ29 DNA polymerase.

In the independent experiment (not shown), the corresponding seedlings from this cross were assessed for the saponin-deficient phenotype via their screening on the presence or absence of saponin in the root tips by fluorescent procedure, as described by Papadopoulou *et al.* (1999). In a complete agreement with the SNP analysis, it was found that 61 F_2 progeny contained saponins in their roots, while 22 offspring were saponin-deficient ones. These results explicitly prove that *sad1* mutation is co-segregated with *AsbAS1* gene. Importantly, all SNP scores were confirmed using a mini-sequencing approach based on the single-primer chain extension (Syvänen *et al.*, 1990).

Concluding Remarks

Ligation-mediated detection of SNPs can be achieved by separation of the ligation products on denatured polyacrylamide gels followed by silver staining (Qi *et al.*, 2001). However, the gel-based SNP detection is labour-intensive and time-consuming, and is also problematic to adapt for automation to achieve high throughput. Presented here non-gel L-RCA protocol readily circumvents these difficulties allowing to detect SNPs in about 7 hours in numerous samples in the microplate wells under UV light directly after the amplification step. All experimental procedures including the RCA and PCR amplification, thermoligation and signal detection can be conducted in the same tube or microplate well using the compatible buffers described in the protocol. The methodology can easily be applied to the analysis of multiple analytes using 384-well microplates and liquid handling systems.

Note that the use of amplified DNA as a reporter followed by the UV detection does not allow a high-multiplexity assay to be performed in a single tube. However, in case of the bi-allelic detection, PCR amplification and subsequent splitting of PCR products for separate ligation of allele-specific padlock probes will help to identify the PCR failures, which are probably the major causes of false-negatives in this detection system. Also the ligation of two allele-specific padlock probes can be carried out simultaneously, followed by two separate probe-specific RCA/HRCA reactions. Multiplexed assay for several loci or alleles in the whole genome can be achieved with the L-RCA method followed by detection of the RCA products with multicolor probes using appropriate laser/filter resources as a comprehensive fluorescence system (Qi *et al.*, 2001).

Acknowledgements

I am indebted to M. Salas for kindly providing ϕ29 DNA polymerase. This work was supported by grants from the Biotechnology and Biological Sciences Research Council (BBSRC) and the Department for International Development (DFID), UK.

References

Alsmadi, O.A., Bornarth, C.J., Song, W., Wisniewski, M., Du, J., Brockman, J.P., Faruqi, A.F., Hosono, S., Sun, Z., Du, Y., *et al.* (2003). High accuracy genotyping directly from genomic DNA using a rolling circle amplification based assay. BMC Genomics *4*, 21.

Banér, J., Nilsson, M., Mendel-Hartvig, M., and Landegren, U. (1998). Signal amplification of padlock probes by rolling circle replication. Nucleic Acids Res. *26*, 5073-5078.

Barany, F. (1991). Genetic disease detection and DNA amplification using cloned thermostable ligase. Proc. Natl. Acad. Sci. USA *88*, 89-193.

Blanco, L., Bernad, A., Lázaro, J.M., Martín, G., Garmendia, C., and Salas M. (1989). Highly efficient DNA synthesis by the phage ϕ29 DNA polymerase. J. Biol. Chem. *264*, 8935-8940.

Brookes, A.J. (1999). The essence of SNPs. Gene *234*, 177-186.

Bryan, G.J., Stephenson, P., Collins, A., Kirby, J., Smith, J.B., and Gale, M.D. (1999). Low levels of DNA sequence variation among adapted genotypes of hexaploid wheat. Theor. Appl. Genet. *99*, 192-198.

Cargill, M., Altshuler, D., Ireland, J., Sklar, P., Ardlie, K., Patil, N., Lane, C.R., Lim, E.P., Kalyanaraman, N., Nemesh, J., *et al.* (1999). Characterization of single-nucleotide polymorphisms in coding regions of human genes. Nat. Genet. *22*, 231-238.

Faruqi, A.F., Hosono, S., Driscoll, M.D., Dean, F.B., Alsmadi, O., Bandaru, R., Kumar, G., Grimwade, B., Zong, Q., Sun, Z., *et al.* (2001). High-throughput genotyping of single nucleotide polymorphisms with rolling circle amplification. BMC Genomics *2*, 4.

Griffin, T.J. and Smith, L.M. (2000). Single-nucleotide polymorphism analysis by MALDI-TOF mass spectrometry. Trends Biotechnol. *18*, 77-84.

Haralampidis, K., Bryan, G., Qi, X., Papadopoulou, K., Bakht, S., Melton, R., and Osbourn, A. (2001). A new class of oxidosqualene cyclases directs synthesis of antimicrobial phytoprotectants in monocots. Proc. Natl. Acad. Sci. USA *98*, 13431-13436.

Hardenbol, P., Banér, J., Jain, M., Nilsson, M., Namsaraev, E.A., Karlin-Neumann, G.A., Fakhrai-Rad, H., Ronaghi, M., Willis, T.D., Landegren, U., and Davis R.W. (2003). Multiplexed genotyping with sequence-tagged molecular inversion probes. Nat. Biotechnol. *21*, 673-678.

Kuhn, H., Demidov, V.V., and Frank-Kamenetskii, M.D. (2002). Rolling-circle amplification under topological constraints. Nucleic Acids Res. *30*, 574-580.

Lizardi, P.M., Huang, X., Zhu, Z., Bray-Ward, P., Thomas, D.C., and Ward, D.C. (1998). Mutation detection and single-molecular counting using isothermal rolling-circle amplification. Nat. Genet. *19*, 225-232.

Luo, J., Bergstrom, D.E., and Barany, F. (1996). Improving the fidelity of *Thermus thermophilus* DNA ligase. Nucleic Acids Res. *24*, 3071-3078.

Nilsson, M., Malmgren, H., Samiotaki, M., Kwiatkowski, M., Chowdhary, B.P., and Landegren, U. (1994). Padlock probes: circularizing oligonucleotides for localized

DNA detection. Science *265*, 2085-2088.

Papadopoulou, K., Melton, R.E., Leggett, M., Daniels, M.J., and Osbourn, A.E. (1999). Compromised disease resistance in saponin-deficient plants. Proc. Natl. Acad. Sci. USA *96*, 12923-12928.

Pickering, J., Bamford, A., Godbole, V., Briggs, J., Scozzafava, G., Roe, P., Wheeler, C., Ghouze, F., and Cuss S. (2002). Integration of DNA ligation and rolling circle amplification for the homogeneous, end-point detection of single nucleotide polymorphisms. Nucleic Acids Res. *30*, e60.

Qi, X., Bakht, S., Devos, K.M., Gale, M.D., and Osbourn, A. (2001). L-RCA (ligation-rolling circle amplification): a general method for genotyping of single nucleotide polymorphisms (SNPs). Nucleic Acids Res. *29*, e116.

Syvänen, A.C., Aalto-Setälä, K, Harju, L., Kontula, K., and Söderlund, H. (1990). A primer-guided nucleotide incorporation assay in the genotyping of Apolipoprotein E. Genomics *8*, 684-692.

Wang, D.G., Fan, J.B., Siao, C.J., Berno, A., Young, P., Sapolsky, R., Ghandour, G., Perkins, N., Winchester, E., Spencer J., *et al.* (1998). Large-scale identification, mapping, and genotyping of single-nucleotide polymorphisms in the human genome. Science *280*, 1077-1082.

3.4

Rolling-Circle Amplification of Duplex DNA Sequences assisted by PNA Openers

Heiko Kuhn and Vadim V. Demidov

Abstract

Peptide nucleic acid (PNA) oligomers can be employed as site-specific openers of the DNA double helix to locally expose a designated marker sequence inside duplex DNA. The opened DNA site is then hybridized to a circularizable oligonucleotide probe, which is subsequently closed by DNA ligase. This way, the marker sequence from the DNA duplex of interest can be isothermally amplified by a variety of DNA polymerases via the rolling-circle amplification (RCA) mechanism. An alternative strategy for the PNA-assisted RCA exploits a restriction enzyme (and an auxiliary linear oligonucleotide) to selectively introduce a nick within the exposed marker sequence. As a result, a single-stranded DNA segment with a free 3' end is obtained, which can serve as a primer in a subsequent RCA reaction, if hybridized to a circular probe oligonucleotide. Besides DNA polymerase, only one extra enzyme is required in these new promising RCA formats and the two examples presented here demonstrate their robust practical potential for DNA diagnostics.

Brief Introduction

Owing to its simplicity, high specificity, sensitivity and multiplexity, rolling-circle amplification (RCA) attracted significant attention in basic and applied research and is now turning into a customary technique for molecular diagnostics (Schweitzer and Kingsmore, 2001; Demidov, 2002, 2004a; Kingsmore and Patel, 2003; Zhang and Liu, 2003). As analytical tool, RCA is generally based on the isothermal enzymatic rolling replication of DNA minicircles hybridized to single-stranded (ss) DNA or RNA targets. Since DNA samples are normally obtained in the double-stranded (ds) form, a denaturation is needed to separate the DNA complementary strands for probe hybridization. Given that fact, we have recently developed two approaches allowing to perform RCA directly on dsDNA (Kuhn *et al.*, 2002a, 2003).

Both methods are based on the local opening of a designated site within the target DNA duplex by a pair of bisPNA openers with subsequent hybridization of an oligonucleotide to the exposed DNA strand, yielding finally a so-called PD-loop (Bukanov *et al.*, 1998; Demidov, 2001, 2004b; Demidov and Frank-Kamenetskii, 2002; Figure 1). This strategy not only makes it possible to avoid the DNA denaturation step but also results in a much higher sequence specificity of DNA targeting with a hybridization/amplification probe since 1) multiple recognition events are involved and 2) most of DNA remains in the duplex state, hence being inaccessible for probe binding.

One of these methods includes the target-directed assembly of the earring-like label (see Figure 1A), which is firmly – *i.e.* topologically – linked to a chosen DNA site. This assembly requires DNA ligase to covalently close a circularizable probe following the hybridization of its linear precursor (Kuhn *et al.*, 1999a, 2000; Demidov *et al.*, 2001; Demidov, 2003). Thus obtained DNA minicircle can then be engaged in single- or double-primed RCA reactions to generate long DNA repeats of the original circle and/or its complement, resulting in up to 10^{10}-fold probe sequence amplification, depending on the particular RCA format (Kuhn *et al.*, 2002a).

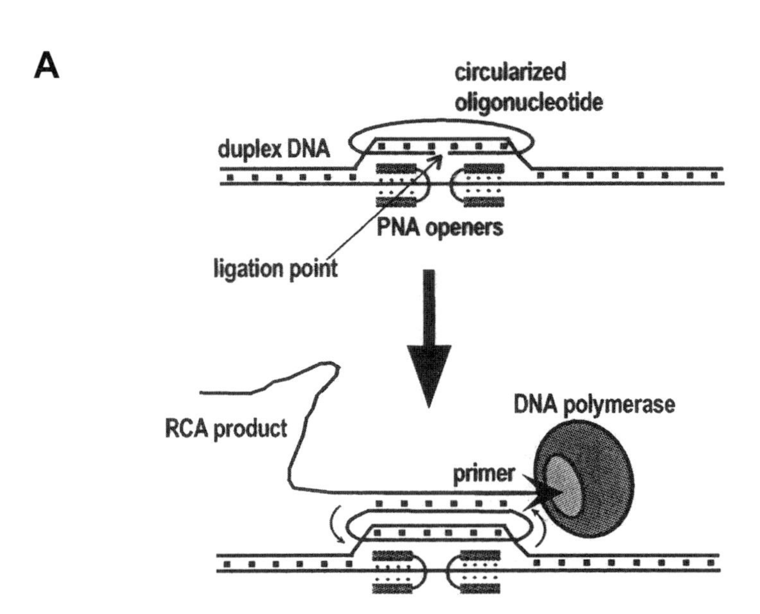

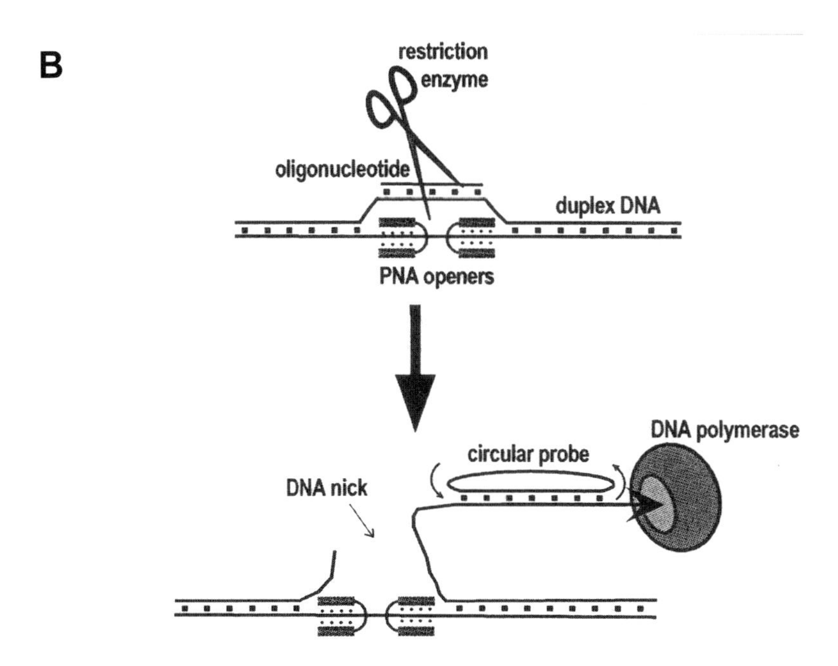

Figure 1. Schematics of the design used for PNA-assisted RCA reactions performed either on earring probe (A) or with employment of artificial nickase system (B). Both schemes are based on the formation of PD-loop consisting of the locally open short dsDNA segment, a pair of bisPNA openers and an oligodeoxyribonucleotide (Bukanov *et al.*, 1998).

Another approach we developed for the PNA-assisted RCA of dsDNA sequences employs the design of site-specific artificial nickase system (Figure 1B; Kuhn *et al.*, 2003). In this case, the relevant PD-loop forming site should comprise a recognition site of some restriction enzyme. Then, a short secondary duplex formed between the auxiliary oligonucleotide and the PNA-opened target sequence is used as a substrate for that restriction enzyme, which ultimately creates a nick in the parent DNA. Using prior circularized hybridization probe, such a nick can subsequently be engaged in a RCA reaction (Kuhn *et al.*, 2003), as it is shown schematically in Figure 1B. Importantly, only one supplementary enzyme is required in both protocols besides DNA polymerase to selectively initiate the RCA reaction on DNA duplexes.

Protocol 1: RCA with *In Situ* Circularizable Probe (Earring Label)

Materials

DNA Target Site

For this protocol, a specific site within the DNA duplex of interest has to be chosen. This site (the PD-loop-forming site; Figure 1) should consist of two 7-8-bp-long homopurine tracts (to bind the all-pyrimidine bisPNA openers) that are closely located on the same DNA strand, being separated by ≤10 bp of any random nucleobase sequence. Examples are: 5'-$\underline{A_3GA_2G_2}CTG_2\underline{A_2G_2A_3}$ (*S. cerevisiae* chromosome IX; the PNA-binding sites are underlined) and 5'-$\underline{AGAG_2A_2G}CTACT\underline{G_2AG_2AGA}$ (HIV-1 *nef* gene; this site was used in the case given below).

PNA Openers

A pair of appropriate [7+7]- and/or [8+8]-mer peptide nucleic acid "clamps" (or bisPNAs) with pseudoisocytosine (J base) instead of cytosine in one half (Egholm *et al.*, 1995) is needed. The triplex-forming J/T-containing PNA strand should be proper arranged so that its N-terminus faces the 5'-end of the target DNA strand (so-called "parallel" PNA-DNA orientation); then the other, duplex-invading, C/T-containing PNA strand with mirror-symmetrical sequence will be consequentially "antiparallel" to the same DNA strand, as required for the most stable and essentially pH-independent bisPNA binding (Egholm *et al.*, 1995). Incorporation of two-three terminal lysines and use of a three-unit linker (made of O-units or eg1; Egholm *et al.*, 1995) is recommended for better stability and specificity of PNA-DNA complexes (Kuhn *et al.*, 1998).

Several biotech companies (*e.g.* Applied Biosystems, Panagen and Metabion) commercially offer custom synthesis of bisPNA oligomers using automated protocols. Both manual and automated "in-house" PNA synthesis on 2-20 μM scale can also be performed in research laboratories with synthetic experience (Braasch and Corey, 2001). In case of difficulties with obtaining the J base-containing bisPNAs (due to a limited availability of the corresponding monomer reagent), PNA oligomers with only C and T nucleobases can be used instead, taking into account that this will result in a much stronger pH-dependence of PNA binding to dsDNA, hence careful choice of optimal pH may then be necessary (Kuhn *et al.*, 1999b). For the study presented here, bisPNAs $HLys_2$-$TCTC_2TC_2$-$(eg1)_3$-J_2TJ_2TJT-$LysNH_2$ and $HLys_2$-$TJTJ_2T_2J$-$(eg1)_3$-CT_2C_2TCT-$LysNH_2$ were purchased from PE Biosystems.

Oligonucleotides

A 5'-phosphorylated circularizable oligodeoxynucleotide (precursor probe) is required having ~10-nt-long end sequences complementary to the PNA-exposed target DNA strand and a linker segment of an arbitrary sequence/length (~20 nt or longer). In the hybridized state, the termini of this oligo will be juxtaposed, thus allowing the probe circularization by enzymatic ligation (see Figure 1A). The ligation yield somewhat depends on the length of oligo employed for circularization, with 45-90 nt total serving quite well for efficient circularization (40-80% yield). The resulting circular, earring-like probe forms roughly two turns around the complementary DNA strand to be topologically linked to the target site. With the use of corresponding primers (and a proper DNA polymerase; see below), such a probe can be involved in linear (single-primed) or branched (double-primed) RCA reactions. Here we use: 5'-CTG_2AG_2AGAT_4GTG_2TATCGAT_2CGTCTCT_2AG A$G_2$$A_2$GCTA (circularizable oligo), 5'-GACGA_2TCGATAC_2AC (primer 1), 5'-GAGACGA_2TCGATAC_2ACA_2 (primer 2), and 5'-GA$G_2$$A_2$GCTACT$G_2AG_2$AGA (primer 3).

Enzymes

DNA Ligase

T4 DNA ligase was used by us for the earring probe circularization. Based on the published data (Zhang and Liu, 2003), we assume that a variety of other DNA ligases can also be employed for this purpose to adjust the ligation efficiency to specific experimental conditions, if necessary.

DNA Polymerase

A number of DNA polymerases, including the thermophilic ones, are able to effectively perform the RCA reactions on earring probes despite certain topological constraints (Kuhn *et al.*, 2002a). They include Sequenase 2.0, Vent exo$^-$ and *Bst* large fragment DNA polymerases. In experiments shown below, the first two polymerase enzymes were used.

Step 1: DNA Targeting with PNA

Incubate the requisite amount of DNA with a pair of PNA openers (0.5-5 μM depending on the binding affinity of bisPNAs) in 10-100 μl of an appropriate buffer solution (*e.g.* 10 mM sodium-phosphate buffer, pH 6.8) for ~1-2 hr at 37°C or other relevant temperature. Typically, PNA openers are taken in a great molar excess over the target DNA yielding pseudo-first-order binding kinetics (Demidov *et al.*, 1995, 1997; for optimization of DNA targeting with PNA openers, see: Demidov and Frank-Kamenetskii, 2001; Demidov, 2004b). To avoid binding of free openers with complementary parts of a probe, remove the surplus of unbound PNA by gel filtration of the sample through Sephadex G-50. In case of long, agarose-embedded DNA templates, PNA-DNA complexes are formed by shaking incubation of the gel sample with the buffered solution of PNA openers for several hours. The gel-filtration procedure is replaced here by thorough washing of the gel sample. Higher PNA concentrations may be required for the in-gel PNA binding as compared with solution-based procedures.

Step 2: Probe Circularization

Add ~10 pmol of circularizable oligonucleotide to ~0.5 pmol of PNA-DNA complex (freshly prepared or stored at -20°C until use). Incubate this mixture with 1-5 U of DNA ligase in a proper ligation buffer (*e.g.* 40 mM Tris-HCl, 10 mM $MgCl_2$, 10 mM DTT, 0.5 mM ATP; pH 7.8) for 1.5 hr at ≤20°C, 15 min at 45°C, and 2 hr at ≤20°C. For agarose-embedded DNA, longer incubation and higher probe concentration may be required. Prior DNA dephosphorylation could be necessary for more efficient probe ligation in case of DNA samples with excess of unrelated DNA. Exonuclease VII (ExoVII) or gel filtration/sample washing can be employed to remove the excess of unprocessed linear precursor (optional).

Step 3: RCA Reactions

Single-Primed RCA

Preincubate the earring-labeled dsDNA with 1 μM oligo primer (primer 1 in our case) for 30 min at 37°C in 10 μl of buffer (20 mM $MgCl_2$, 50 mM NaCl, 40 mM Tris-HCl, pH 7.2). Then, add 5.7 μl of H_2O, 1.0 μl of 100 mM DTT, 3.2 μl of all four dNTPs (25 mM each), 0.5 μl of SSB protein (2.2 mg/ml) and 2.0 μl of Sequenase 2.0 (1.6 U/μl) and incubate the reaction mixture for 3.5 hr at 37°C. For kinetic experiments, the RCA reaction can be stopped at different incubation times by addition of 0.5 M EDTA (pH 8.0) to a final concentration of 100 mM followed by removal of unused primer and

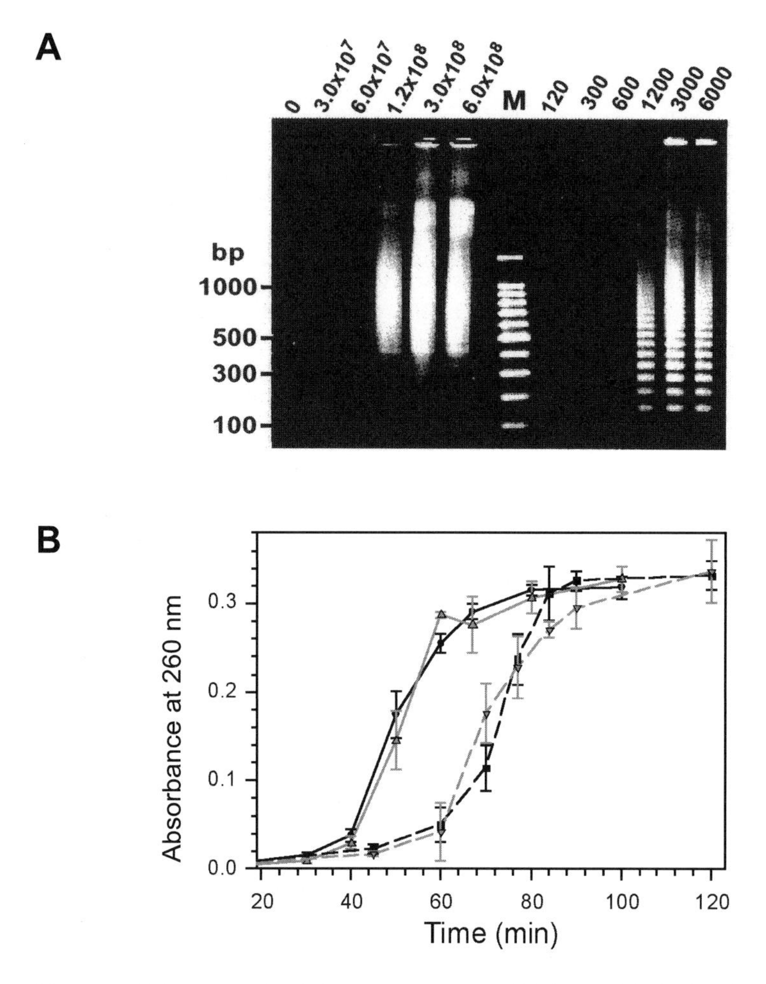

Figure 2. PNA-assisted RCA initiated on the HIV-1 dsDNA site by the earring-type labeling (see Materials/Protocol 1 section for sequences of the DNA target site, PNA openers and circularizable probe, and Kuhn *et al.*, 2002a for experimental details/conditions).
A: Electrophoretic analysis of RCA products in 2% agarose gels stained by ethidium bromide. Single-primed (left part; Sequenase 2.0 DNA polymerase) and double-primed (right part; Vent exo$^-$ DNA polymerase) RCA reactions were performed on 350 bp dsDNA target fragments carrying the marker HIV-1 site in the middle. The RCA amplicons were generated starting from different initial inputs of the target DNA fragment (given in number of molecules above each lane). Lane M corresponds to a 100 bp size marker.
B: Triplicate-averaged time courses of the amplicons' accumulation for the single-primed RCA on a free DNA minicircle (black curves) and the earring probe (grey curves), as assessed by UV absorbance measurements. Numbers of input molecules, the DNA minicircle or earring-labeled DNA target fragment (~100 bp DNA dumbbell with the centrally located HIV-1 marker site), were 3.6×10^9 (solid lines) or 1.2×10^8 (dash lines), respectively.

dNTPs by Sephadex G-50 gel filtration, if necessary. Note that although the single-primed RCA typically results in single-stranded products, dsDNA amplicons could also be obtained in some cases (Fire and Xu 1995; Sabanayagam *et al.*, 1999; Kuhn *et al.*, 2002a).

Double-Primed RCA

Incubate the earring-labeled dsDNA at 62°C for 1.5 hr in a 35 μl reaction volume containing 20 mM Tris-HCl (pH 8.8 at 25°C), 2.5 mM $MgCl_2$, 10 mM KCl, 10 mM $(NH_4)_2SO_4$, 0.1 % Triton-X-100, 5% v/v DMSO, 500 μM each of the four dNTPs, 1.33 μg T4 gene-32 protein, 1 μM each of two oligo primers (primers 2 and 3 in our case), and 10 U of Vent exo^- DNA polymerase. The double-primed RCA (Lizardi *et al.*, 1998; Zhang *et al.*, 1998; not shown in our schematics for simplicity) is characterized by complex, exponentially-branching kinetics, yields higher amplification than the single-primed RCA, and always results in dsDNA amplicons.

Example

Figure 2 presents typical results of RCA experiments with rolling replication of earring probes (Kuhn *et al.*, 2002a), which demonstrate the zeptomolar sensitivity of the double-primed RCA format. Indeed, it makes possible to reliably detect down to several hundred copies of HIV-1 DNA marker in the duplex form (see the double-primed RCA reactions in Figure 2A). Figure 2B shows that the efficiency of the earring-directed RCA is not compromised by topological constraints. We have also demonstrated that in case of large, agarose-embedded DNAs, the in-gel earring-based RCA reactions are feasible (Demidov *et al.*, 2001). Furthermore, we have found that the earring probe can be formed with high yield on the dsDNA target fragment in the presence of equimolar amounts of human genomic DNA followed by successful RCA (not shown).

Not less important for applications is that the earring assembly has virtually zero tolerance to the single-base probe-target mismatches

(Kuhn *et al.*, 1999a, 2000). Hence, only the correct dsDNA site will be recognized by the corresponding earring probe essentially independent on how much unrelated DNA is present in a sample to be analyzed, thus providing with a highly sequence-specific RCA signal. All these data hold promise for the earring-based detection of specific sequences within intact genomes and native chromosomes with the complete protocol requiring ≤10 hr.

Protocol 2: Nick-Induced RCA with Pre-Circularized Probe

Materials

DNA Target Site, PNA Openers and Oligonucleotides

As in the previous method, the dsDNA target site should be capable of forming a PD-loop (see above for corresponding requirements). This site has also to embody a recognition site for a restriction enzyme. One example is the HIV-1 DNA site described in the preceding protocol, which contains in the middle the recognition sequence 5'-AG↓CT for the restriction enzyme AluI (the arrow marks the cleavage position) and which was used in the exemplary application below. Another example is the *B. anthrax* site (*lef* gene), 5'-$\underline{AG_2A_2GAG}CAT_3{\downarrow}\underline{A_3G_2A_3}$, containing the recognition sequence T_3A_3 for the restriction endonuclease DraI (the octapurine stretches for binding the bisPNAs are underlined). For choice and synthesis of PNA openers, see protocol above; in the prototypal study of the nick-induced RCA presented below, the same pair of bisPNAs as before was employed.

For efficient DNA nicking, sufficiently long (≥15 nt) auxiliary oligonucleotide is required as a cleavage-directing template to make the secondary duplex (see Figure 1B) readily digestible by the restriction endonuclease. For the same reason, this oligo should carry the enzyme's recognition sequence well in the middle (15-mer 5'-AGAGGAAGCTACTGG was employed here as the AluI template).

The RCA-reactive circular DNA probe can be prepared from a circularized linear precursor with the use of a splint oligo (post-assembly digested by ExoVII exonuclease), as described by Kuhn *et al.*, 2002a. These three oligonucleotides were used in the RCA experiment presented below:

5'-GTGTATCATCCTCGCATCCGTAAGAAGAAAATCACAAGTCGTTC TCGTACACACTACTGGAGGAGACTATATTGTATTCATCACACTCAG TATCAATC (98-nt-long circularized probe),
5'-GAGGATGATACACGATTGATACTGAG (splint oligonucleotide),
5'-ATCAATCGTGTATCATCCTCGCATC (RCA primer).

Enzymes

Restriction Enzyme

Essentially any type II restriction enzyme can be converted into an artificial nickase with designated sequence specificity to selectively initiate the RCA reaction on dsDNA (Kuhn *et al.*, 2003; AluI was used in the nick-induced RCA example below). Note that some exceptional restriction enzymes, such as DdeI, HhaI and HapII can cut ssDNA, although with lower efficiency. These enzymes may not only yield nicks but also double-stranded breaks at the PD-loop site. Still, such a side effect obviously does not impair the ability of the cleaved PD-loop (no matter nicked or completely cut) to be afterward involved in a RCA reaction with the hybridized DNA minicircle.

Note in addition that a restriction enzyme will cut dsDNA not only at the PD-loop but also at all other corresponding recognition sites resulting in fragmentation of sufficiently long DNA duplexes. For many RCA-based diagnostic assays, this will obviously constitute no problem, as the primary task there is to sequence-specifically introduce a detectable nick within a chosen dsDNA target. Furthermore, the enzymatic DNA methylation could be employed to avoid the byproduct DNA fragmentation, if required. Such a treatment, if performed directly after the PNA binding and prior to oligonucleotide hybridization and DNA cleavage, will block the DNA digestion by a restriction enzyme at essentially all recognition sites. Besides the locally opened designated

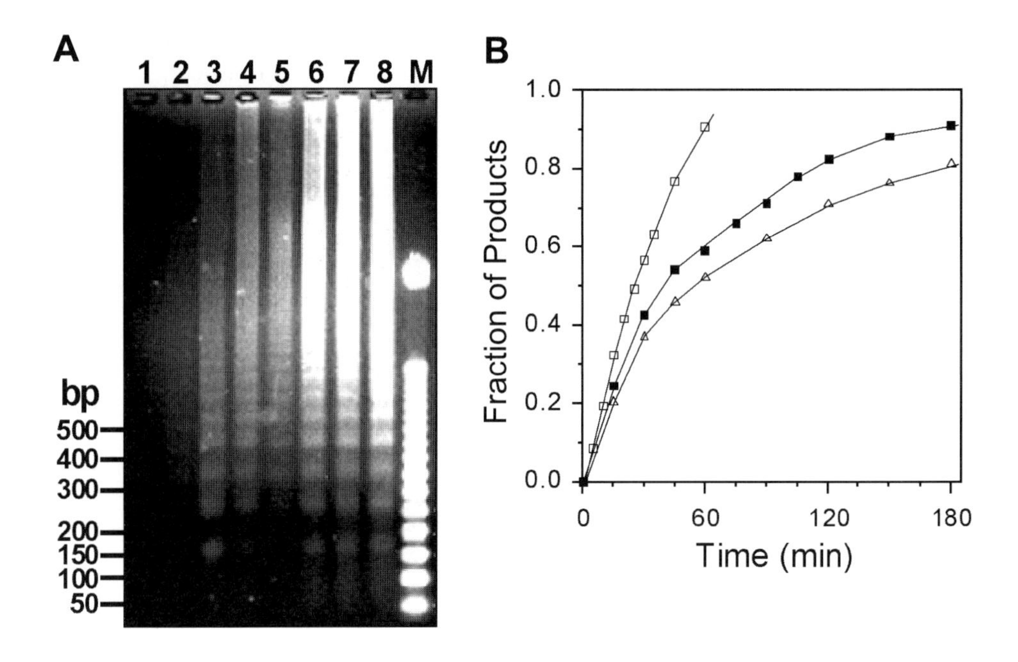

Figure 3. Use of PNA-based artificial DNA nickases in the site-specific dsDNA nicking and subsequent nick-induced RCA reactions (see Protocol 2 section and Kuhn *et al.*, 2003 for details).

A: PNA-directed, nick-induced RCA on the HIV-1 DNA site centrally located within 350 bp dsDNA fragments. Input numbers of AluI-nicked DNA targets: lane 2, 5×10^3; lane 3, 10^4; lane 4, 5×10^4; lane 5, 5×10^5; lane 6, 5×10^6; lane 7, 5×10^7; lane 8, 5×10^8. Lane 1, control with non-nicked DNA (5×10^8 molecules). M is a 50-bp dsDNA ladder.

B: Comparative kinetic analysis of the restriction enzyme cleavage efficiency on regular DNA duplex (open squares) and PD-loops formed by 16-mer (open triangles) or 25-mer (closed squares) oligonucleotides within ~350 bp DNA target fragment with the assistance of PNA openers. In these experiments, the BglII restriction endonuclease was used to site-specifically generate nicks (PD-loop samples) or double-stranded breaks (intact DNA fragment) within the artificially designed sequence 5'-$\mathbf{A_2G_2AGAGA}\underline{\mathbf{A}{\downarrow}\mathrm{GATCT}}\mathbf{A_2GA_2GA_4}$ (PNA-binding sites are in boldface; restriction enzyme recognition sequence is underlined). PNA openers: H-$T_2JT_2JT_4$-$(eg1)_3$-$T_4CT_2CT_2$-$LysNH_2$, $HLys_2$-$T_2J_2TJTJT_2$-$(eg1)_3$-$T_2CTCTC_2T_2$-$LysNH_2$; PD-loop-forming oligo templates: 5'-$AG_2AGAGA_2GATCTA_2$ (16-mer), 5'-$A_2G_2AGAGA_2GATCTA_2GA_2GA_4$ (25-mer).

dsDNA target, only very few other recognition sites that overlap with binding sites for any of the employed PNA openers will be protected from methylation and subsequently cut (Veselkov *et al.*, 1996ab).

DNA Polymerase
Any convenient RCA-active DNA polymerase could be used in this protocol. In the experiment shown below, Sequenase 2.0 was used.

Step 1: PD-Loop Assembly

The DNA targeting with bisPNAs is performed essentially as described in the previous protocol. The resulting DNA-PNA complex is precipitated with ethanol to be re-dissolved in 10 μL of TPE buffer (10 mM Tris-phosphate, 0.1 mM EDTA, pH 6.8). To form PD-loops, a mixture of 7 μL H_2O, 1 μL of DNA-PNA complex (0.15-0.3 μg), 1 μL of corresponding 10×buffer for the restriction enzyme used next, and 2 μL of 10 μM solution of the template oligo was incubated for 10-15 min at 37°C.

Step 2: DNA Nicking

For DNA nicking, typically 1 μL with 5-10 U of the requisite restriction enzyme is added to the PD-loop containing samples, followed by incubation for 2-3 hr at 37°C. If desired, the yield of DNA nicks can be checked by electrophoresis according to Kuhn *et al.*, 2002b, 2003. For this analysis, it is necessary to dissociate the PNA openers from the analyzed DNA sample at 70°C for ~1 hr (important: the DNA-PNA complex must be retained for the RCA reaction!).

Step 3: Nick-Induced RCA

The PNA-bound, nicked DNA is preincubated with 0.1 μM of circular probe and 0.5 μM of primer for 30 min at 37°C in 20 μL of DNA polymerase reaction buffer. Then, 3.5 μL of H_2O, 1.0 μL of 100 mM DTT, 1.0 μL of a mixture containing all four dNTPs (25 mM each),

0.5 µL of SSB protein (2.2 mg/ml) and 1.0 µL of DNA polymerase (1-2 U) are added, and the reaction mixture is incubated for several hours at 37°C. This will result in the double-primed, branched RCA reaction. Omitting extra primer oligonucleotide at this stage will result in the single-primed RCA. Aliquots of the amplified samples can be analyzed by electrophoresis in an agarose gel.

Example

Figure 3A shows the results of the PNA-assisted nick-induced RCA obtained with the unique HIV-1 dsDNA site capable of PD-loop formation (Kuhn *et al.*, 2003). One can see that the intact DNA control did not generate RCA products, while the nicked DNA samples yielded, after the gel electrophoresis, the distinct, ladder-like bands typical to the hyperbranched-type RCA reaction used here (Lizardi *et al.*, 1998; Zhang *et al.*, 1998; Demidov, 2002), if the input number of target molecules was $\geq 10^4$. Note that the goal of these experiments was to demonstrate the principal workability of this RCA format. Its sensitivity may be much higher under optimized conditions.

It is also worth mentioning that the nick-induced RCA reaction proceeds on dsDNA without any restrictions imposed by topology (in contrast to the topology-constrained design of the previous protocol) and that in case of linear RCA reactions the entire amplicon will be linked to the target site, which may be advantageous in some applications. Also of importance is the fact that the cleavage efficiency of restriction enzymes on PD-loops is comparable to that on regular DNA duplexes (Figure 3B), thus making it possible to quantitatively nick specific sites on linear DNA duplexes in ~2 hr at 37°C and to complete the entire protocol in ≤10 hr.

Summarizing Discussion

Protocols and examples presented in this chapter demonstrate the potent and unique ability of PNA openers to generate local DNA structures suitable for subsequent RCA reactions directly with duplex

DNA, the major DNA form. The PNA-assisted RCA approaches could be used for site-specific dsDNA labeling/detection (*e.g.* with immunoreactive or fluorescent tags) and offer to this end several advantages. First, they allow to bypass the prior DNA denaturation step intrinsic in all other RCA protocols applicable to dsDNA, which could be convenient for the assay miniaturization/automation. Second, they generally result in a much higher sequence specificity of DNA targeting with a hybridization/amplification probe than assays involving DNA denaturation. Third, the PNA-assisted approaches may result in a stable attachment of the RCA product to a target DNA site, if highly localized amplification signal for *in situ* gene-diagnostic assays is necessary.

Although DNA sites of special sequences are required for the approaches described here (they must form the PD-loop), simple statistical estimations supported by analysis of known DNA sequences show that such DNA sites are quite frequent in genomes: they should be met on average every 1-2 kb of a random DNA sequence. Hence, virtually every prokaryotic and eukaryotic gene will contain at least one PD-loop-forming site to be employed into the RCA reaction on dsDNA. Note that each DNA site of this type will normally be unique in the whole genome as a typical PD-loop spans more than 20 bp. Accordingly, these sequences may serve as useful and selective DNA markers in the RCA-based DNA diagnostics featuring potential for highly selective, ultrasensitive and site-positioned detection of specific DNA analytes.

References

Braasch, D.A. and Corey, D.R. (2001). Synthesis, analysis, purification, and intracellular delivery of peptide nucleic acids. Methods *23*, 97-107.

Bukanov, N.O., Demidov, V.V., Nielsen, P.E., and Frank-Kamenetskii, M.D. (1998). PD-loop: a complex of duplex DNA with an oligonucleotide. Proc. Natl. Acad. Sci. USA *95*, 5516-5520.

Demidov, V.V. (2001). PD-loop technology: PNA openers at work. Expert Rev. Mol. Diagn. *1*, 343-351.

Demidov, V.V. (2002). Rolling-circle amplification in DNA diagnostics: the power of simplicity. Expert Rev. Mol. Diagn. *2*, 542-548.

Demidov, V.V. (2003). Earrings and padlocks for the double helix: topological labeling of duplex DNA. Trends Biotechnol. *21*, 148-151.

Demidov, V.V. (2004a). Rolling-circle amplification (RCA). In: Encyclopedia of Diagnostic Genomics and Proteomics. J. Fuchs and M. Podda, eds. Marcel Dekker, New York (in press).

Demidov, V.V. (2004b). PNA openers for duplex DNA: basic facts, fine tuning and emerging applications. In: Peptide Nucleic Acids: Protocols and Applications (2nd Edition). P.E. Nielsen, ed. Horizon Scientific Press, Wymondham, p. 207-226.

Demidov, V.V., Yavnilovich, M.V., Belotserkovskii, B.P., Frank-Kamenetskii, M.D., and Nielsen, P.E. (1995). Kinetics and mechanism of polyamide ("peptide") nucleic acid binding to duplex DNA. Proc. Natl. Acad. Sci. USA *92*, 2637-2641.

Demidov, V.V., Yavnilovich, M.V., and Frank-Kamenetskii, M.D. (1997). Kinetic analysis of specificity of duplex DNA targeting by homopyrimidine PNAs. Biophys. J. *72*, 2763-2769.

Demidov, V.V. and Frank-Kamenetskii, M.D. (2001). Sequence-specific targeting of duplex DNA by peptide nucleic acids via triplex strand invasion. Methods *23*, 108-122.

Demidov, V.V., Kuhn, H., Lavrentyeva-Smolina, I.V., and Frank-Kamenetskii, M.D. (2001). PNA-assisted topological labeling of duplex DNA. Methods *23*, 123-131.

Demidov, V.V. and Frank-Kamenetskii, M.D. (2002). PNA openers and their applications. In: Peptide Nucleic Acids: Methods and Protocols. P.E. Nielsen, ed. Humana Press, Totowa, New Jersey. p. 119-130.

Egholm, M., Christensen, L., Dueholm, K.L., Buchardt, O., Coull, J., and Nielsen, P.E. (1995). Efficient pH-independent sequence-specific DNA binding by pseudoisocytosine-containing bis-PNA. Nucleic Acids Res. *23*, 217-222.

Fire, A. and Xu, S.-Q. (1995). Rolling replication of short DNA circles. Proc. Natl. Acad. Sci. USA *92*, 4641-4645.

Kingsmore, S.F. and Patel, D.D. (2003). Multiplexed protein profiling on antibody-based microarrays by rolling circle amplification. Curr. Opin. Biotechnol. *14*, 74-81.

Kuhn, H., Demidov, V.V., Frank-Kamenetskii, M.D., and Nielsen, P.E. (1998) Kinetic sequence discrimination of cationic bis-PNAs upon targeting of double-stranded DNA. Nucleic Acids Res. *26*, 582-587.

Kuhn, H., Demidov, V.V., and Frank-Kamenetskii, M.D. (1999a). Topological links between duplex DNA and a circular DNA single strand. Angew. Chem. Int. Ed. *38*, 1446-1449.

Kuhn, H., Demidov, V.V., Nielsen, P.E., and Frank-Kamenetskii, M.D. (1999b). An experimental study of mechanism and specificity of peptide nucleic acid (PNA) binding to duplex DNA. J. Mol. Biol. *286*, 1337-1345.

Kuhn, H., Demidov, V.V., and Frank-Kamenetskii, M.D. (2000). An earring for the double helix: assembly of topological links comprising duplex DNA and a circular oligodeoxynucleotide. J. Biomol. Struct. Dyn. *Sp. Iss. S2*, 221-225.

Kuhn, H., Demidov, V.V., and Frank-Kamenetskii, M.D. (2002a). Rolling-circle amplification under topological constraints. Nucleic Acids Res. *30*, 574-580.

Kuhn, H., Protozanova, E., and Demidov, V.V. (2002b). Monitoring of single nicks in duplex DNA by gel electrophoretic mobility-shift assay. Electrophoresis *23*, 2384-2387.

Kuhn, H., Hu, Y., Frank-Kamenetskii, M.D., and Demidov, V.V. (2003). Artificial site-specific DNA-nicking system based on common restriction enzymes assisted by PNA openers. Biochemistry *42*, 4985-4992.

Lizardi, P.M., Huang, X., Zhu, Z., Bray-Ward, P., Thomas, D.C., and Ward, D.C. (1998). Mutation detection and single-molecule counting using isothermal rolling-circle amplification. Nat. Genet. *19*, 225-232.

Sabanayagam, C.R., Berkey, C., Lavi, U., Cantor, C.R., and Smith, C.L. (1999). Molecular DNA switches and DNA chips. In: Micro- and nanofabricated structures and devices for biomedical and environmental applications II. M. Ferrari, ed. Proc. SPIE *3606*, 90-97.

Schweitzer, B. and Kingsmore, S. (2001). Combining nucleic acid amplification and detection. Curr. Opin. Biotechnol. *12*: 21-27.

Veselkov, A.G., Demidov, V.V., Frank-Kamenetskii, M.D., and Nielsen, P.E. (1996a). PNA as a rare genome-cutter. Nature *379*, 214.

Veselkov, A.G., Demidov, V.V., Nielsen, P.E., and Frank-Kamenetskii, M.D. (1996b). A new class of genome rare cutters. Nucleic Acids Res. *24*, 2483-2488.

Zhang, D.Y., Brandwein, M., Hsuih, T.C.H., and Li, H. (1998). Amplification of target-specific, ligation-dependent circular probe. Gene *211*, 277-285.

Zhang, D.Y. and Liu, B. (2003). Detection of target nucleic acids and proteins by amplification of circularizable probes. Expert Rev. Mol. Diagn. *3*, 237-248.

3.5

Phi29 DNA Polymerase Based Rolling Circle Amplification of Templates for DNA Sequencing

John C. Detter, John R. Nelson
and Paul M. Richardson

Abstract

The generation of DNA sequencing templates is typically an inconsistent and labor-intensive procedure, especially in a high-throughput facility. Purification of recombinant plasmids from *E. coli* and PCR amplification of inserts have generally been employed, but require many laborious/time-consuming steps and do not always yield suitable amounts of high-quality templates to be used in downstream applications. Replication by a rolling-circle mechanism is common among bacteriophages in nature. Recently, rolling-circle amplification (RCA) with Phi29 DNA polymerase has been applied *in vitro* to marker DNA sequences (using specific primers) and to circular cloning vectors (using random hexamer primers) to achieve their exponential amplification via DNA strand displacement. The US DOE Joint Genome Institute has successfully implemented random-primed RCA into their high-throughput process for production of sequencing templates. Here, we describe the RCA-based plasmid amplification protocol, as well as several practical applications for using amplified DNA in sequencing and related procedures.

Background

Strand-displacement process allows a DNA polymerase to synthesize a new complementary DNA strand while displacing the previous one lying upstream. In contrast to 'DNA-unproductive' nick translation, which is accomplished during the reparative replication of double-stranded DNA by DNA polymerases having associated 5'-3' exonuclease activity (*e.g. E. coli* DNA Pol I), replication by strand-displacing DNA polymerases results in the net DNA synthesis. This is particularly useful when the polymerase is being used for rolling-circle DNA replication (Kornberg and Baker, 1992). Once the enzyme has completed replication of the circular single-stranded DNA template, strand-displacement activity would be required to begin the 'rolling' mode of amplification.

Indeed, DNA circles smaller than 100 nt can readily be 'rolled' by most DNA or RNA polymerases due to strong flexional stress created by requiring the newly made DNA duplex to bend sharply. This stress causes the 5' end of the replicated DNA strand to unwind and dangle at some distance behind the polymerase; hence no actual strand displacement is necessary in case of small circular DNA templates (Kool, 1996; Demidov, 2002). But as the size of the input template circle exceeds ~100 nt, the strand-displacement activity of DNA polymerase becomes an absolute requirement for the RCA reaction to proceed (Kool, 1996). To this end, Phi29 DNA polymerase has no upper limit on the size of DNA being replicated, and in fact performs the strand-displacement replication on circular DNA templates as efficiently as with linear DNAs (Blanco *et al.*, 1989; Lizardi *et al.*, 1998; Dean *et al.*, 2001). This robust ability can be used both for signal amplification in DNA diagnostics with the use of DNA minicircles (Lizardi *et al.*, 1998; Demidov, 2002) and for generation of DNA sequencing templates from plasmids, as described in this chapter.

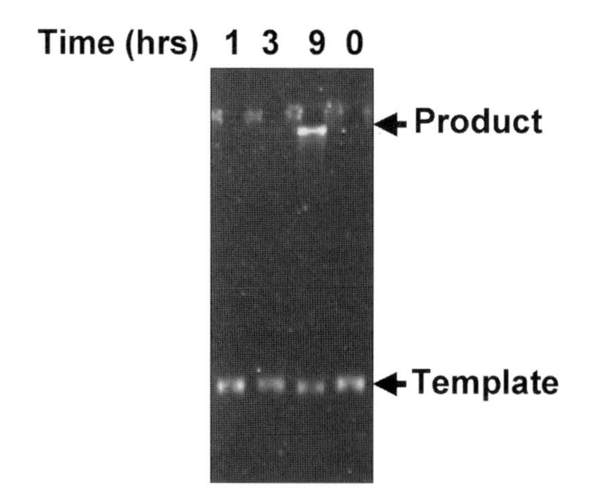

Figure 1. Strand-displacement activity and processivity of Phi29 DNA polymerase in the single-primed RCA reaction with single-stranded circular M13 DNA template. Reaction samples containing 25 ng of M13mp18 DNA and 5 pmoles of –40 universal primer in 10 μl TE buffer were heated to 95°C for 3 min and slow cooled to room temperature for annealing the primer. These were then chilled, reaction buffer and a sub-saturating amount of Phi29 DNA polymerase was added, and reactions were incubated at 30°C for the indicated times. The RCA products were then resolved by electrophoresis in 0.6% agarose gel (1× TBE buffer).

Phi29 DNA polymerase is a highly processive enzyme that can incorporate ≥70,000 nucleotides per binding event (Blanco *et al.*, 1989; see also the chapter by Salas and Blanco in this book). In a processivity assay performed by us using M13 single-stranded DNA as circular template, the single-primed DNA replication effectively proceeds via rolling-circle mechanism even with limiting amounts of Phi29 DNA polymerase (Figure 1). Phi29 DNA polymerase does not apparently pause when encountering double-stranded template. Thus, it can perform the strand-displacement synthesis in the absence of DNA helicase or SSB protein, which normally assist DNA polymerase in performing this reaction. In Figure 1, high-molecular-weight amplification products are observed after 9 hrs. These products are so large that they fail to migrate into the gel and it is difficult to size them. Note that fragments of intermediate size are not observed, indicating the product has been synthesized during a single binding event. Phi29 DNA polymerase is known to incorporate nucleotides at 25-50 nt/second (Blanco *et al.*, 1989; Lizardi *et al.*, 1998), which would suggest the potential size of these RCA amplicons is actually over 800,000-nt-long.

Accordingly, a high-throughput strategy for isothermal DNA amplification that employs Phi29 DNA polymerase and rolling-circle DNA synthesis has been developed to generate high-quality templates for DNA sequencing. The RCA-based method, called TempliPhi™ DNA Amplification (Amersham Biosciences), utilizes the random-primed replication by this enzyme to exponentially amplify circular DNA templates consisting of plasmid vector with cloned DNA inserts (Dean *et al.*, 2001; Detter *et al.*, 2002). One can see that using a random set of hexamer primers for initiation of multiple simultaneous replication forks, tandem (concatenated) double-stranded copies of the circular DNA could be generated in a hyperbranched reaction, as depicted schematically in Figure 2. Phosphorothioate modification between the last three nucleotides at the 3' end of the primer prevents the DNA polymerase from degrading the primers via intrinsic 3'-5' exonuclease activity and dramatically accelerates the amplification kinetics (Dean *et al.*, 2001).

This RCA format is a very powerful approach that requires little starting material. Typically there is a brief (~0.5 hr) lag period followed by rapid synthesis of DNA products until the supply of nucleotides is exhausted (Figure 3). The method can isothermally produce the microgram quantities of DNA from the picograms inputs in a few hours, allowing greater than $5 \cdot 10^6$-fold amplification. We utilized the electron microscopy to visualize and examine the amplification products (Figure 4). With this technique, products of ≥100 kb have been observed after 4 hr of amplification using a 2.7-kb circular DNA template and certain RCA amplicons consisted of >50 end-to-end tandem repeats of the input material.

Importantly, our method of DNA amplification should replicate DNA with high fidelity: Phi29 DNA polymerase has an associated 3'-5' exonuclease 'proofreading' activity (Blanco and Salas, 1996) and has a reported error rate of $5 \cdot 10^{-6}$ (Esteban *et al.*, 1993), about 100-fold lower than that of Taq DNA polymerase (Dunning *et al.*,

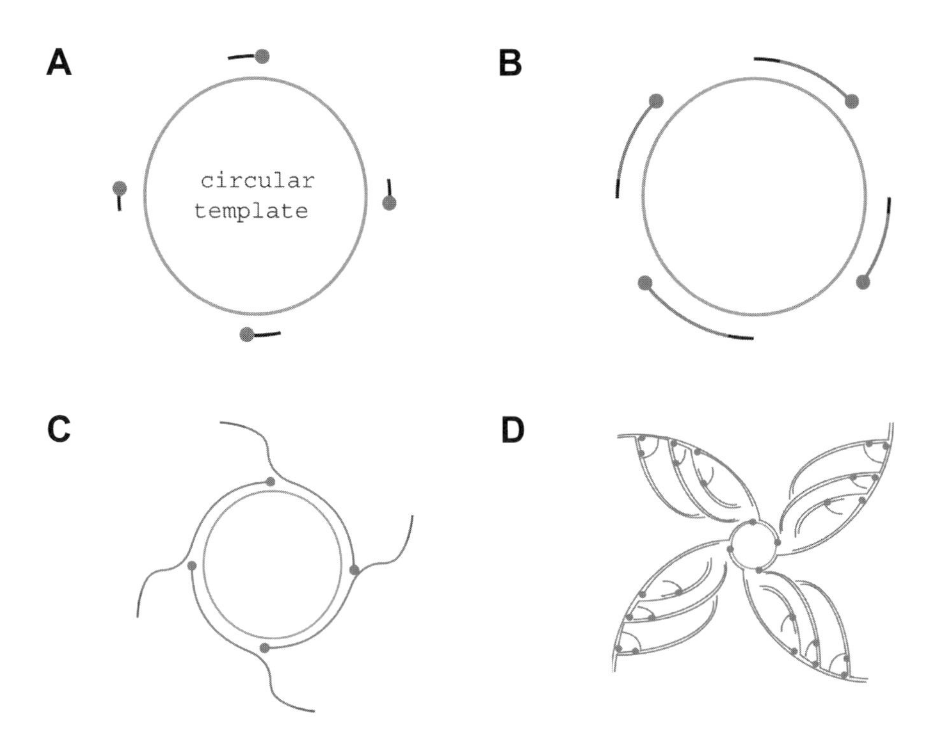

Figure 2. Schematics of the random-primed RCA reaction with circular vector template employing strand displacement. A: Random primers hybridize to multiple locations on DNA circle (single-stranded or denatured plasmid or phage DNA). B: Phi29 DNA polymerase (black sphere) initiates replication from many primers simultaneously synthesizing multiple end-to-end copies. C: Strand displacement allows the polymerase to continue the RCA-type replication. D: Random-priming events subsequently occur on the displaced DNA strands yielding the highly-branched amplification products. This series of displacement and replication events continues until the nucleotide pool is depleted. **The colour version of this figure is located in the colour plate section at the back of this book.**

1988). In agreement with this expectation, an error rate of $3 \cdot 10^{-6}$ was determined for TempliPhi DNA amplification following a modified Kunkel method (Kunkel, 1985; Nelson *et al.*, 2002). In addition, multiply repeated sequencing experiments indicated that Phi29 DNA polymerase faithfully replicates input DNA as no amplification errors were detected. In these experiments (data not shown), DNA sequence data were obtained from primer walks for TempliPhi-amplified M13mp18 and pUC18 DNA (10,000-fold amplified; approximately 20,000 nt of sequence) using 24 pairs of forward and reverse vector-

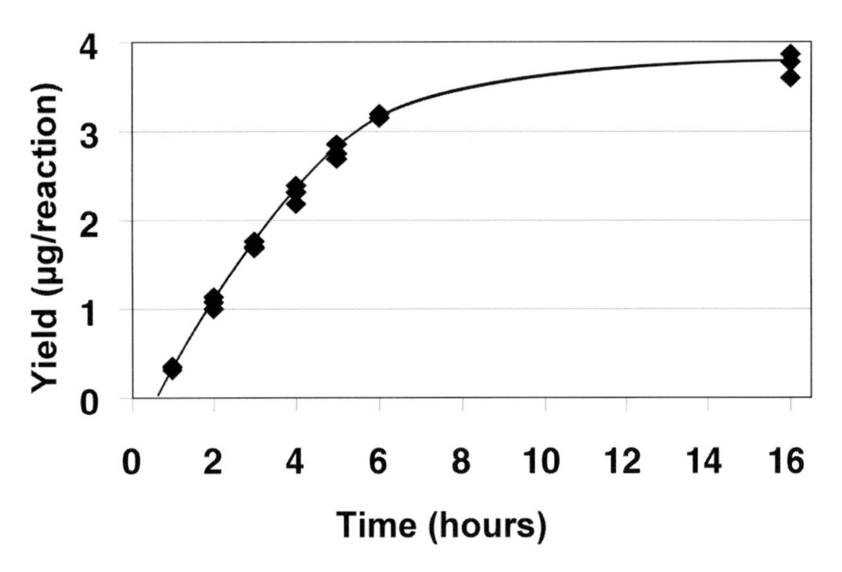

Figure 3. Exemplary TempliPhi DNA amplification kinetics. Supercoiled pUC18 DNA (1 ng) was added to 10 µL of Sample Buffer (TempliPhi™ DNA Sequencing Template Amplification Kit; Amersham Biosciences), heated at 95°C for 3 min to denature the template and chilled to 4°C. Then, 10 µL of TempliPhi Premix from the kit was added and reaction tubes were held at 30°C for the indicated times. To stop the amplification reaction, samples were heated to 65°C for 10 min to inactivate the Phi29 DNA polymerase and then quantified.

specific primers in a total of eleven 96-well-plate sequencing runs on MegaBACE™ 1000 sequencing complex.

The actual steps of the method are quite simple and it can be used to amplify the recombinant plasmid or M13 phage DNA directly from cell culture, colonies or plaques. Interestingly, while chromosomal DNA isolated from a bacterial host can also be efficiently amplified by this method, TempliPhi reactions performed on bacterial cells containing medium to high copy number of plasmid DNA show no evidence for the amplification of host DNA, yet contain up to 2-3 µg of newly made plasmid DNA. The presumed reason for this is that the initial heat treatment of bacterial cells (95°C for 3 min) is harsh enough to release plasmid DNA but it does not totally lyse the bacterial cells. Therefore, only plasmid DNA is free to be amplified in this reaction. There is also some evidence that a difference in amplification kinetics of plasmids versus chromosomal DNA (the former could be more

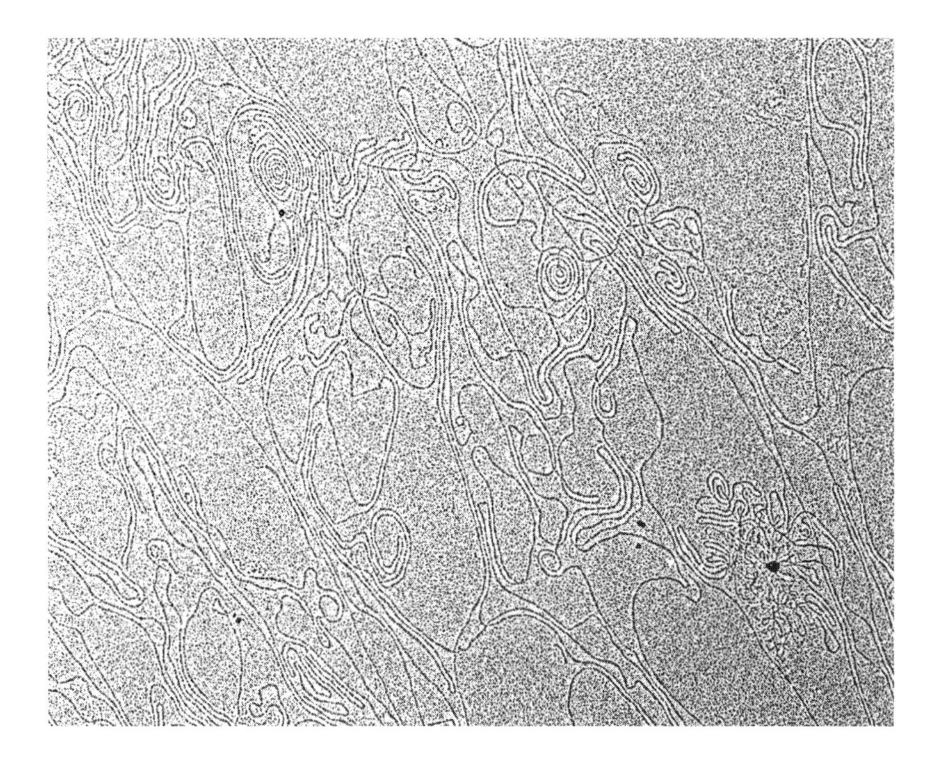

Figure 4. Electron microscopy of the amplification product (with Daniel Collins assistance). Plasmid DNA (pUC18) was amplified for 4 hr using the TempliPhi method. An aqueous basic protein film technique was used (Kleinschmidt method) to analyze the products by transmission electron microscopy using a JEOL JEM 1230 equipment. The sample was shadowcast under vacuum using palladium/platinum wire as the source metal for deposition and the image was captured on Kodak SO-163 film. Exact product length measurements were not possible to make as long DNA strands could not be found, which were entirely contained in the picture. However, many linear double-stranded DNA products greater than 50-120 kb long, representing RCA products of 20-50 concatenated repeats of the input template circles were identified.

rapidly amplified) may result in a bias for plasmid amplification over small amounts of chromosomal DNA, which can be present at the start of the reaction.

The RCA reaction products obtained do not need to be processed prior to DNA sequencing. As a matter of fact, the completed amplifications are depleted of nucleotides and residual random hexamers will not anneal under typical DNA sequencing conditions. Consequently, the amplified samples can be sequenced directly after the amplification procedure without any purification using any of the commercially available dideoxynucleotide DNA sequencing kits. In addition, since the amplification self-terminates once the supply of nucleoside triphosphate is exhausted, each reaction contains approximately the same concentration of DNA product. This inherent normalization provides a consistency that can improve the DNA sequencing success rate, which is strongly influenced by variability in template DNA concentration. This is especially true in high-throughput settings with capillary sequencers where even slight differences can negatively affect pass rates and read lengths, thus reducing throughput and increasing costs. Besides, the sequenced samples that are processed after the sequencing reaction by G-50 spin columns or precipitation on carboxylate modified magnetic beads (detailed in Step 2a) to remove reaction components typically have a significant portion of the template DNA selectively removed. This can also lead to improved results on capillary sequencing instruments, which are known to be less tolerant to high concentrations of template DNA than slab gel instruments.

Experimental Protocols and Exemplary Applications

Materials

Amplification

Source of double-stranded plasmid DNA (either purified recombinant plasmid or *E. coli* colony and saturated culture (with or without glycerol) containing plasmid); denaturation buffer (10 mM Tris-HCl/pH 8.2; 0.1 mM EDTA); RCA reaction mixture (reaction buffer, dNTPs, thiophosphate protected random hexamers and phi29 DNA polymerase; all from the Amersham Biosciences TempliPhi DNA Sequencing Template Amplification Kit).

Sequencing

Aliquot of amplified plasmid as sequencing template; sequencing primer; dH_2O; DYEnamic™ ET terminator sequencing kit for MegaBACE sequencers (Amersham Pharmacia Biotech) or BigDye™ sequencing kit for ABI sequencers (Applied Biosystems).

Clean-Up of Sequencing Reaction

The microbead-based method uses the following: dH_2O, 100% ethanol, tetraethyleneglycol (Aldrich), and magnetic beads (Seradyn); standard ethanol-precipitation method uses the following: dH_2O, 100% ethanol, 7.5 M ammonium acetate.

Restriction Digestion

Aliquot of amplified plasmid as digestion template; desired restriction endonuclease with corresponding buffer.

Core Procedure: Plasmid Amplification

Plasmid DNA present in bacterial colonies can be used directly as template for TempliPhi amplifications. This is a great convenience since clones can be screened at the colony stage without having to wait for liquid cultures to grow and without the need for DNA plasmid preparation. As little as one part in 10,000 of a single colony can act as a good template for RCA using the high copy-number plasmid pUC18 and generating more than 1 μg of product in 4-6 hr at 30°C (data not shown). In case of the lower copy-number plasmids, such as pBR322, pCUGIblu21 and pMCL200, amplifications should be performed for 8-16 hr to achieve maximal yields. Note that presence of host chromosomal DNA does not affect the efficiency of plasmid amplification, as mentioned above (see the concluding section of Background).

M13-derived plaques are also excellent starting material for amplification. Post-lytic supernatants from M13-infected bacterial culture may be used, too. We have taken 0.02-1 μL of these supernatants in 4-hr RCA reactions to successfully generate DNA sequencing templates (data not shown). Since the resultant amplicons are double stranded, both strands of M13 clones can be sequenced.

Liquid cultures provide good templates for RCA reactions, as well. This is particularly useful to researchers who prefer to grow small overnight cultures for archival storage. There is no significant difference in amplification yield when using 0.001-1.0 μL of a saturated overnight *E. coli* culture as initial template for RCA reactions. A small portion of frozen glycerol stock can be amplified directly using this method. We use this process at the Joint Genome Institute in our high-throughput sequencing production line and in doing so have significantly increased throughput, read lengths, and quality of data, while reducing costs and manpower (Detter *et al.*, 2002).

Below, detailed plasmid amplification steps are presented followed by an example of the process along with quantification results. For additional information, including high-throughput scaling of the process, see research protocols at www.jgi.doe.gov.

Step 1: Sample Heat Denaturation

The sample can be in the form of bacterial liquid culture (typically 2 μL), 7.5% glycerol stock (typically 2 μL), colony (use small to medium picked piece with pipette tip, not tooth pick) or purified plasmid DNA (use <100 ng, down to as little as 1 pg). Mix the above amounts of plasmid-containing sample with denaturation buffer (1 mM Tris-HCl/pH 8, 0.05 mM EDTA) in 10 μL volume total. Heat sample(s) to 95°C for 3-5 min (to avoid release of bacterial chromosomal DNA when amplifying from bacterial cells, do not exceed the 5 min limit). Place samples on ice or keep them at 4°C for no less than 5 min.

Step 2: RCA Reaction

Add 10 μL of RCA reaction mixture (*e.g.* RCA mix from TempliPhi kit) to 10 μL of denatured sample(s). Carry out the RCA reaction at 30°C for 12-24 hr. In our experience, an overnight incubation (~16 hr) yields more consistent and reliable results in a high-throughput setting although shorter (down to 4-6 hr) incubation may also give the adequate results. Note that alternative (larger or smaller) reactions volumes can be used with success.

Step 3: Enzyme Inactivation

Heat inactivate the DNA polymerase at 65°C for 10 min and place sample on ice or keep it at 4°C for at least 5 min. This step inactivates the exonuclease activity of the enzyme and prevents it from potentially interfering with the sequencing reaction. The final product can be stored for ~7 days at 4°C, for ~30 days at –20°C or for much longer time at –80°C before being used as template for sequencing without prior purification.

Exemplary Results

The first use of the RCA-based TempliPhi technology at the Joint Genome Institute (JGI) was to replace the existing plasmid DNA isolation method (Hawkins *et al.*, 1997; Elkin *et al.*, 2001) used in our high-throughput sequencing production process. Conditions were optimized for amplification of arrayed shotgun libraries in a 384-well plate format from 2 μL of bacterial glycerol stocks containing randomly cloned fragments in the pUC18 cloning vector. This amplification procedure (detailed in the protocol section above) produces a uniform quantity of DNA for each well of a plate (Figure 5a; see also Detter *et al.*, 2002). When run out on an agarose gel, most of the amplified DNA remains in the gel-loading wells due to the high-molecular-weight nature of the highly-branched and concatenated amplification products. A portion of the amplified DNA co-migrates with the 23 kb λ/HindIII fragment regardless of the size of the initial starting material. This material is of high enough molecular weight so that it migrates at the limit of resolution of the gel system. When these samples are heated to 95°C and then quickly cooled, the amplified concatemeric DNA forms a complex crosslinked structure that cannot migrate in a non-denaturing gel system and remains in the wells of the agarose gel (data not shown). Using a Fluor-S™ MultiImager with Quantity One 4.2.3 software (BioRad, Hercules, CA), we quantified the relative abundance of the amplified DNA for each 20 μL sample to average 137 ng/μL with a standard deviation of ±20 ng/μL. In contrast, relative abundances in 50 μL DNA samples isolated according to our previous, magnetic bead-based protocol (solid-phase reversible immobilization or SPRI; Hawkins *et al.*, 1997; Elkin *et al.*, 2001) were half as concentrated, averaged 72 ng/μL with a two times higher standard deviation of ±46 ng/μL, as can be seen in Figure 5b.

Application 1: Plasmid Sequencing

The RCA-based generation of templates for plasmid sequencing provides the user with high-quality and mass-uniform DNA samples. Here, sequencing and clean-up steps will be outlined followed by exemplary results obtained with both DYEnamic ET and BigDye v3.1 terminator sequencing chemistries. If necessary, see research protocols at www.jgi.doe.gov for additional information.

Step 1a: DYEnamic ET Terminator Chemistry Sequencing

Each reaction contains 1 μL amplified product, 4 pmoles of primer, 5 μL of dH_2O and 4 μL DYEnamic ET terminator sequencing mix for a 10 μL total reaction volume (smaller volumes can also be used with success). The reaction(s) are cycled 30 times: 95°C for 25 s, 50°C for 10 s, 60°C for 2 min with a hold at 4°C.

Step 1b: BigDye Terminator Chemistry Sequencing

Each reaction contains 1 μL amplified product, 4 pmoles of primer, 2 μL of dH_2O, 1 μL 5× sequencing buffer and 1 μL BigDye v3.1

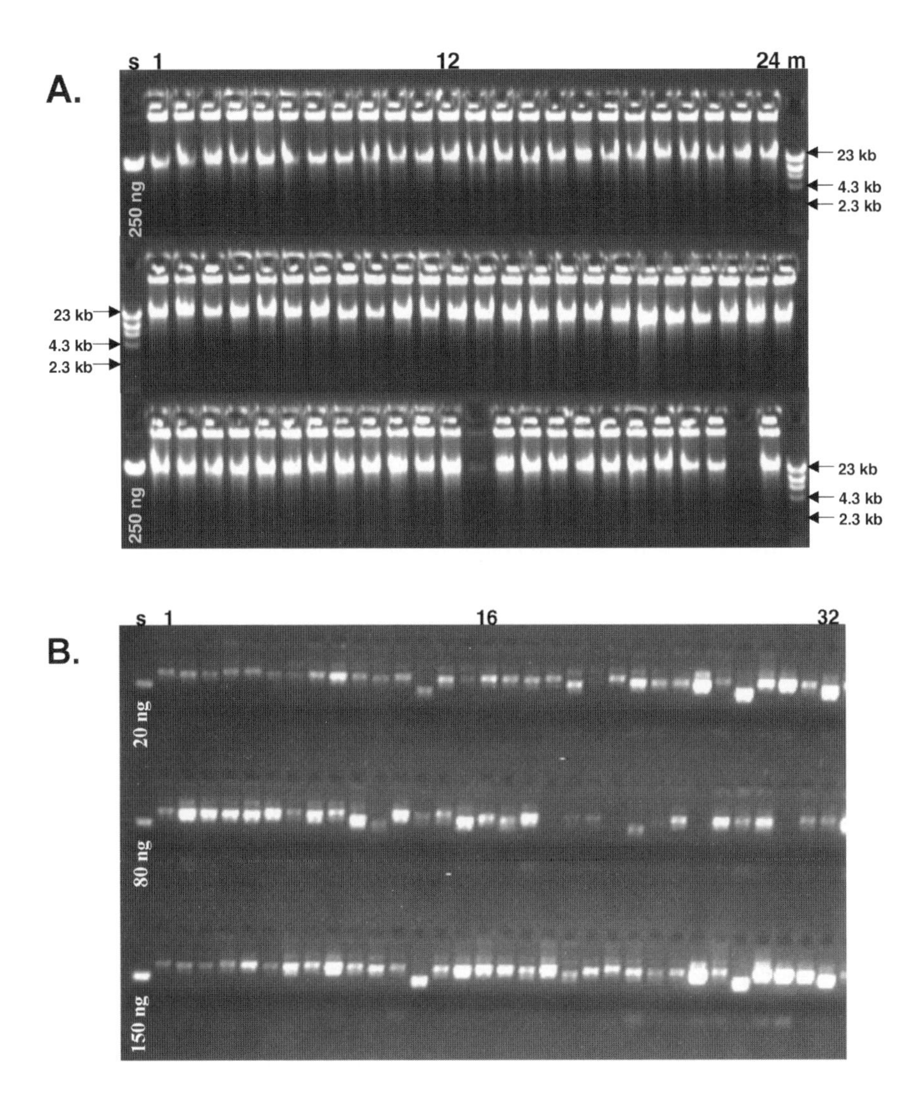

Figure 5. Comparison of the TempliPhi-amplified plasmid subclones and SPRI (solid-phase reversible immobilization)-isolated plasmids.
A: Strand-displacement RCA of plasmids in production. 2 μL aliquots of saturated *E. coli* culture were heat lysed and amplified with TempliPhi for 12 hr at 30°C (see experimental protocol). 3 μL of products from 96 wells of a 384-well plate were run on a 1% agarose gel and stained with ethidium bromide (to conserve space, only 72 wells are shown). Lanes: (s) 250-ng λ mass standard; (1-72) amplified products; (m) λ/HindIII size marker with 23, 4.3, 2.3 kb marked.
B: SPRI isolation of plasmid DNA in production. Plasmid DNA was isolated from 185 μL aliquots of saturated *E. coli* cultures in 96-well format using SPRI protocol (Elkin *et al.*, 2001; Hawkins *et al.*, 1997). 5 μL of the isolated DNA from 96 wells was run on a 1% agarose gel and stained with ethidium bromide. Lanes: (s) 80- and 150-ng λ mass standard; (1-96) isolated products.

terminator sequencing mix for a 5 μL total reaction volume (smaller volumes can also be used with success). The reaction(s) are started at 95°C for 1 min, then they are cycled 25 times: 95°C for 30 s, 50°C for 20 s, 60°C for 4 min with a hold at 4°C.

Step 2a: SPRI-Based Clean-Up of Sequencing Reaction

Make stock of washed Seradyn beads as follows. Add 15 mL of thoroughly resuspended Seradyn beads to a 50 mL Falcon tube. Fill tube with dH_2O and place it on a magnet for ~15 min or until almost clear. With tube on magnet, pour off water. Perform the wash step thrice. Then add 15 mL dH_2O to the Falcon tube. Invert the tube until its content completely mixed (keep well suspended before aliquoting for use).

Prepare fresh BET (bead, ethanol, tetraethyleneglycol) mixture just before use. To process ten 384-well plates make the following. In a 50 mL flask or bottle, mix 32 mL 100% ethanol, 3.5 mL dH_2O, and 3.2 mL tetraethyleneglycol. Cover and mix by inverting several times. Then add 1.0 mL of washed beads from above, cover and mix. Additionally, prepare a 70% ethanol wash solution.

To clean up the sequencing reaction(s) do the following: Transfer 5 μL of the sequencing reaction(s) to a clean PCR plate with either 96 or 384 wells. Add 10 μL of the properly mixed BET solution (made as above). Thoroughly mix each well on a plate with a pipette or cover this plate and shake it gently on a vortex. Incubate covered with a lid plate in dark area (closed drawer, *etc.*) for 15 min at room temperature. Place PCR plate on a high-field plane magnet for 1 min. Aspirate out all liquid with a pipette or Hydra device. Remove plate from a magnet and add 15 μL 70% ethanol. Once again, place PCR plate on a magnet for 1 min and aspirate out all liquid. Air-dry samples in a closed drawer or cabinet (dark, ventilated area) for 10 min.

Resuspend purified sequencing reaction with 15 μL dH_2O. Seal PCR plate and mix its wells with medium vortexing for ~2 min on vortex plate. Quick spin PCR plate and incubate at room temperature for

10 min. If needed, purified reaction samples can be stored at –20°C for up to two weeks before loading onto sequencer.

Step 2b: Reaction Clean-Up by Ethanol

To each sequenced well add the following. For a 10 μL reaction, add 1/10th volume 7.5 M ammonium acetate (1 μL) and 2.5 volumes 100% ethanol (28 μL). Cut amount added in half for 5 μL sequencing reactions. Seal reaction plate and vortex it at medium setting for 2 min on vortex plate. Centrifuge reaction plate for 30-60 min at 3000×*g*. If spinning at lower relative centrifugal force, increase time. Dump waste and invert spin on paper towel for 1 min at ~ 400×*g*. Repeat process with 70% ethanol to wash pellet if desired. Dump waste and tap lightly on paper towel. Place plate in dark drawer to air dry for ~15 min.

Resuspend purified sequencing reaction with 15 μL dH_2O. Seal reaction plate and mix wells with medium vortexing for ~2 min on vortex plate. Quick spin reaction plate and incubate it at room temperature for 10 min. If needed, purified reaction samples can be stored at –20°C for up to two weeks before loading onto sequencer.

Exemplary Results

Using the uniformly amplified plasmid DNA, sequencing conditions were optimized for capillary electrophoresis instruments, including the MegaBACE 1000 and 4000 (Amersham Biosciences, Piscataway, NJ) and the ABI PRISM® 3700 (Applied Biosystems, Foster City, CA) sequencers (Detter *et al.*, 2002). We have found that such a high level of uniformity in DNA template concentration we may achieve (see Figure 5a) significantly increases success rates for capillary sequencing. As described in the protocol section above, the protocol we designed utilizes 1 μL of the amplified product and consistently gives the high-quality (Phred Q20 score) sequence reads >600 bases (Ewing *et al.*, 1998a,b). Figures 6a and 6b illustrate the sequencing results we commonly observe using either the DYEnamic ET or the BigDye terminator chemistries.

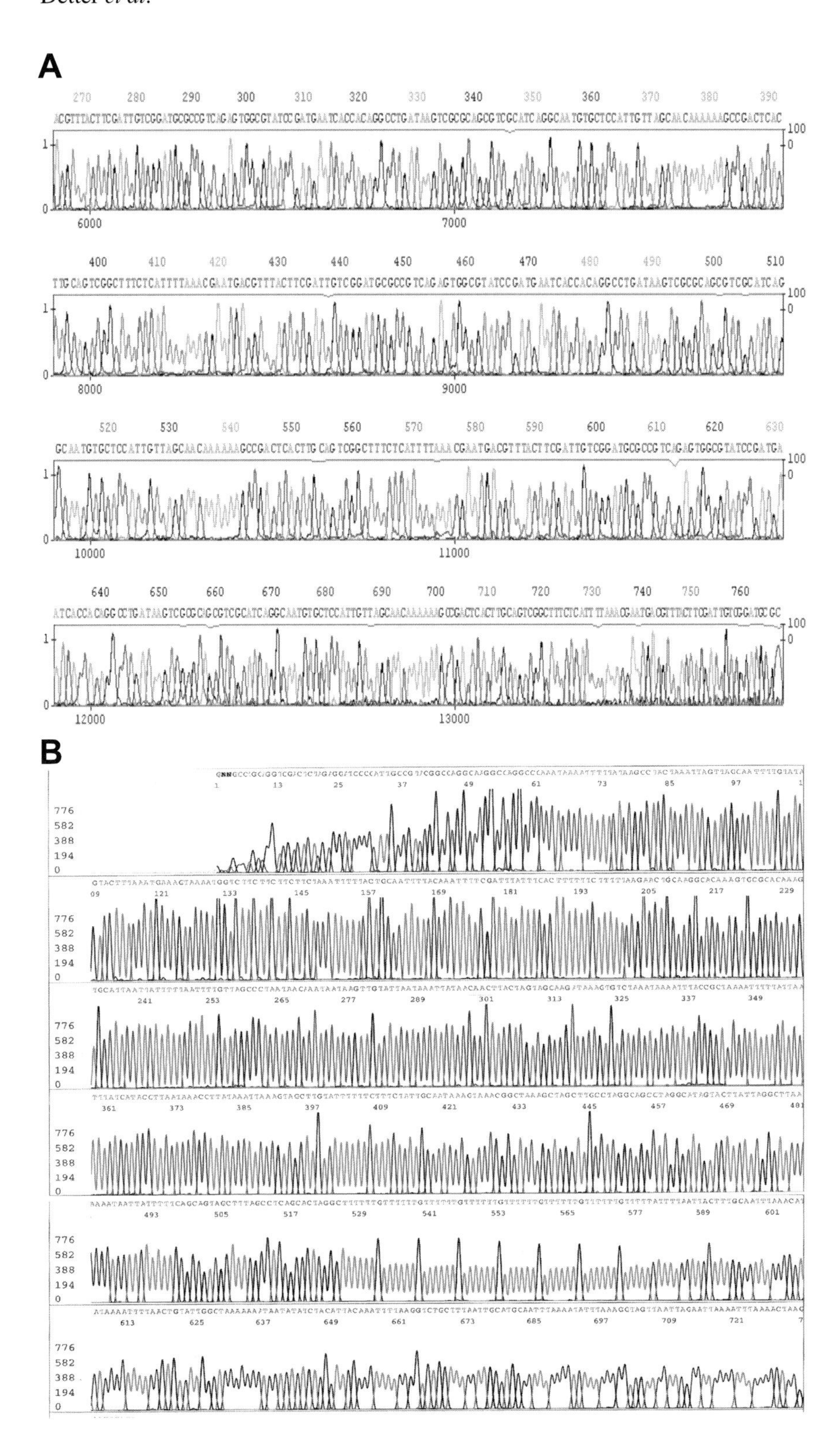
A
B

Most DNA isolation processes rely heavily on manpower and reagents, while producing large quantities of waste material and significant lag times between inoculation and sequencing. In contrast, our new RCA-based process has been largely automated, requires fewer reagents and process steps, and produces little to no waste (Detter *et al.*, 2002). The major advantages of this technology have been the complete elimination of labor-intensive steps of bacterial lysis and DNA purification as well as streamlining of the production process from 10 steps down to 4, which has lead to cutting the time it requires to go from inoculation to sequencing in half. This process also facilitated moving from a 96-well plate format to 384-well plates, and opens the prospect for even higher density and alternative formats.

Remarkably, the production results from the RCA-based approach have been extremely encouraging. Now that the amplification method has been fully implemented, the average pass rate (% reads/96 with >50 Q20 bases) for this process has been 92% with read-lengths greater than 600 Phred Q20 bases per passed lane. Since being phased into production at the JGI in early 2001, the process has yielded over 25 billion high quality raw bases from a variety of different Prokaryotic and Eukaryotic organisms (including Human, Mouse, Chicken, Xenopus, Fugu, Ciona, Poplar, as well as many microbes). Furthermore, an additional 1-2 billion high quality bases of various genomic sequences are added each month using this process.

Application 2: Plasmid Restriction Digestion

Digestion of plasmids has many uses, including insert size verification, subclone library quality control and insert excision/manipulation. Standard plasmid-prep methods are satisfactory for small scale, but in a larger scale they become laborious, time consuming and

Figure 6. Sequencing traces obtained from the RCA-generated DNA templates (3 kb inserts from the TempliPhi amplified pUC18 plasmids). Both DYEnamic™ ET terminator chemistry (A) and BigDye terminator chemistry (B) were used. Sequencing reactions were cleaned up using the BET SPRI method and run on either the MegaBACE 1000/4000 (A) or ABI3730 (B) capillary sequencer. **The colour version of this figure is located in the colour plate section at the back of this book.**

expensive. Since use of amplified plasmids overcomes all these issues, we employed the RCA-based high-fidelity TempliPhi amplification for this purpose. Here, the general steps will be outlined (see plasmid amplification protocol) followed by results.

Plasmid Digestion Protocol

Aliquot 5 μL of amplified plasmid into tube or well of a plate (depending on throughput desired). Add 1 μL of 10× enzyme buffer (use buffer that goes with enzyme of choice). Add ~1 μL of desired enzyme that cuts your plasmid and/or your insert. Bring total volume up to 10 μL with dH_2O. Incubate at recommended enzyme reaction temperature (usually 37°C) for ~2-4 hr. Heat kill the enzyme at 65°C (usually) for 20 min. Store between +4 and –20°C.

Volumes of the above reaction are dependant on end use and concentration of enzyme. The above conditions are standard for our purposes. We use high concentrations of enzymes and are digesting the plasmid to check the insert size of 8-10 kb libraries. Reaction volumes can be customarily scaled to fit end user's needs.

Exemplary Results

For our purposes, we became interested in using the RCA-based approach for shotgun library quality control to qualify sizes of our large insert (8-10 kb) plasmid libraries and our high GC-content small insert (3 kb) plasmid libraries that were poorly amplified by PCR due the nature of their inserts. We generally test between 24 and 96 colonies per library in order to determine the average insert size. Standard DNA isolation procedures were not feasible for our quality control digest because of the time and effort they require. We were already using the in-house RCA-based process for our plasmid sequencing line, so we investigated its usefulness for creating template for restriction digests. Typical plasmid preps require an overnight growth period followed by several hours of cell lysis, centrifugation and DNA purification. Our approach utilizes a similar overnight incubation but has no additional steps. Furthermore, the resulting RCA products are very consistent in concentration, contrary to the plasmid-prep procedure (see Figures 5a and 5b for comparison). This makes the digestions very reliable and easy to qualify. As seen in Figure 7, the vector-related band as well as the insert-related band(s) thus obtained can easily be identified after the gel electrophoresis. As this is typical with restriction digests, rare-cutting enzyme (*Swa*I in Figure 7a) allows more discriminative size estimates than do frequent cutters (*Bam*HI and *Hin*dIII in Figure 7b).

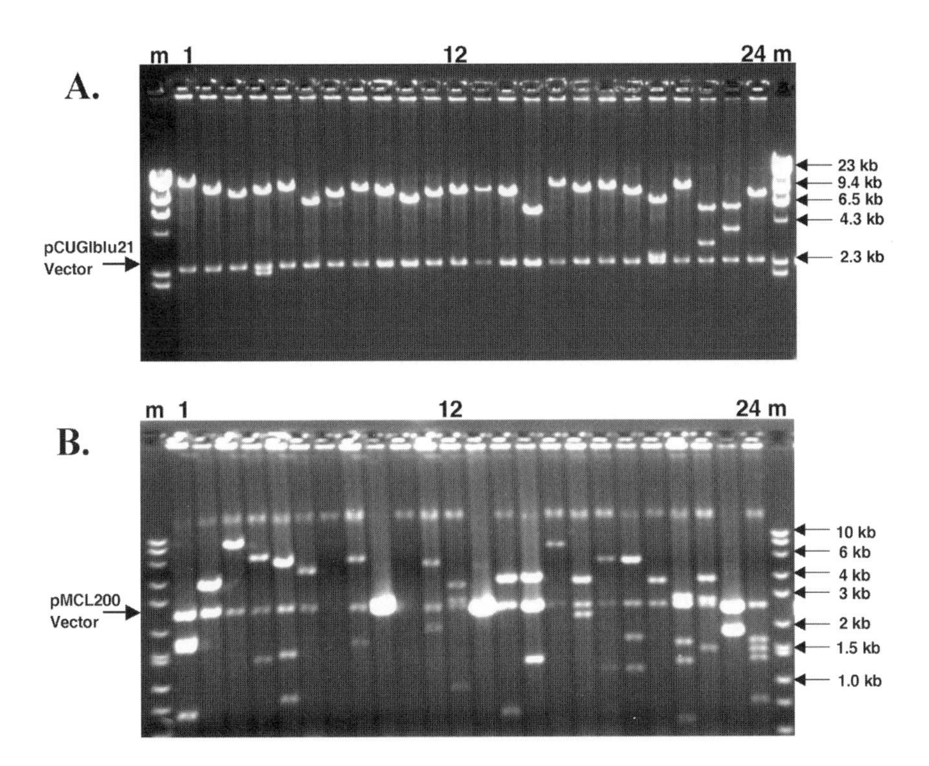

Figure 7. Quality control of shotgun library by restriction digest of RCA products. *E. coli* colonies containing low-copy plasmids with 8-10 kb inserts were amplified with the use of TempliPhi. 5-μL aliquot of amplified products was taken for digestion with either *Swa*I enzyme (A), which cuts DNA insert out of pCUGIblu21 vector, or *Bam*HI and *Hin*dIII enzymes (B), which cut DNA insert out of pMCL200. Both the vector-specific band and the insert-specific bands are clearly seen. The DNA marker (m) identifies the size of inserts.

In addition to digestion of amplified plasmids, the products of this high-fidelity amplification can also be used for cloning purposes, *in vitro* transcription and translation, the products can been spotted on microarray and can be used in both Southern and Northern blot analysis. The only applications, in which these amplification products cannot be used interchangeably with intact plasmid DNA, are those that require supercoiled plasmids (*e.g.* in transformation procedures). However, we have found that decatenation/linearization of amplification products by the restriction enzyme digestion and their subsequent ligation/ circularization can be used to make the transformation-proficient DNA from the amplification products.

Summarizing Discussion

The use of random hexamer-primed isothermal strand-displacement amplification with Phi29 DNA polymerase offers many benefits in a variety of molecular biology applications compared to traditional techniques. We have utilized this robust RCA format to optimize conditions for amplification of plasmids from standard glycerol stocks to increase the production efficiency of the sequencing process at the US DOE Joint Genome Institute (JGI) (Detter *et al.*, 2002). Our plasmid amplification method has significant advantages over traditional template generation by greatly simplifying the process and will be helpful for genome centers as well as for core labs, service centers and academic researchers that isolate and sequence DNA from a variety of vectors. This approach has few simple steps, can be highly automated, is completely scalable and is already giving excellent, cost-effective results at our production genomics facility. The collection of plasmid amplification and sequencing protocols we developed based on this approach are available on the web (www.jgi.doe.gov) and are included in the technical literature with the TempliPhi kit.

Our amplification technology was extended for use in the end-sequencing of a variety of other vector constructs. Most high-copy *E. coli* plasmids (pGEM, pET, pSMART, *etc.*) that were attempted produced results similar to those obtained in production with pUC18. Additionally, several low copy plasmids (pCUGIblu21 and pMCL200) containing inserts up to 10 kb also produce results similar to those obtained in production with pUC18. In fact, these low-copy, medium size-insert constructs now work well enough to have been phased into the standard production amplification sequencing line at the JGI. We have also demonstrated that this process can be used to sequence plasmids cloned into yeast cells (*e.g.* pGADT7 was cloned into yeast cells Y187 for the purpose of a yeast two-hybrid assay; CLONTECH). We have found that a simple lysis procedure similar to that used for amplification of plasmids from *E. coli* also can be used for amplification and sequencing of plasmids from yeast colonies or cultures.

Preliminary experiments with the large-insert vectors, such as cosmids, fosmids and BACs, have been less successful. In our experience, low-copy, large-insert clones do not sequence robustly using this amplification process. Amplified DNA from these clones often contain variable amounts of contaminating *E. coli* DNA and, as a result, yield poor sequencing results (data not shown). Experiments with improved lysis and specific primers are currently underway to reduce the *E. coli* chromosomal DNA amplification and optimize the RCA-based process for these large-insert constructs.

Acknowledgements

We would like to thank members of the Cloning Technology Group, as well as many others, at the Joint Genome Institute for their assistance with this work. Thanks to Daniel Collins for the electron microscopy. This work was performed under the auspices of the U.S. Department of Energy, Office of Biological and Environmental Research, by the University of California, under Contracts No. W-7405-Eng-48, No. DE-AC03-76SFOO098, and No. W-7405-ENG-36.

References

Blanco, L., Bernad, A., Lazaro, J.M., Martin, G., Garmendia, C., and Salas, M. (1989). Highly efficient DNA synthesis by the phage phi 29 DNA polymerase. Symmetrical mode of DNA replication. J. Biol. Chem. *264*, 8935-8940.

Blanco, L. and Salas, M. (1996). Relating structure to function in phi29 DNA polymerase. J. Biol. Chem. *271*, 8509-8512.

Dean, F.B., Nelson, J.R., Giesler, T.L., Lasken, R.S. (2001). Rapid amplification of plasmid and phage DNA using Phi 29 DNA polymerase and multiply-primed rolling circle amplification. Genome Res. *11*, 1095-1099.

Demidov, V.V. (2002). Rolling-circle amplification in DNA diagnostics: the power of simplicity. Expert Rev. Mol. Diagn. *2*, 542-548.

Detter, J.C., Jett, J.M., Lucas, S.M., Dalin, E., Arellano, A.R., Wang, M., Nelson, J.R., Chapman, J., Lou, Y., Rokhsar, D., Hawkins, T.L., and Richardson, P.M. (2002). Isothermal strand-displacement amplification applications for high-throughput genomics. Genomics *80*, 691-698.

Dunning, A.M., Talmud, P., and Humphries, S.E. (1988). Errors in the polymerase chain reaction. Nucleic Acids Res. *16*, 10393.

Esteban, J.A., Salas, M., and Blanco, L. (1993). Fidelity of phi 29 DNA polymerase. Comparison between protein-primed initiation and DNA polymerization. J. Biol. Chem. *268*, 2719-2726.

Elkin, C.J., Richardson, P.M., Fourcade, H.M., Hammon, N.M., Pollard, M.J., Predki, P.F., Glavina, T., and Hawkins, T.L. (2001). High-throughput plasmid purification for capillary sequencing. Genome Res. *11*, 1269-1274.

Ewing, B., Hillier, L., and Green, P. (1998a). Base-calling of automated sequencer traces using Phred. I. Accuracy probabilities. Genome Res. *8*, 175-185.

Ewing, B. and Green, P. (1998b). Base-calling of automated sequencer traces using Phred. II. Error probabilities. Genome Res. *8*, 186-194.

Hawkins, T.L., McKernan, K.J., Jacotot, L.B., MacKenzie, J.B., Richardson, P.M., and Lander, E.S. (1997). A magnetic attraction to high-throughput genomics. Science *276*, 1887-1889.

Kool, E.T. (1996). Circular oligonucleotides: new concepts in oligonucleotide design. Annu. Rev. Biophys. Biomol. Struct. *25*, 1-28.

Kornberg, A. and Baker, T.A. (1992). *DNA Replication*. WH Freeman and Co, New York, pp. 502-503.

Kunkel, T.A. (1985). The mutational specificity of DNA polymerase-beta during *in vitro* DNA synthesis. Production of frameshift, base substitution, and deletion mutations. J. Biol. Chem. *260*, 5787-5796.

Lizardi, P.M., Huang, X., Zhu, Z., Bray-Ward, P., Thomas, D.C., and Ward, D.C. (1998). Mutation detection and single-molecule counting using isothermal rolling-circle amplification. Nat. Genet. *19*, 225-232.

Nelson, J.R., Cai, Y.C., Giesler, T.L., Farchaus, J.W., Sundaram, S.T., Ortiz-Rivera, M., Hosta, L.P., Hewitt, P.L., Mamone, J.A., Palaniappan, C., and Fuller, C.W. (2002). TempliPhi, phi29 DNA polymerase based rolling circle amplification of templates for DNA sequencing. Biotechniques *Suppl.*, 44-47.

3.6

Multiple-Displacement Amplification (MDA) of Whole Human Genomes from Various Samples

Roger S. Lasken, Seiyu Hosono
and Michael Egholm

Abstract

Methods for whole genome amplification (WGA) are becoming increasingly important to generate the large amounts of DNA required for genetic testing. Obtaining high-quality human genomic DNA (gDNA) samples can be a limiting factor for many applications. For example, most DNA collections used for association and linkage studies are generally a nonrenewable precious resource available to only a limited number of laboratories. Preparation of gDNA from clinical samples is also a bottleneck in high-throughput genetic testing and DNA sequencing, and is frequently limited by the amount of specimen available. In this chapter, protocols are presented for WGA of human DNA from various sources by multiple-displacement amplification (MDA). MDA can be used for extensive DNA generation directly from blood, buccal swabs and other clinical specimens, as well as for amplification of extracted archival gDNA. The MDA reaction has several advantages over older WGA methods, including more complete and unbiased coverage of the human genome, higher DNA yields and much longer DNA products ranging from 10 to 100 kb.

Background

Since its introduction, the MDA approach has been used to amplify gDNA from a wide range of clinical samples, including blood, plasma, buccal swabs and tissues. MDA is also used to recover depleted DNA collections, often being the only simple and reliable method available to extend the usefulness of valuable DNA samples obtained from population and epidemiological studies. For association and linkage studies, MDA can generate the hundreds of micrograms of gDNA now frequently required for newer, high-throughput SNP assays. Reliable whole genome amplification by MDA has also made it possible to generate an abundant supply of DNA for widespread distribution of samples that were previously available to only a few laboratories.

Principles of the MDA Reaction

MDA is the first method that can accurately and comprehensively amplify whole genomes with minimal amplification bias and without loss of essential sequences (Dean *et al.*, 2002; Hawkins *et al.*, 2002; Lasken and Egholm, 2003). Initially proposed by Paul Lizardi of Yale University, MDA differs from older WGA methods in not being based on the polymerase chain reaction (PCR). MDA achieves exponential amplifications entirely at 30°C with random hexamer primers by a polymerase-driven strand-displacement reaction (Dean *et al.*, 2001). The DNA template is replicated repeatedly by a "hyperbranching" mechanism (Lizardi *et al.*, 1998) with DNA polymerase laying down a new copy as it concurrently displaces previously made copies (Figure 1).

Use of exonuclease-resistant primers is required for 10^4-10^6-fold amplifications (Dean *et al.*, 2001). MDA generates a relatively constant

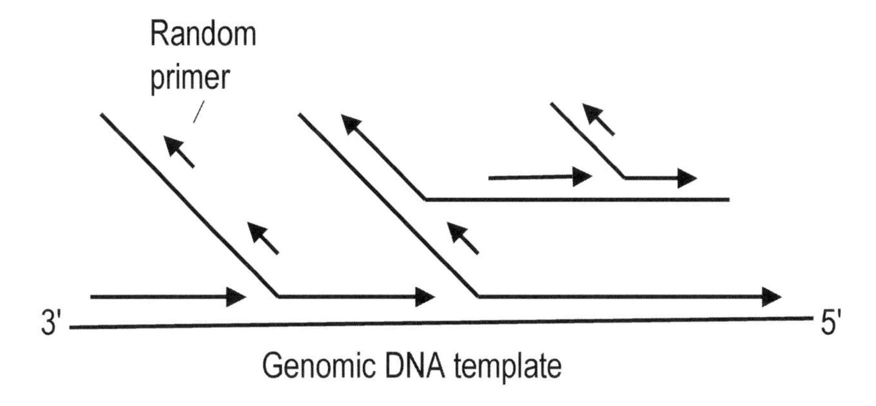

Figure 1. Multiple-displacement amplification reaction. DNA synthesis is primed by random hexamers. Exponential amplification occurs by a "hyperbranching" mechanism. Unlike PCR, which requires thermal cycling to repeatedly melt template and anneal primers, the ϕ29 DNA polymerase acts at 30°C to concurrently extend primers as it displaces downstream DNA products.

yield of approximately 80 μg DNA/100 μL reaction, which does not depend on the amount of input DNA template. Since MDA is performed at a fixed temperature, it can be conveniently used as a preparative method scaled to any volume. For example, more than a milligram of genomic sequence is generated from a 1.5 mL reaction.

MDA is carried out with a DNA polymerase from the bacteriophage ϕ29 that has exceptionally tight binding to DNA and is able to continuously add about 70,000 nucleotides every time it binds to the primer (Blanco *et al.*, 1989; see also the corresponding chapter about this enzyme here). This is the highest processivity reported for any DNA polymerase in the absence of cellular multi-subunit complexes and accounts for its ability to generate extremely long DNA products in MDA averaging 12 kb in length and ranging up to more than an estimated 100 kb. Long products are an advantage in many applications, including restriction fragment length polymorphism (RFLP), generation of probes and DNA sequencing (Dean *et al.*, 2002).

Accurate amplification results from the high fidelity of ϕ29 DNA polymerase, which has an error rate of only 1 in 10^6-10^7 bp (Esteban *et al.*, 1993). The total accumulation of errors in an MDA reaction, over the entire course of the amplification, is about 3 per 10^6 bp (Nelson *et al.*, 2002) compared to an estimated one mutation per 900 bp after 20 PCR cycles (Saiki *et al.*, 1988). The accuracy of genotyping assays using the MDA-generated DNA confirms reliable representation of alleles (Hosono *et al.*, 2003). MDA also achieves even representation of different genes throughout the genome (see Example 1). Amplification of whole bacterial genomes by MDA and subsequent DNA sequencing also indicates accurate and unbiased coverage (Detter *et al.*, 2002; Hawkins *et al.*, 2002).

DNA Template Requirements

As little as 1 ng of gDNA template can be used for successful MDA. However, if the template is degraded to an average length of less than two to five kb, somewhat uneven coverage of the genome may result. Yet, even in these cases, MDA may generate amplified DNA of adequate quality for accurate genotyping analysis, so that this method should be considered as a means to salvage older, partially degraded samples, where no other alternative is available. The amount of DNA synthesized in an MDA reaction is independent of the starting amount of DNA: inputs between 1 ng and 100 ng of DNA template provide the same yield of high quality DNA (we recommend a standard input of 10-100 ng). In our experience, stored DNA is often partially degraded and has a considerably lower concentration than estimated at the time of collection. In general, smaller amounts of biological samples are required compared to use of previously extracted and stored DNA samples. This is thought to result from the high quality of DNA released from gently lysed cells, which has suffered less damage than extracted DNA samples. For example, less than 1 μL of blood is required to generate 100+ μg of gDNA from whole blood.

Protocols

I. Sample Preparation

The proper handling of DNA template is important to achieve optimum whole genome amplification. Two different types of samples are encountered (Table 1): 1) Previously purified DNA samples, such as those stored in DNA sample archives. Amplification is carried out in order to replenish supplies. 2) Clinical and other biological specimens. In this case, the reaction serves two purposes. First, it is a rapid method for preparing DNA from cell-based sources, bypassing the need for phenol extraction or bead-/cartridge-based DNA purification methods. Second, the amplification can generate large quantities of DNA from very small specimens.

For biological samples, an alkaline treatment with KOH (Table 1) serves both to lyse cells, releasing the DNA, and also to denature the DNA template converting it to the single-stranded form required for MDA. Neutralization with a Tris buffer completes the sample preparation steps. In the case of previously isolated DNA samples, KOH treatment is also employed for preparing the single-stranded DNA template. Alternatively, heating the DNA solution to 95°C has been used to denature the DNA (Dean *et al.*, 2001). However, heating can damage the DNA template and should be limited to no more than 3 minutes at 95°C. In this chapter, all cell lysis and DNA denaturation is carried out with KOH treatment following prescribed protocols (Molecular Staging Inc., 2003).

Denaturation of Purified gDNA

1. Required reagents:
    - Assemble reactions at room temperature
    - 10 – 100 ng of gDNA in 2.5 μL volume
    - Denaturation solution: 50 mM potassium hydroxide, 1 mM EDTA (pH 8.0)
    - Neutralization buffer: 80 mM Tris-hydrochloride, pH ~4 (not adjusted)
2. Denaturing reaction: mix together 2.5 μL of gDNA and 2.5 μL of denaturation solution and incubate at room temperature for 3 min.
3. Neutralization reaction: stop the denaturation reaction by adding 5 μL of neutralization buffer and proceed directly to the amplification protocol below (Section II, MDA amplification

Table 1. DNA sources employed in MDA[a]

Sample Types	Common Uses	MDA Advantages	Lysis/Denaturation Condition	MDA reaction vol.	Comments
Stored genomic DNA	Population genetics, epidemiology, clinical trials	Replenish depleted archived DNA samples from ≥10 ng of DNA.	Denatured with KOH for 3 min	≥ 50 μL	Suboptimal recovery if DNA template degraded to < 2-5 kb average length
Whole blood	Population genetics, epidemiology, clinical diagnosis, clinical trials	Only 1 µL of blood required per amplification. Amplified DNA is assay ready without further purification.	Lysed/denatured with KOH for 10 min	≥ 50 μL	Fresh or frozen whole blood may be used. No inhibitory effect from common anticoagulant and preservative.
Buffy coat	Population genetics, epidemiology, clinical trials	Only 1 µL of buffy coat required per amplification. Amplified DNA is assay ready without further purification.	Lysed/denatured with KOH for 10 min	≥ 50 μL	Fresh or frozen buffy coat may be used. No inhibitory effect from common anticoagulants and preservatives.
Tissue culture	Cancer genetics, molecular pathology, population genetics	Only few hundred cells required per amplification.	Lysed/denatured with KOH for 10 min	≥ 50 μL	Fresh or frozen tissue culture cells may be used.
Frozen tissue (biopsy, laser dissected tissue)	Cancer genetics, molecular pathology	Only few hundred cells required per amplification. Amplified DNA is assay ready without further purification.	Lysed/denatured with KOH for 30 min	≥ 500 μL	No homogenization or sonication required.

Dried blood spot	Population genetics, epidemiology, clinical diagnosis, clinical trials	Only 3 mm filter punch required per amplification. Ease of long-term storage. Amplified DNA is assay ready without further purification.	Lysed/denatured with KOH for 10 min	≥ 500 μL	Blood spots as old as 8 years have been tested for amplification
Buccal swab and cytobrush	Population genetics, epidemiology, clinical diagnosis, clinical trials	Low cost, non-invasive, and self-administered procedure. Amplified DNA is assay ready without further purification.	Lysed/denatured with KOH for 10 min	≥ 500 μL	Air-drying of the swab/cytobrush or immediate freezing of the swab/cytobrush at -20°C recommended
Plasma and serum	Population genetics, epidemiology, clinical trials	May obtain gDNA from minute amount of DNA present in plasma or serum. Amplified DNA is assay ready without further purification.	Lysed/denatured with KOH for 10 min	≥ 750 μL	At least 50 μL of serum or plasma required per amplification

[a] Paraffin embedded tissue currently cannot be used for MDA due to heavy fragmentation of cross-linked genomic DNA

reaction). The amplification protocol was always carried out within 1 hr of completing the DNA denaturation and neutralization steps above. Longer times have not been tested.

Processing of Blood, Buffy Coat or Tissue Culture Cells

Drawn whole blood or finger stick blood and its subsequent preparation of buffy coat are still the most common methods used in collecting clinical samples (see reviews by Gunter, 1997; Landi and Caporaso, 1997; Holland *et al.*, 2003). The MDA reaction works efficiently from both fresh and frozen blood samples. No inhibitory effect on the MDA reaction was observed with the four most common anti-coagulants and preservative used in blood storage–EDTA, citrate, acid-citrate-dextrose and heparin. The MDA reaction requires less than 1 μL of whole blood, finger stick blood or buffy coat (Table 1), which is ideal for rare and precious collections.

Tissue culture cells are another common source of gDNA. The time consuming process of growing a large number of cells to obtain sufficient quantity of gDNA can be eliminated with MDA as only a few hundred cells are required here.

1. Required reagents:
   - Assemble reactions on ice
   - 0.5 μL of whole blood, finger stick blood and buffy coat or approximately 300 tissue culture cells all diluted to 3 μL with 1X phosphate-buffered saline (PBS)
   - Lysis solution: 400 mM potassium hydroxide, 100 mM dithiothreitol (DTT), 10 mM EDTA (pH 8.0)
   - Neutralization buffer: 800 mM Tris-hydrochloride, pH ~4 (not adjusted)
2. Lysis/denaturation reaction: mix 3 μL of PBS-diluted biosamples with 3.5 μL of lysis solution and incubate on ice for 10 min. Avoid exceeding 10 min as this may damage the DNA.
3. Neutralization reaction: Stop the lysis/denaturation reaction by adding 3.5 μL of neutralization buffer. Proceed to the amplification protocol below (Section II, MDA amplification reaction). The

amplification protocol was always carried out within 1 hr of completing the lysis, denaturation and neutralization steps above. Longer times have not been tested.

Processing of Buccal Cells Collected by Swab or Cytobrush

The low-cost and non-invasive exfoliated buccal cell collection method has become popular since many sample donors are reluctant to provide blood or tissue samples (Lench *et al.*, 1988; Richards *et al.*, 1993; Garcia-Closas *et al.*, 2001). Mishandling during collection and improper storage of the buccal swab or cytobrush leads to cell disintegration followed by the degradation of gDNA due to apoptosis or the release of nucleases from buccal cells, as well as mold and bacteria that may grow in the sample (Steinberg *et al.*, 2002). DNA extracted from a single buccal swab can range from 4 to 60 μg depending on the extraction method. However, the median percentage of human DNA present in the DNA extracted from buccal cells is estimated to be only about 11.5% (Garcia-Closas *et al.*, 2001). MDA is ideal for obtaining larger quantities of DNA from these small samples. For optimal MDA performance using buccal cells, the swab or brush must be air-dried for at least 2 hr before long-term storage at -20°C (see Table 1). Alternatively, the air-dried swab or brush can be kept at room temperature for up to a week, allowing enough time for a subject to self-administer the collection and send the sample through the first class mail.

1. Required reagents:
   - Assemble the reaction on ice
   - One buccal swab or cytobrush. For swab, the Catch-All™ sample collection swab (Epicentre, Madison, WI) was used. For cytobrush, the CytoSoft™ cytology brush (Medical Packaging Corporation, Camarillo, CA) was used. Equivalent products from other manufactures are also available.
   - 1X phosphate-buffered saline (PBS)
   - Buccal cell lysis solution: 400 mM potassium hydroxide, 10 mM EDTA (pH 8.0). (Note: DTT is not included here, as it is for the lysis solutions of other sample types described in this

chapter. In some of our observations, performance for buccal swabs was better with DTT omitted.)
- Neutralization buffer: 800 mM Tris-hydrochloride, pH ~4 (not adjusted)

2. Buccal cell collection: rinse out the subject's mouth twice with water. The subject should abstain from drinking coffee for 1 hr before this process. Collect the buccal cells by rolling and dragging the swab or brush firmly on the inside of the cheek 20 times on each cheek, making sure to scrub across the entire cheek. Note: when storing the swab or cytobrush for future use, let it air-dry for 2 hr prior to storage. After air-drying, both swabs and brushes can be stored at room temperature for up to 7 days. However, best results are obtained when used fresh or stored at -20°C.

3a. Lysis/denaturation reaction for swabs: cut off the swab from the stick with a sterile razor blade and place it into a 1.5 mL microcentrifuge tube containing 100 μL of 1X PBS and vortex the swab for 10 sec. Then lift the swab out of the solution and recover any solution still remaining in the swab by pipetting to obtain as much liquid as possible before discarding the swab. The swab must be removed before proceeding to the cell lysis step since gDNA appears to be trapped by the swab. Mix together 100 μL of PBS-diluted buccal swab cells and 100 μL of buccal cell lysis solution, then incubate on ice for 10 min. Avoid exceeding 10 min as this may damage the DNA.

3b. Lysis/denaturation reaction for cytobrush: cut off the cytobrush from the stick into a 1.5 mL microcentrifuge tube containing 30 μL of 1X PBS. For cytobrushes it is recommended to leave the brush head in the tube throughout the lysis and MDA steps. Mix 30 μL of PBS-diluted buccal swab cells with 35 μL buccal cell lysis solution, vortex hard for 30 sec, spin down the cell debris for few seconds and incubate on ice for 10 min. Avoid exceeding 10 min as this may damage the DNA.

4a. Neutralization reaction for swabs: stop the lysis/denaturation reaction by adding 100 μL of neutralization buffer. Take 100 μL and proceed to the amplification protocol below (Section II, MDA amplification reaction) for a 500 μL volume. The amplification protocol was always carried out within 1 hr of completing the lysis, denaturation and neutralization steps above. Longer times have not been tested.

4b. Neutralization reaction for cytobrush: stop the lysis/denaturation reaction by adding 35 μL of neutralization buffer and mix by vortexing for 10 sec. Use all 100 μL total, with the brush head still in the reaction tube, and proceed to the amplification protocol below (Section II, MDA amplification reaction) for the 500 μL volume. The amplification protocol was always carried out within 1 hr of completing the lysis, denaturation and neutralization steps above. Longer times have not been tested.

Use of Dried Blood Spots on Guthrie Cards

Stored dried blood spots on Guthrie cards are a vast resource for population and clinical genetic studies (McEwen *et al.*, 1994; Clayton *et al.*, 1995). For example, blood spot samples have been collected for the majority of newborns in many countries and potentially provide a DNA source for all newborns in the United States although the policy varies from state to state (Therrell *et al.*, 1996). MDA has been validated for amplification directly from blood spots including spots that have been stored for many years. Only a 3 mm diameter punch of a blood spot is required for amplification.

1. Required reagents:
   - Assemble reactions on ice
   - Dried blood spot (~3 mm punch)
   - 1X phosphate-buffered saline (PBS)
   - Lysis solution: 400 mM potassium hydroxide, 100 mM dithiothreitol (DTT), 10 mM EDTA (pH 8.0)
   - Neutralization buffer: 800 mM Tris-hydrochloride, pH ~4 (not adjusted)
2. Suspend dried blood with 10 μL of 1X PBS, soak the spot at room temperature for 30 min, with 5 sec of vortexing every 10 min..
3. Lysis/denaturation reaction: mix together 10 μL of blood spot punch suspended in PBS and 10 μL of lysis solution, then incubate on ice for 10 min with 5 sec of vortexing every 2 min. Avoid exceeding 10 min as this may damage the DNA.
4. Neutralization reaction: stop the lysis/denaturation reaction by adding 10 μL of neutralization buffer and proceed to the

amplification protocol below (Section II, MDA amplification reaction) for a 500 μL reaction with the blood spot punch still in the reaction tube. The amplification protocol was always carried out within 1 hr of completing the lysis, denaturation and neutralization steps above. Longer times have not been tested.

Processing of Flash Frozen Tissue Sections

Examples of tissue samples include flash frozen tissue sections, laser capture microdissection samples (Emmert-Buck *et al.*, 1996) and needle aspirate biopsies (Euhus *et al.*, 2002). Conceptually, MDA will be ideal for large population studies such as the Cancer Genome Project. For example, genome-wide screens have been used to identify genes involved in the disease process (Davies *et al.*, 2002). However, this approach utilizes DNA from cultured tumor cells. Some of these cells, such as those from prostate and pancreatic cancers, are difficult to culture (Pollock *et al.*, 2002). In addition, cultured cell lines accumulate mutations over time and may not truly reflect the genetic make-up of the original tumor cells. MDA would allow amplification directly from tumor cells without the costly and labor-intensive process of generating cell lines. Furthermore, minimal processing of the tissue section is required since no homogenization or sonication of the tissue is needed.

1. Required reagents:
   - Assemble reactions on ice
   - Flash frozen tissue section (about 2 mm in diameter)
   - 1X phosphate-buffered saline (PBS)
   - Lysis solution: 400 mM potassium hydroxide, 100 mM dithiothreitol (DTT), 10 mM EDTA (pH 8.0)
   - Neutralization buffer: 800 mM Tris-hydrochloride, pH ~4 (not adjusted)
2. Suspend flash frozen tissue section with 10 μL of 1X PBS, incubate at room temperature for about 10 min, with 5 sec of vortexing every 2 min.
3. Lysis/denaturation reaction: mix together 10 μL of tissue section suspended in PBS and 10 μL of lysis solution, then incubate on ice

for 30 min with 5 sec of vortexing every 10 min. Avoid exceeding 30 min as this may damage the DNA.

4. Neutralization reaction: stop the lysis/denaturation reaction by adding 10 μL of neutralization buffer and proceed to the amplification protocol below (Section II, MDA amplification reaction) for a 500 μL reaction with the tissue section in the reaction tube. The amplification protocol was always carried out within 1 hr of completing the lysis, denaturation and neutralization steps above. Longer times have not been tested.

Use of Cells in Plasma or Serum

Vast collections of plasma and serum samples exist in hospitals and laboratories around the world. In many epidemiological studies, sample collection has been limited to plasma and serum (Holland *et al.*, 2003). However, the amount of DNA that can be extracted from plasma or serum is extremely limited and is preventing large-scale genetic analysis based on these samples. MDA amplification has been shown to work consistently from only 50 μL of plasma or serum, which contains small amounts of cells or DNA. Plasma and serum samples have also been used to detect gDNA that is released from cells in many pathologic conditions (Lee *et al.*, 2001) and MDA may also be of value in amplifying this DNA.

1. Required reagents:
   - Assemble reactions on ice
   - 50 μL of plasma or serum in any anticoagulant or preservative
   - 1X phosphate-buffered saline (PBS)
   - Lysis solution: 400 mM potassium hydroxide, 100 mM dithiothreitol (DTT), 10 mM EDTA (pH 8.0)
   - Neutralization buffer: 800 mM Tris-hydrochloride, pH ~4 (not adjusted)
2. Lysis/denaturation reaction: mix together 50 μL of plasma or serum and 50 μL of lysis solution, then incubate on ice for 10 min. Avoid exceeding 10 min as this may damage the DNA.

3. Neutralization reaction: stop the lysis/denaturation reaction by adding 50 μL of neutralization buffer and proceed to the amplification protocol below (Section II, MDA amplification reaction) for a 750 μL reaction. The amplification protocol was always carried out within 1 hr of completing the lysis, denaturation and neutralization steps above. Longer times have not been tested.

II. MDA Reaction

MDA is a fully scalable reaction and has been tested between 50 μL to 10 mL. It generates a constant DNA yield of approximately 0.8 μg/μL ± 15%. Primers are rendered exonuclease resistant by incorporation of thiophosphates at the 3'-terminus to protect them from degradation by the 3'-5' exonuclease proofreading activity of ϕ29 DNA polymerase (Dean *et al.*, 2001). At the reaction temperature of 30°C the reaction time is between 6 to 16 hr. The yield of DNA product plateaus in about 6 hr with little additional synthesis occurring beyond this time. Sixteen hours is frequently used because it is convenient to set up the reaction at the end of the day and retrieve the amplified samples the following morning.

The MDA reaction is terminated by heating to 65°C for 3 min, which inactivates the ϕ29 DNA polymerase. A thermocycler can conveniently be used to maintain the 30°C reaction temperature, the 65°C reaction termination step, and then 4°C until the samples are retrieved. Alternatively, reactions can be carried out in an oven or water bath, particularly when many samples and/or large volume reactions are required. For larger reaction volumes, longer incubation time at 65°C may be needed to completely inactivate the polymerase. For example, for a 10 mL preparative reaction carried out in a Falcon tube, 10 min in a 65°C water bath was required to effectively terminate the polymerase and associated 3'-5' exonuclease activities. After the DNA polymerase is inactivated, the amplified DNA is ready for storage and will have a shelf-life comparable to that of gDNA samples. For long-term storage, we recommend that the amplified DNA be aliquoted into multiple tubes to minimize degradation from freeze/thaw cycles (Steinberg *et al.*, 1997) and stored at -80°C.

Table 2. Typical MDA reaction volumes

Reagent (μL)	50 μL reaction	500 μL reaction	750 μL reaction
DNA or lysate	10	100 [b]	150
Sterile, distilled water	27	270	405
4X REPLI-g mix [a]	12.5	125	187.5
ϕ29 DNA polymerase	0.5	5	7.5
Total volume	50	500	750

[a] 4X REPLI-g mix (see Table 3) and ϕ29 DNA Polymerase are components from REPLI-g Whole Genome Amplification Kit (Molecular Staging Inc, New Haven, CT).
[b] For Guthrie card or tissue section, 30 μL of lysate should be used, therefore 340 μL of sterile, distilled water must be added for the 500 μL reaction.

Selecting the Appropriate Reaction Volume

The optimal and recommended amplification reaction volume varies with the source of gDNA (see Table 2 for reaction set up). For DNA template sources such as purified gDNA, whole blood, buffy coat and tissue culture cells, there is a high degree of confidence in the amount of DNA present in the sample, and therefore a reaction volume as small as 50 μL can be used successfully. For tissue section, buccal cell and dried blood spot lysate, varying quality and amounts of gDNA is obtained, and therefore a reaction volume of at least 500 μL should be used. This allows for enough of the cell lysis reaction to be added, thus assuring that a sufficient number of genome copies will be present to give the complete and unbiased amplification coverage. For plasma and serum lysate, only trace amounts of gDNA is present; therefore a reaction volume of 750 μL should be used.

1. MDA components: ϕ29 DNA polymerase and exonuclease-resistant hexamer primers in reaction buffer (Dean *et al.*, 2001) are available at Molecular Staging Inc. in the REPLI-g kit. When assembling the reactions, refer to Table 2 for reaction volumes.

2a. For single tube reactions, always add the components in the order shown in Table 2 and assemble the reaction on ice. Briefly mix after addition of the 4X REPLI-g mix (Table 3) prior to addition of the ϕ29 DNA polymerase. Mix again when the entire reaction is assembled with polymerase present.

Table 3. Contents of the 4X REPLI-g mix[a]

Tris-HCl, pH 7.5	150 mM
KCl	200 mM
$MgCl_2$	40 mM
$(NH_4)_2SO_4$	80 mM
dNTPs	4 mM of each nucleotide
Exonuclease-resistant hexamer	0.2 mM

[a] Molecular Staging Inc, 2003

2b. For multiple reactions, a master reaction mix can be used, consisting of 4X REPLI-g mix, distilled water, and ϕ29 DNA polymerase. In that case, 4X REPLI-g mix and distilled water are combined and briefly mixed before the polymerase is added. This master reaction mix can be scaled for any number of reactions according to Table 2. In cases where the volume of DNA or lysate added differs from that shown in Table 2, the amount of distilled water added should be adjusted accordingly. When the reaction is fully assembled, immediately transfer the tubes to a 30°C incubator.

Note that polymerase is added to master reaction mix prior to the presence of DNA template or lysate. Therefore, it is important to add polymerase to the master reaction mix on ice just prior to starting reactions. Having reaction tubes numbered and dispensing the denatured and neutralized DNA template or lysate (Section I, Sample preparation) to the tubes in advance allows prompt addition of the Master Reaction Mix after it is completed by mixing in the ϕ29 DNA polymerase.

III. Quantification of MDA Amplicons

Since MDA amplification products contain unused reaction primers and dNTPs, it is important to utilize a DNA quantification method that is specific for double-stranded DNA such as PicoGreen® (Molecular Probes).

PicoGreen Protocol for MDA-Amplified Product

1. Follow the instructions for PicoGreen reagent preparation (Molecular Probes). Prepare 50 μl of genomic DNA standards at 1.6 μg/mL, 0.8 μg/mL, 0.08 μg/mL, 0.008 μg/mL and 0 μg/mL in a 1X TE buffer to create a reference five-point standard curve in duplicates.
2. Perform the measurement of MDA amplified DNA in duplicates. The MDA reaction typically generates about 0.8 μg/μL of DNA. Test an amount of DNA that falls within the linear range of the standard curve. In our experience, diluting 2 μL of the MDA amplicon in 1.998 mL of 1X TE buffer giving a final dilution of 1/1000 results in a final concentration of an approximately 0.8 μg/mL, which is within the linear range of the standard curve. However, care must be taken to assure this, as failure to measure DNA within the linear range of the standard curve is a common source of error.
3. Add 50 μL of the working solution of PicoGreen reagent to the standard and the test samples at room temperature and mix. For microtiter plates, mix by tapping the side of the plate several times. Spin down the plate briefly using a table-top centrifuge. In our work, the sample fluorescence is measured using a fluorescence microplate reader as recommended by Molecular Probes.
4. For sample analysis, subtract the fluorescence value of the reagent blank from that of each of the samples. Use corrected data to generate a final standard curve (between 1.6 μg/mL to 0.0 μg/mL) of fluorescence versus DNA concentration. We conveniently use Microsoft Excel to calculate the DNA concentration of the sample from the fluorescence value using the slope equation obtained from the gDNA standard curve. The correction factor for dilution is 1/1000. For a typical MDA product, the concentration range should be approximately 0.8 μg/μL (±15%).

Note: This procedure should also be used for determining the template gDNA amount prior to amplification.

IV. Qualitative Analysis of MDA Amplicons

The TaqMan® approach can determine the yield of a specific gDNA sequence. Evaluation of a few representative loci by this quantitative PCR method will give a reliable assessment of the overall quality of the MDA-amplified DNA, and by inference, the starting DNA template quality as well. If the selected loci are well represented, and are present at similar copy number, this indicates a generally successful amplification free of loci drop out or large amplification bias between different genes. Quantification of a given locus can be performed using the TaqMan-based system of assays, developed by Applied Biosystems Inc. For example, two representative TaqMan assays measuring c-*JUN* and human RFLPO (large ribosomal protein) have been used by us for copy number determination in evaluation of MDA amplification performance. For RFLPO and c-*JUN*, the locus representation, defined as the ratio of the copy number in the amplified DNA to the copy number in the unamplified template, should typically be between 40% and 200%. The preservation of copy number is a great advantage of the MDA method compared to older, PCR-based WGA amplification methods, which introduce large bias between different regions (Hosono *et al.*, 2003).

When older DNA samples of lower quality are used as template for the MDA reaction, some increase in amplification bias may be observed. However, even in these cases, the MDA-amplified DNA often performs with high accuracy in most genotyping assays. By carrying out genotyping studies only on those samples that pass this initial locus representation screen, even older DNA collections having some degradation of DNA can be recovered for use. Performance for older samples will vary depending on overall quality and must be evaluated for each DNA collection. Examples of the locus representation analysis and SNP genotyping with MDA-amplified DNAs are given below.

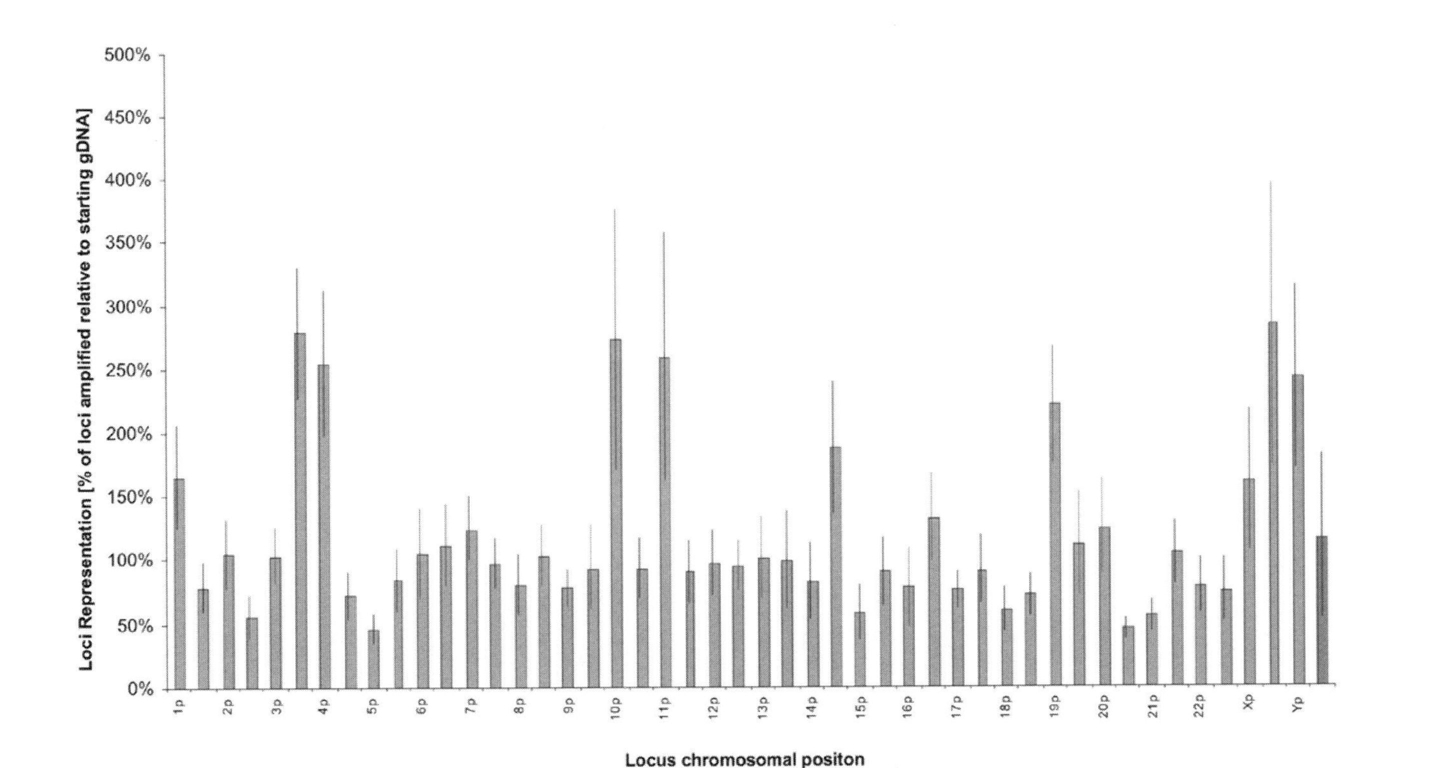

Figure 2. Amplification bias analysis for MDA by the quantitative TaqMan PCR assay for 47 human loci. Locus representation expressed as the percent of locus copies in the amplified DNA relative to the starting DNA template. Each bar represents the average of amplifications of 44 different individual's DNA. The average of all of the 47 loci was 117% (far right bar). Reprinted from (Hosono *et al.*, 2003) with permission of *Genome Research*, Cold Spring Harbor Laboratory Press.

Example 1: Even Amplification of Human Chromosomal DNA

The robust ability of φ29 DNA polymerase to replicate through difficult sequence as well as the extensive coverage afforded by 10-100 kb DNA products results in even amplification of the genome. In the test study (Hosono *et al.*, 2003), 47 different single copy human genetic loci were found to be represented at between 0.5 and 3 copies per genome in the amplified DNA from 44 individuals (Figure 2). In contrast, PCR-based WGA methods generated between 10^3 and 10^6-fold differences in copy number across the same arbitrarily selected loci (Dean *et al.*, 2002). These data demonstrate the unique potential of MDA to uniformly represent different human genes throughout the genome (reprinted with permission from *Genome Research*).

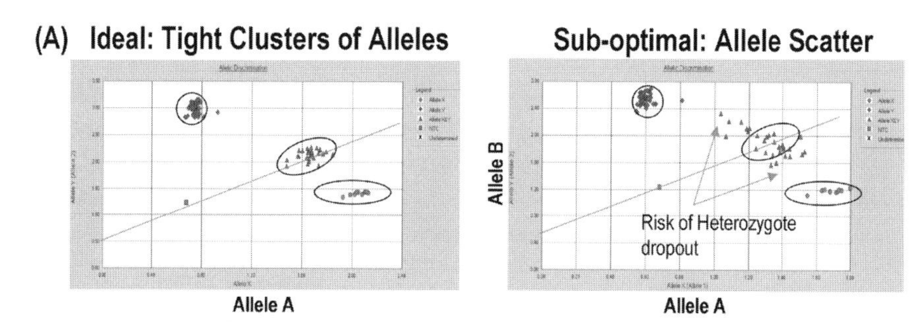

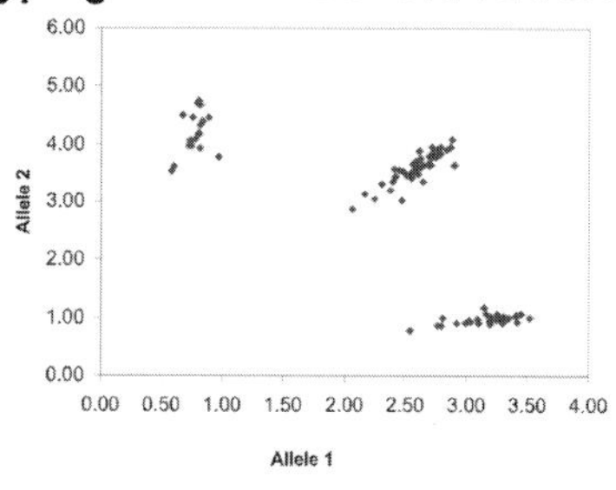

Figure 3. MDA amplification quality assessment by genotyping of SNP using TaqMan assays. (A) Analysis of the quality of a particular set of MDA-amplified gDNA samples by endpoint analysis using a TaqMan SNP assay. If allele bias is introduced in the amplification, heterozygotes may be difficult to score as they deviate from the diagonal (arrows). (B) Actual genotyping of MDA amplicons from 84 different gDNA samples.

Example 2: SNP Genotyping Using MDA-Generated DNA as Template

The amplified DNA has been tested for performance in most genotyping methods commonly in use. Genotyping from MDA product was validated by using the manufacturer's recommended amount of gDNA in performing the PCR-based SNP TaqMan genotyping assay (Applied-Biosystems, 2002; Hosono *et al.*, 2003) and the AmpFLSTR Profiler Plus™ assay (Applied-Biosystems, 2000; Hosono *et al.*, 2003). For Southern analysis of the restriction fragment length polymorphism (RFLP), a standard amount of 10 to 20 μg of MDA product was used for restriction digestion (Southern, 1975; Dean *et al.*, 2002). No DNA purification step is required prior to the use of MDA product in these assays.

SNP loci with a high frequency of heterozygotes should be used for the analysis. If both alleles of a SNP are well represented in the amplified DNA, then three separate populations will be clearly distinguished representing heterozygotes and each homozygote. The tightness of the cluster along the diagonal determines the accuracy of genotyping that will be achieved, and is a good indicator of allele bias in the amplified product (Figure 3A).

Allele bias is a stochastic event that occurs when only a limited number of gDNA templates are present to initiate the amplification. By using the starting template amounts recommended above, the equal amplification of both alleles is established early in the amplification reaction. The accurate SNP genotyping with the TaqMan system using MDA amplicons is shown in Figure 3B. This technique also serves as a good assay for checking the overall quality of the MDA amplified product.

Example 3: Visualization of MDA Amplicon

MDA amplicons have an average product length exceeding 10 kb, as determined by agarose gel electrophoresis (Figure 4). To this end, the double-stranded DNA products of the MDA reaction were denatured

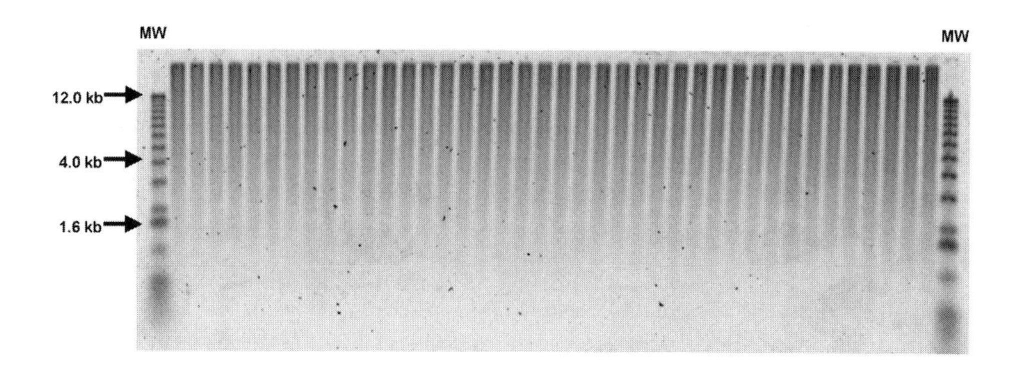

Figure 4. Denaturing gel analysis of MDA amplification product size. 100 ng of MDA amplicons from 40 different gDNA samples were denatured by addition of a 2X loading buffer containing 100 mM NaOH and 2 mM EDTA and electrophoresed through a 1% agarose gel using solution of 50 mM NaOH and 1 mM EDTA as a running buffer. After electrophoresis, the gel was soaked in water for 30 min and stained by SYBR Green® (Molecular Probes). The stained DNA in gel was visualized by PhosphorImager (Molecular Dynamics).

by NaOH and resolved on the denaturing agarose gel in order to reveal the true length of the single-stranded polymerase products free of nicks. Importantly, the yield and average length of amplified DNA was reproducibly uniform among 40 different DNA templates amplified.

References

Applied-Biosystems (2000). *AmpFLSTR® Profiler Plus™ Product User Manual.*

Applied-Biosystems (2002). *TaqMan SNP Genotyping Product User Manual.*

Blanco, L., Bernad, A., Lázaro, J.M., Martin, G., Garmendia, C., and Salas, M. (1989). Highly efficient DNA synthesis by the phage phi 29 DNA polymerase. Symmetrical mode of DNA replication. J. Biol. Chem. *264*, 8935-8940.

Clayton, E.W., Steinberg, K.K., Khoury, M.J., Thomson, E., Andrews, L., Kahn, M.J., Kopelman, L.M., and Weiss, J.O. (1995). Informed consent for genetic research on stored tissue samples. JAMA *274*, 1786-1792.

Davies, H., Bignell, G.R., Cox, C., Stephens, P., Edkins, S., Clegg, S., Teague, J., Woffendin, H., Garnett, M.J., Bottomley, W., *et al.* (2002). Mutations of the BRAF gene in human cancer. Nature *417*, 949-954.

Dean, F.B., Nelson, J.R., Giesler, T.L., and Lasken, R.S. (2001). Rapid amplification of plasmid and phage DNA using Phi 29 DNA polymerase and multiply-primed rolling circle amplification. Genome Res. *11*, 1095-1099.

Dean, F.B., Hosono, S., Fang, L., Wu, X., Faruqi, A.F., Bray-Ward, P., Sun, Z., Zong, Q., Du, Y., Du, J., *et al.* (2002). Comprehensive human genome amplification using multiple displacement amplification. Proc. Natl. Acad. Sci. USA *99*, 5261-5266.

Detter, J.C., Jett, J.M., Lucas, S.M., Dalin, E., Arellano, A.R., Wang, M., Nelson, J.R., Chapman, J., Lou, Y., Rokhsar, D., *et al.* (2002). Isothermal strand-displacement amplification applications for high- throughput genomics. Genomics *80*, 691-698.

Emmert-Buck, M.R., Bonner, R.F., Smith, P.D., Chuaqui, R.F., Zhuang, Z., Goldstein, S.R., Weiss, R.A., and Liotta, L.A. (1996). Laser capture microdissection. Science *274*, 998-1001.

Esteban, J.A., Salas, M., and Blanco, L. Fidelity of phi 29 DNA polymerase. (1993). Comparison between protein-primed initiation and DNA polymerization. J. Biol. Chem. *268*, 2719-2726.

Euhus, D.M., Cler, L., Shivapurkar, N., Milchgrub, S., Peters, G.N., Leitch, A.M., Heda, S., and Gazdar, A.F. (2002). Loss of heterozygosity in benign breast epithelium in relation to breast cancer risk. J. Natl. Cancer Inst. *94*, 858-860.

Garcia-Closas, M., Egan, K.M., Abruzzo, J., Newcomb, P.A., Titus-Ernstoff, L., Franklin, T., Bender, P.K., Beck, J.C., Le Marchand, L., Lum, A., *et al.* (2001). Collection of genomic DNA from adults in epidemiological studies by buccal cytobrush and mouthwash. Cancer Epidemiol. Biomarkers Prev. *10*, 6878-6896.

Gunter, E.W. (1997). Biological and environmental specimen banking at the Centers for Disease Control and Prevention. Chemosphere *34*, 1945-1953.

Hawkins, T.L., Detter, J.C., and Richardson, P.M. (2002). Whole genome amplification–applications and advances. Curr. Opin. Biotechnol. *13*, 65-67.

Holland, N.T., Smith, M.T., Eskenazi, B., and Bastaki, M. (2003). Biological sample collection and processing for molecular epidemiological studies. Mutation Res. *543*, 217-234.

Hosono, S., Faruqi, A.F., Dean, F.B., Du, Y., Sun, Z., Wu, X., Du, J., Kingsmore, S.F., Egholm, M., and Lasken, R.S. (2003). Unbiased whole-genome amplification directly from clinical samples. Genome Res. *13*, 954-964.

Landi, M.T. and Caporaso, N. (1997). Sample collection, processing and storage. IARC Sci. Publ. *142*, 223-236.

Lasken, R.S. and Egholm, M. (2003). Whole genome amplification: abundant supplies of DNA from precious samples or clinical specimens. Trends Biotechnol. *21*, 531-535.

Lee, T.H., Montalvo, L., Chrebtow, V., and Busch, M.P. (2001).Quantitation of genomic DNA in plasma and serum samples: higher concentrations of genomic DNA found in serum than in plasma. Transfusion *41*, 276-282.

Lench, N., Stanier, P., and Williamson, R. (1988). Simple non-invasive method to obtain DNA for gene analysis. Lancet *1*, 1356-1358.

Lizardi, P.M., Huang, X., Zhu, Z., Bray-Ward, P., Thomas, D.C., and Ward, D.C. (1998). Mutation detection and single-molecule counting using isothermal rolling-circle amplification. Nat. Genet. *19*, 225-232.

McEwen, J.E. and Reilly, P.R. (1994). Stored Guthrie cards as DNA "banks". Am. J. Hum. Genet. *55*, 196-200.

Molecular Staging Inc. (2003). *REPLI-g Whole Genome Amplification Kit: Instruction of Use.* New Haven, CT.

Nelson, J.R., Cai, Y.C., Giesler, T.L., Farchaus, J.W., Sundaram, S.T., Ortiz-Rivera, M., Hosta, L.P., Hewitt, P.L., Mamone, J.A., Palaniappan, C., and Fuller, C.W. (2002). TempliPhi, phi29 DNA polymerase based rolling circle amplification of templates for DNA sequencing. Biotechniques *Suppl.*, 44-47.

Pollock, P.M. and Meltzer, P.S. (2002). Lucky draw in the gene raffle. Nature *417*, 906-907.

Richards, B., Skoletsky, J., Shuber, A.P., Balfour, R., Stern, R.C., Dorkin, H.L., Parad, R.B., Witt, D., and Klinger, K.W. (1993). Multiplex PCR amplification from the CFTR gene using DNA prepared from buccal brushes/swabs. Hum. Mol. Genet. *2*, 159-163.

Saiki, R.K., Gelfand, D.H., Stoffel, S., Scharf, S.J., Higuchi, R., Horn, G.T., Mullis, K.B., and Erlich, H.A. (1988). Primer-directed enzymatic amplification of DNA with a thermostable DNA polymerase. Science *239*, 487-491.

Southern, E.M. (1975). Detection of specific sequences among DNA fragments separated by gel electrophoresis. J. Mol. Biol. *98*, 503-517.

Steinberg, K.K., Sanderlin, K.C., Ou, C.Y., Hannon, W.H., McQuillan, G.M., and Sampson, E.J. (1997). DNA banking in epidemiologic studies. Epidemiol. Rev. *19*, 156-162.

Steinberg, K., Beck, J., Nickerson, D., Garcia-Closas, M., Gallagher, M., Caggana, M., Reid, Y., Cosentino, M., Ji, J., Johnson, D., *et al.* (2002). DNA banking for epidemiologic studies: a review of current practices. Epidemiology *13*, 246-254.

Therrell, B.L., Hannon, W.H., Pass, K.A., Lorey, F., Brokopp, C., Eckman, J., Glass, M., Heidenreich, R., Kinney, S., Kling, S., *et al.* (1996). Guidelines for the retention, storage, and use of residual dried blood spot samples after newborn screening analysis: statement of the Council of Regional Networks for Genetic Services. Biochem. Mol. Med. *57*, 116-124.

Section 4

DNA Amplification in Detection of Non-DNA Analytes

4.1

Enhanced Protein Detection Using Real-Time Immuno-PCR

Michael Adler and Christof M. Niemeyer

Abstract

A method for the ultra-sensitive protein detection in the range of 10^{-20} to 10^{-14} M is described, using a combination of Immuno-PCR (IPCR) with real-time detection of PCR products as well as a combination of IPCR with Enzyme Linked Oligonucleotide Sorbent Assay (ELOSA). The antigen is immobilized on microplates and coupled with an antigen-specific primary antibody. In a second step, a commercially available DNA-labelled species-specific antibody is added, and finally the DNA-marker is amplified in PCR in the presence of a TaqMan probe for on-line detection. Alternatively, PCR is performed in the presence of a hapten-coupled nucleotide for subsequent PCR-ELOSA. In the PCR-ELOSA, the labelled PCR-product is immobilized by hybridization

Abbreviations:
mIgG, Mouse-IgG; **αm-rIgG,** Rabbit-anti-mouse IgG; **αr-gIgG(b),** Biotinylated goat-anti-rabbit IgG; **AP,** Alkaline phosphatase; **ELISA,** Enzyme linked immuno sorbent assay; **ELOSA,** Enzyme-linked oligonucleotide sorbent assay; **IPCR,** Immuno-PCR; **HP,** Horseredish peroxidase; **PCR,** Polymerase chain reaction; **rtPCR,** Real-time PCR; **rtIPCR,** Real-time immuno-PCR; **RSR,** "Anti-rabbit IgG secondary reagent", goat-anti rabbit-IgG coupled with DNA; **STV,** Streptavidin.

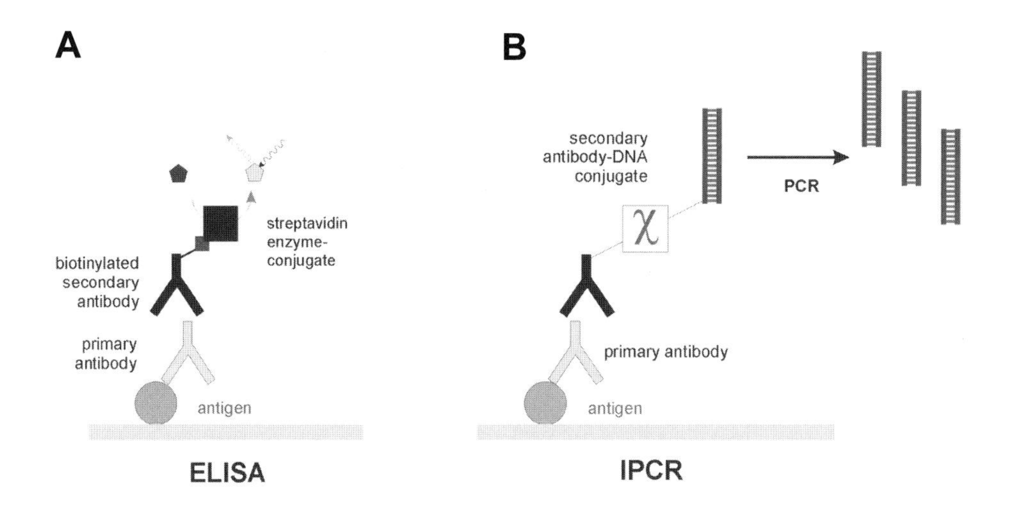

Figure 1. Schematic outline of the indirect ELISA (A) and IPCR (B).
A: The antigen mouse-IgG is immobilized on a microplate surface, coupled with an antigen-specific primary rabbit-anti-mouse antibody and subsequently with a biotinylated species-specific secondary goat-anti-rabbit antibody. The detection is carried out with a streptavidin-enzyme conjugate, which binds to the biotinylated antibody and changes a colourless substrate into a coloured product. The assay could be used for the detection of either the immobilized antigen or the primary antibody. The protocol given in the text describes an assay for the detection of the antigen.
B: The assay is carried out similar to the previously described ELISA, but instead of a secondary species-specific biotinylated antibody, a secondary antibody-DNA conjugate is used for detection. This polyvalent chimeric DNA-protein conjugate (designated as χ) contains a multiplicity of antibodies and DNA-marker. During PCR amplification of the DNA-marker, biotin- and hapten-labels may be incorporated in the amplicon.

to capture oligonucleotide-coated microplates and detected with an antibody-enzyme conjugate. The protocol could easily be generalized and adapted for the detection of other antibodies or antigens by using corresponding antigen and the antigen-specific primary antibody.

Background

The polymerase chain reaction (PCR, Mullis *et al.*, 1987) is a powerful tool for detection of nucleic acids at the levels far below those accessible for other biomolecules. The combination of the exponential amplification power of PCR with antibody-based immunoassays, such as the Enzyme Linked Immuno Sorbent Assay (ELISA, *e.g.* Crowther 1995), allows for detection of proteins at a level of a few hundred molecules (Figure 1). Immuno-PCR (IPCR), first introduced in 1992 by Sano *et al.*, (Sano *et al.*, 1992), is a very demanding technique due to its enormous sensitivity. Therefore, it is not yet broadly applied in routine clinical assays, despite the fact that there is a huge interest in ultra-sensitive protein detection methods. Nonetheless, successful applications of the modified "Universal-IPCR" protocol introduced by Zhou (Zhou *et al.*, 1993), have been reported in research studies (Furuya *et al.*, 2001; Komatsu *et al.*, 2001; Ozaki *et al.*, 2001; Ren *et al.*, 2001; Wu *et al.*, 2001; Adler *et al.*, 2003a; Liang *et al.*, 2003). Further advances in the IPCR, especially by improving the antibody-DNA coupling (Niemeyer *et al.*, 1999b) and by developing commercially available pre-conjugated DNA-antibody reagents (Adler *et al.*, 2003a,b), nowadays allow even an inexperienced researcher to successfully use this technique. Additionally, the introduction of multiplex-IPCR for simultaneous detection of several antigens (Joerger *et al.*, 1995), *in situ* IPCR (Cao 2002), competetive IPCR for small molecules (Niemeyer *et al.*, 2001) or DDI-IPCR (Niemeyer *et al.*, 2003) has substantially broadened the application area of this method.

The development of real-time PCR (Heid *et al.*, 1996) has led to a breakthrough in the PCR-based analyses. The on-line amplicon detection with fluorescent hybridization probes provides reliable quantitative data in each amplification experiment. The power of this technology is impressively demonstrated by the rapidly increasing number of published real-time PCR applications (Jung *et al.*, 2000; Walker 2001). This chapter describes an adaptation of the real-time PCR to the IPCR (Figure 2A; Sims *et al.*, 2000; Adler *et al.*, 2003b) as well as "classical" combination of IPCR with a PCR-ELOSA (Figure 2B; Maia *et al.*, 1995; Niemeyer *et al.*, 1997) and discusses advantages and drawbacks of both methods.

As an example we report here on the detection of mouse IgG in an indirect IPCR-assay, as well as on the use of the PCR-ELOSA for quantitation of PCR amplicons. The IPCR method can easily be adapted for the detection of other antigenes, while the PCR-ELOSA technique

A

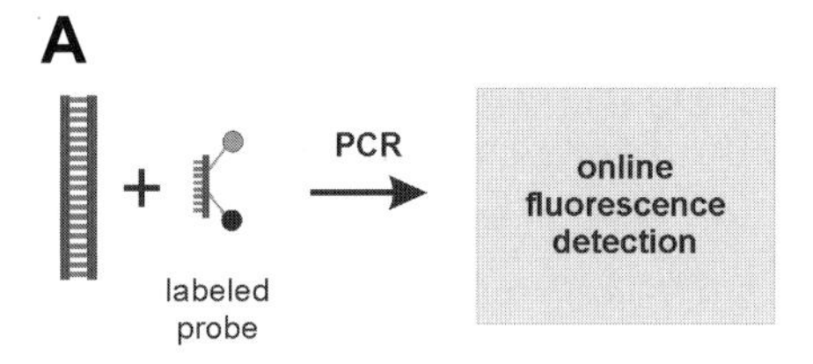

B

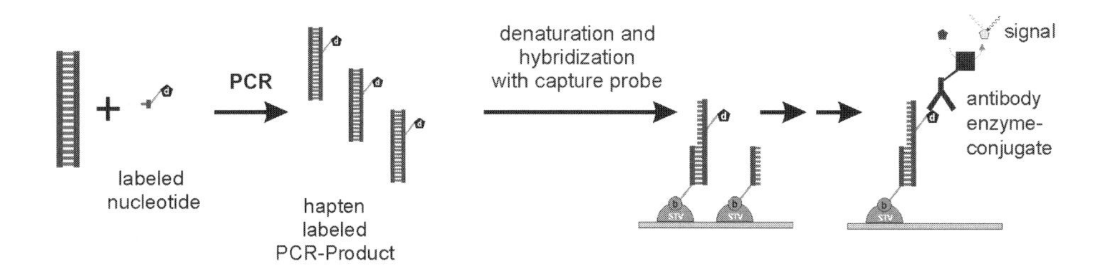

Figure 2. Comparison of real-time (A) and PCR-ELOSA (B) detection methods.
A: An oligonucleotide TaqMan probe with a fluorescent dye and a quencher is a part of the PCR-mix. During PCR amplification, the fluorescent dye is set free and the increase in fluorescence with increasing amounts of DNA amplicon is detected on-line (see also Figure 3).
B: Prior to PCR, digoxigenin-dUTP is added to the PCR amplification mix. The hapten-labeled amplicon is then denaturated by heat and immobilized on a streptavidin-coated microplate by hybridization with biotinylated capture-oligonucleotides. Subsequently, amplicon is coupled with an antibody-enzyme conjugate and detected by the enzymatic reaction.

might be combined with standard PCR assays for quantitative detection of nucleic acids as well.

Protocol

We present the IPCR protocol consisting of four steps: in step 1 microplates needed for the assay are prepared, in step 2 they are coupled with immuno-reagents containing DNA marker, and in step 3 DNA marker is PCR amplified. PCR is performed either in real-time setup (step 3a), or conventional PCR is followed by ELOSA (step

3b), or conventional ELISA is performed without PCR amplification (step 3c). The last step 4 includes quantitative analysis of the data.

Materials

For performing real-time immuno-PCR we suggest using an ABIPrism 7000 thermocycler (Appled Biosystems), however other thermocyclers able to detect fluorescence (FAM and/or VIC labels) can also be used. Additionally the following devices are needed: digital pipettes 1-10, 10-100 and 100-1000 μL (Eppendorf), multi-channel pipettes 5-50 μL and 50-300 μL (Finnpipette), and orbital shaker for microtiter plates (Heidolph). For PCR-ELOSA, a PCR thermocycler with heated lid compatible with microtiter plate format (*e.g.* MWG, MJ Research, PerkinElmer) and a microtiter plate reader for fluorescence detection (*e.g.* Victor, PerkinElmer) should be used.

For IPCR amplification, microplate surfaces should combine protein binding ability and compatibility with PCR thermocyclers. Polypropylen, a common material for PCR tubes, has insufficient protein binding capacity, whereas most ELISA plates are not compatible with PCR thermocyclers. We found that TopYield™ microplates (TopYield™ starter kit, Nunc) is the material of choice for IPCR (Niemeyer *et al.*, 1999a).

Siliconized tubes (1.5 mL., *e.g.* Biozym, Starlab) are needed for sample dilutions. We strongly suggest using pipette tips with filters (*e.g.* from Eppendorf, or Biozym) to avoid cross-contamination. The following consumables are also needed for PCR-ELOSA: polypropylen microtiter plates (Nunc); black polystyrol microtiter plates (*e.g.* FluoroNunc™ MaxiSorp™ Nunc); 8-well 300 μl polypropylen strip-tubes with caps (*e.g.* Biozym, Starlab) and adhesive foil (Dynex). For real-time PCR, optical adhesive covers (Applied Biosystems) are needed, too.

The following reagents and buffer solutions are needed: borate buffer for immobilization of the antigen (50 mM, pH 9,5); mouse-IgG as antigen (Sigma); rabbit-anti-mouse IgG as primary antibody

(Sigma); biotinylated goat-anti-rabbit IgG for control ELISA (Sigma; note that other providers for these commercially available antibodies are possible). The protocol as described is not limited to the model system employing mouse IgG. By simply choosing other available anti-species antibody-DNA conjugates, other IgGs, *e.g.* from rabbit, human or goat could be detected using the same robust protocol in combination with anti-rabbit, anti-human or anti-goat reagents, respectively (all available from Chimera Biotec GmbH). Several other components of the assay can be purchased from Chimera Biotec: conjugate dilution buffer; assay buffer A, assay buffer B, blocking solution; amplification buffer, primer mix, PCR supplement mix, real-time IPCR accessory set and/or IPCR-ELISA accessory set. Taq DNA polymerase including 10x buffer and PCR-enhancer stock solution were from Molzym GmbH. For ELISA and ELOSA, digoxigenin dUTP, anti-digoxigenin alkaline phosphatase conjugate, attophos® substrate, recombinant streptavidin, and streptavidin-alkaline phosphatase conjugate can be purchased from Roche. We used biotin from Sigma; use of avidin is not recommended due to potential high non-specific binding.

Step 1: Coating of Microtiter Plate Surfaces

Prior to assay, Top-Yield™ modules are coated with the antigen for IPCR and black MaxiSorp™ modules are coated with STV and capture-oligonucleotide for the PCR-ELOSA.

Coating with Antigen for IPCR

Immobilize a 10-fold dilution series of mIgG in borate buffer ranging from 0.01 to 10,000 amol/30μL/well (e. g., 50 fg/mL to 50 ng/mL) on TopYield microplates (see Table 1). Generally, whenever possible, novel IPCR-assays should be performed in triplicates. Well established and validated assays could be carried out in duplicates. Include one well containing borate buffer as a negative control. Also prepare 3 "spiking" samples with 0.3, 30 and 300 amol/30μL of antigen. All dilution steps should be carried out in siliconized tubes and prepared

Table 1. Typical IPCR setup. Row 1 and 2, duplicates of "known samples" serial dilutions for generation of the calibration curve; row 3, "unknown" samples in duplicates. It is recommended to leave two wells blank during the complete assay for a PCR no-substrate control ("PCR-NC") to test possible contamination of the PCR reagents.

	A	B	C	D	E	F	G	H
1	0.01 amol	0.1 amol	1 amol	10 amol	100 amol	1000 amol	NC IPCR	10,000 amol
2	0.01 amol	0.1 amol	1 amol	10 amol	100 amol	1000 amol	NC IPCR	10,000 amol
3	0.3 amol	30 amol	300 amol	NC PCR	0.3 amol	30 amol	300 amol	NC PCR

in duplicates. Incubate microplates at 4°C for 12-48 h. Subsequently, wash the IgG-coated modules three times for three minutes each with 240μL/well of buffer A. Incubate microplates at 4°C with 240μL/well of blocking solution for at least 12 h. The blocked modules are stable at 4°C for about one week.

Coating with STV and Capture-Oligonucleotide for PCR-ELOSA

For PCR-ELOSA, immobilize 50μL/well of 10 μg/mL solution of STV on black MaxiSorp™ F16 modules at 4°C for 72 h for homogenous coating. A shorter incubation time (at least 12h at 4°C) is possible but it results in the increase of the well-to-well error. Subsequently, wash the STV-coated plates three times for three minutes each with 240 μL/well of buffer A. Then incubate the modules with 150 μL/well of blocking solution at 4°C for at least 12 h. The blocked modules could be sealed with adhesive foil and stored for about 4 weeks at 4°C. As an alternative to in-house prepared STV-coated plates, ready-to use STV-coated microplates provided by several suppliers can be used. Using STV is favored as compared to avidin because of the generally lower background of the former.

Dilute the biotinylated capture-oligonucleotide 1:500 in buffer B according to the manufacturer's instructions (final concentration of about 1 pmol/μl). Wash the STV-coated modules with 240 μL/well of buffer B two times for 30 sec and two times for 4 min under orbital shaking, and incubate with 50 μl/well of the capture-oligonucleotide solution for 1h at RT under orbital shaking. Prepare a 800 μM solution of biotin in buffer B. Wash the modules with 240 μL/well of buffer B/biotin solution twice for 30 sec and once for 4 min. Use the oligonucleotide –coated modules either immediately (for PCR-ELOSA see step 3b), or add 240 μl/well of buffer B/biotin containing 10% blocking solution and store the plates sealed with adhesive foil at 4°C for up to 48 h.

Step 2: Coupling with Antibody-DNA Conjugate

Wash the antigen-coated modules with 240 μL/well of buffer B two times for 30 sec and two times for 4 min under orbital shaking. Incubate with 30 μg/ml solution of αm-rIgG in buffer B at room temperature for 25 min under orbital shaking. Then wash four times with buffer B as described above. Add 30 μL/well of RSR diluted 1:300 for IPCR-ELOSA or 1:30 for real-time IPCR in conjugate dilution buffer. Incubate again for 25 min as described above. Wash seven times with buffer B (4 x 30 sec, 3 x 3 min) and two times for 1 min with buffer A.

Step 3a: PCR with Real-Time Detection

Prepare a real-time PCR-mix according to the recipe given in Table 2 (for larger amount of wells use multiples of the volumes given in Table 2). Preparation of the PCR mix should be performed during the last washing steps. The fluorescent probe should be protected from light to avoid photobleaching. It is not recommended to often thaw and freeze the probe. Dispense 30 μL of the PCR master-mix in each well. Seal the wells with light-permeable optical adhesive covers and place them in the thermocycler. Due to the multiple washing steps in the protocol, it is likely that the bottom of the modules is

Table 2. PCR-mastermix components calculated for eight wells, 30 μl vol

	Real-time IPCR	PCR-ELOSA
PCR supplement mix	200 μL	214 μL
10x Buffer	25 μl	25 μL
PCR-Enhancer	2.5 μl	2,5 μL
Dig-dUTP (0.3 mM)	-	2 μL
Primer Mix (100 μM each)	5 μl	5 μL
Taq polymerase	1.25 μl	1.25 μL
Real-time supplement mix	16 μl	-
H_20 to final volume	250 μL	250μL

contaminated by dust or powder from the gloves, which may result in unspecific fluorescence signals. Therefore use powder-free gloves for the final washing steps and clean the bottom of the modules carefully with precision wipes (*e.g.* Kimberley-Clark) before placing them into the thermocycler. Use the programme given in Table 3 for performing PCR.

Step 3b. PCR-ELOSA

In PCR-ELOSA, PCR amplicon is labelled with a hapten-coupled nucleotide. The labelled amplicon is immobilized on microtiter plate and coupled with an antibody-enzyme conjugate. Subsequently, the detection is carried out using a fluorescence-generating substrate.

PCR

Prepare PCR master-mix according to the recipe in Table 2, calculated for 8 wells. The preparation of the PCR mix should be performed either during the last washing step or in advance. In the last case store aliquots frozen. Pipette 30 μL of the PCR master-mix in each well. Seal the wells with adhesive foil. Carry out a PCR according to the

Table 3. IPCR programmes.

		Real-Time IPCR	PCR-ELOSA
Time	Temperature	Repeat number	Repeat number
5 min	95°C	1 x	1 x
30 sec	50°C		
30 sec	72°C	40 x	28 x
12 sec	95°C		
5 min	50°C	-	1 x
5 min	72°C	-	1 x

programme in Table 3. After completion of PCR, add 30 μL of sample storage buffer (Chimera biotec, part of the IPCR-ELISA acessory) to each well. Transfer the product in a polypropylen microplate. The plate could be sealed with adhesive foil and stored at 4°C for up to four weeks.

ELOSA-Detection of the Amplicon

Dilute the PCR amplicon 1:40 by adding 6 μL of amplicon to 234 μL of buffer B containing 800 μM biotin. The dilution should be carried out in 300 μl polypropylen strip-tubes, using a multichannel pipette. Close the tubes with caps and denature DNA for 10 min at 96°C followed by 10 min on ice. Apply 50 μL of the diluted and denatured amplicon to each well of a capture oligonucleotide-coated microplate (see step 1) in duplicate. While handling the PCR amplicon, take care not to contaminate your workplace and the pipette. Avoid aerosol formation due to hectic pipetting and use tips with filters for washing and pipetting. Incubate the microplate at room temperature for 45 min under orbital shaking. Wash four times with buffer B as described above. Add 50 μL/well of a 1:5000 dilution of anti-digoxigenin - alkaline phosphatase conjugate in buffer A. Incubate at room temperature for 45 min under orbital shaking. Wash four times with buffer B as described above and three times for 1 min with buffer A. Add 50 μL of AttoPhos™ substrate to each well. Incubate at room temperature for 15 min under orbital shaking. Measure fluorescence at 550 nm. If the signals obtained after 15 min incubation are too low, re-measure fluorescence in 10 min intervals and/or incubate the plates at 37°C. If no fluorescence reader is available, a photometric substrate could be used as an alternative to the fluorescence-generating substrate AttoPhos™. For photometric quantitation, using *e.g. p*-nitrophenylphosphate (pNpp), measurements after 45 min incubation at 37°C are recommended. In this case it is necessary to use transparent microplates for the assay. For comparison of different substrates and protocols see (Niemeyer *et al.*, 1997).

In contrast to the real-time IPCR, IPCR/PCR-ELOSA allows only for endpoint-detection of the PCR amplicon. If too high or too low signals were obtained, the number of cycles during PCR could be adjusted correspondingly. It is preferable to use a lower number of cycles to avoid the increase of the non-specific background.

Step 3c: Control ELISA

Parallel to the IPCR, a conventional ELISA is carried out as a control experiment (see Figure 1a).

Perform immobilization of the antigen and coupling with the primary antibody as described above. After 25 min incubation at room temperature under orbital shaking wash the wells four times with buffer B as described above. Add 30μl/well of 5 nM solution of "αr-gIgG(b)" in conjugate dilution buffer. Incubate again for 25 min as described above. Wash four times with buffer B as described above. Add 30 μl/well streptavidin-alkaline phosphatase conjugate in 1:5,000 dilution in reagent dilution buffer. Incubate again for 25 min as described above. Wash again four times with buffer B as described above and three times for 1 min with buffer A. Add 50 μL of AttoPhos™ substrate to each well. Incubate for 15 min at room temperature under orbital shaking. Measure fluorescence intensity at 550 nm as described in step 3b.

Step 4: Quantitative Analysis of the Data

Real-Time IPCR

The real-time PCR-cycler records the increase of the normalized fluorescence signal (ΔRn) for each cycle. Subsequent to the run, apply a baseline correction and use typically the fluorescence during the first 10-15 cycles as a background signal. The software calculates the threshold cycle (Ct), which represents the first PCR cycle at which the reporter signal exceeds the signal of the baseline ("threshold") determined by the user (see Figure 3). Determine this threshold level by choosing a ΔRn value above the background and set it in the phase where signal increases linearly. Typical threshold values vary between 70 and 150 ΔRn. Choose one threshold and baseline value for all signals to be compared. To render an easy comparison of data obtained from real-time IPCR and conventional IPCR/ELOSA or ELISA, the problem has to be circumvented that Ct values are inversely proportional to antigen concentrations while ELISA signals are directly proportional to antigen concentrations (NC has the smallest numerical value). Therefore, ΔCt values were calculated by subtracting the Ct values obtained for each signal from the total number of cycles carried out in the experiment. As shown in Figure 3, for Ct values of 26.7 and 32.7 and a maximum cycle number of 40, ΔCt values of 13.3 and 7.3 were obtained, respectively. This simple mathematical conversion facilitates the comparison of the data. Alternatively, 1/Ct could be used. Calculate average values and standard deviations of the ΔCt for each experiment performed in duplicate.

Plot ΔCt for control samples against the log concentration and carry out linear regression. The resulting equation will be used for determination of the concentrations of the three "unknown" spiking samples. To evaluate reliability of this quantitation, the recovery rate of the spiked samples is calculated. Typically, recovery rates of 90-100% were obtained. If the recovery rate is lower than 90%, the assay should be checked for systematic errors such as antigen loss during sample handling. We recommend including the recovery rate control by spiked samples in each new quantitation experiment with unknown samples and an alteration of the regression curve.

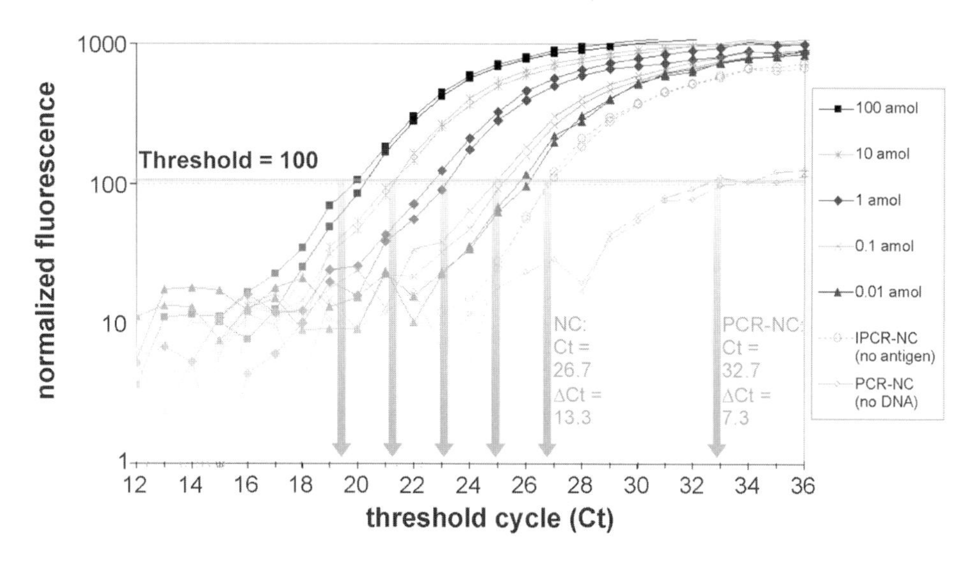

Figure 3. Real-time IPCR, determination of ΔCt. Mouse-IgG immobilized on a microplate was detected with an anti-mouse IgG-DNA conjugate. The fluorescence intensities (Rn) were plotted against the cycle number and normalized against a baseline using fluorescence in the first 15 cycles. Then a uniform threshold was set at the Rn =100 in the linear amplification range. With a higher threshold, the PCR negative control (PCR-NC) could be excluded completely from the Ct values. The Ct value was determined as the intersection of the threshold and the signal curves. Note excellent reproducibility of the method because nearly identical curves were obtained in duplicate experiments. The average standard deviation of the duplicates is 0.7 +/- 0.4 %. The average of the duplicate Ct values was calculated for each protein concentration and subtracted from the total number of cycles (40) to calculate the ΔCt value, proportional to the concentration of the protein in the sample.

In each IPCR experiment, two standard negative controls should be included: an "IPCR-NC" negative control without the antigen and a "PCR-NC" negative control with PCR mix but without DNA in a separate well.

In contrast to typical PCR assays, in the IPCR the DNA to be amplified is added to each well, even to the standard IPCR-NC negative control, as part of the antibody-DNA conjugate. Therefore, even with excessive washing steps, it is not possible to completely remove unspecific bound DNA from each well. Because of this, a high background signal for the "IPCR-NC" without antigen is typical for the IPCR-method and one of the major obstacles in most applications of the IPCR. In contrast, the "PCR-NC" represents the purity of the PCR-mix without DNA.

By comparison of the "IPCR-NC" and the "PCR-NC" it is possible to estimate the amount of nonspecific binding. The systematic difference between the "PCR-NC" and the "IPCR-NC" is also obvious in Figure 3. While it is possible to completely eliminate the signal of the "PCR-NC" with a slightly increased threshold value, the curve of the "IPCR-NC" is similar to the signal curves obtained for samples containing antigens. It is also observed that for very low amounts of the antigen, the distance between the Ct-values is decreasing. For the discrimination between the negative control without antigen and the detection limit of the assay, therefore the standard deviation of the values is important. If an average Ct value minus its standard deviation is higher than the negative control plus its double standard deviation, the signal is positive. Very low standard deviations (see Figure 3) hence allow high sensitivity of the IPCR due to clear difference in signals from low concentration of analyte and the "IPCR-NC".

If background in the "PCR-NC" is high, all reagents, especially the PCR-mix should be substituted with news samples. For lowering the background of unspecific binding in "IPCR-NC", the concentrations of the primary antibodies and/or of the species-specific antibody-DNA conjugate could be lowered in 3-fold increments.

PCR-ELOSA

Prepare duplicate ELOSA samples from each PCR product (which was done in duplicate too), measure fluorescence intensities, and calculate average and standard deviation. To compare different experiments normalize the average intensities by calculating the ratio of signal to IPCR-NC background for each sample. Plot normalized values against the log of spiked antibody concentration and carry out a linear regression. The resulting equation will be used for the determination of the concentrations of the "unknown" samples as described above.

Example : Detection of Mouse IgG with a Secondary Antibody in Ascites Fluid

We applied real-time IPCR for detection of the solid phase-immobilized mouse IgG with anti-mouse IgG-DNA conjugate (Figure 3) and a combination of rabbit-anti-mouse IgG and anti-rabbit IgG-DNA conjugate (Figure 4). In this example we used a commercially available non-purified anti-mouse IgG in rabbit ascite fluid as a primary antibody. Serial dilutions of mIgG and three "unknown" samples were immobilized on the microplate surface and subsequently coupled with the rabbit anti-mouse IgG and the anti-rabbit-DNA conjugate for IPCR. The detection limit was defined as a minimal amount of protein giving rise to a signal with standard deviations not overlapping with the standard deviations of the negative control. In a parallel experiment, a control ELISA with alkaline phosphatase was carried out as described in Step 3c.

The comparison of rtIPCR, IPCR/ELOSA and ELISA shown in Figure 4 revealed a detection limit of 0.1 amol of mouse-IgG for both IPCR-based methods and 100 amol IgG for the control-ELISA, which means that a 1,000-fold increase in sensitivity has been achieved. The linear regression of the log concentration against the ΔCt revealed a correlation coefficient of 0.99. Using this regression for the quantitation of three "unknown" spiked samples, we obtained an average recovery rate 98%. A double logarithmic linear regression of the IPCR/ELOSA signals gives a correlation coefficient of 0.98, with the average recovery rate being 92%. Thus, both IPCR-based methods allow 1000-fold increase in sensitivity as compared with conventional ELISA and robust and reliable quantitation of the protein analyte.

Discussion

The sensitivity of the IPCR-based methods is 1,000-fold higher than sensitivity of conventional ELISA. The comparison between direct and indirect assays shows that increase in the complexity of the assay leads to the decrease of the IPCR absolute detection limit

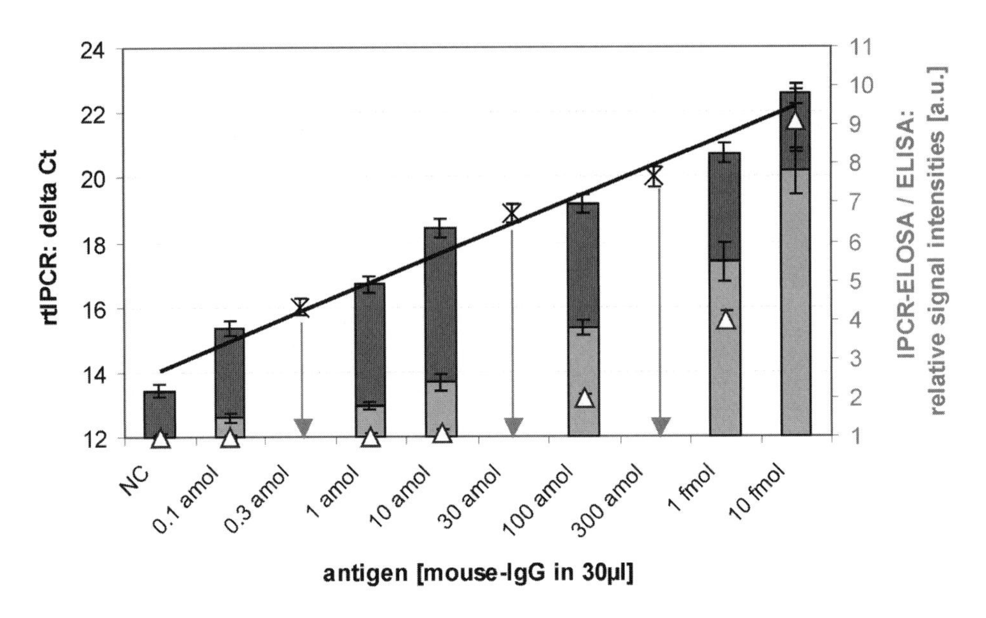

Figure 4. Comparison of rtIPCR, IPCR/ELOSA and conventional ELISA. Mouse-IgG immobilized on a microplate was detected with rabbit-anti-mouse IgG and an anti-rabbit IgG-DNA conjugate. For experimental details see text. Calibration curve obtained in rtIPCR (dark grey columns and black line) was used for concentration determination of the "unknown" samples (crosses). For comparison, IPCR/ELOSA results (light grey columns) and conventional ELISA results (triangles) are shown.

(Figures 3 and 4). However, note that the sensitivity of the control-ELISA is also decreased; therefore the overall 1000-fold sensitivity enhancement of IPCR is maintained. It has been repeatedly proven (Niemeyer *et al.*, 1997; Adler *et al.*, 2003a; Sugawara *et al.*, 2000) that the detection limit of the IPCR is strongly dependent on the performance of the antibodies. Generally, polyclonal antibodies show superior performance compared to monoclonal antibodies, so that, whenever possible, polyclonal antibodies have to be chosen for analysis. Especially in sandwich-IPCR, the use of the same polyclonal antibodies for capture and detection is appropriate for the reduction of the non-specific binding.

The main difference between IPCR and PCR is the additional ELISA-based antigen detection step prior to the signal amplification with

PCR. Generally speaking, the enzyme used for signal amplification in ELISA has been substituted with a DNA-marker in IPCR. When the PCR-step is performed, the protein analyte is already purified in prior antibody-antigen binding steps and subsequent washings. PCR conditions for DNA-marker amplification are optimized in advance. Thus, it is unlikely that the PCR step can fail. However, there may be another trouble with the PCR step. Any ultra-sensitive method is prone to potential errors, such as false positives. Therefore, a successful routine application of the IPCR is possible only with standard reagents, simple and robust protocols and avoidance of unnecessary and potentially error-generating steps. Thus using commercially available DNA-antibody conjugates greatly reduces the number of steps involved and potential error sources.

After the decision is made to use IPCR, a suitable detection technique for analyzing the DNA amplicons has to be chosen. In other words, the choice between real-time detection and ELOSA has to be made (Figure 2). The sensitivities of both methods are usually comparable, which has been shown in a number of experiments. We however, observed that the average standard deviations in IPCR using real-time detection are typically lower (CV < 5%) as compared to the combination of IPCR and ELOSA detection (typically CV > 10%). Additionally, performing PCR-ELOSA requires an additional 4 hours following PCR amplification, while when using real-time detection, the quantitation is completed right after the PCR step. For performing PCR-ELOSA, a lot of additional materials and reagents are needed, such as functionalised nucleotides which are incorporated in the PCR amplicon, capture oligonucleotide-coated microplates and an antibody-enzyme conjugate with substrate for detection. This translates into additional steps and controls which are necessary for quality control of ELOSA assay. Compared to this, real-time IPCR needs only a fluorescent probe in the PCR-mix.

From the other hand, the more work-intensive IPCR/PCR-ELOSA assay requires less antibody-DNA conjugate than real-time IPCR assay (typically dilution 1:300 versus 1:30).

Another potential drawback of the real-time assay is the more complex acquisition of data and data analyses. While PCR-ELOSA gives fluorescent intensities which could directly be used for establishing a standard curve for quantitation, real-time PCR needs a baseline correction and threshold estimation for signal interpretation.

In conclusion, if you have the facilities to apply both methods and the experience in real-time PCR, it is preferable to use the real-time approach as it combines a broad range of advantages. The ELOSA is suggested for all users who have experience and equipment for ELISA applications.

The combination of ELISA and PCR makes IPCR attractive: it combines all the advantages of the well established ELISA technique with the amplification power of the PCR, but without all the typical difficulties of PCR because of amplification of DNA marker under optimised conditions.

An important point beneficial for laboratories engaged in routine diagnostics is that the dramatic increase in sensitivity obtained with the rtIPCR, is not associated with increased handling time and thus, with higher costs for personnel. Hands-on time in the rtIPCR is comparable to the analogous ELISA. A further reduction in assay steps is also possible in combination with novel self assembly-based assay formats, such as the DNA-directed immobilization (DDI) (Niemeyer et al., 2003). The DDI-IPCR is carried out on capture oligonucleotide-coated microtiter plates and allows incubating oligonucleotide-linked capture antibodies, antigen and antibody-DNA conjugates in a single step. Availability of commercially available kits and ready-to-use reagents makes IPCR fully compatible with most of the common ELISA applications, ranging from proteome research to biomedical and clinical diagnostics.

Acknowledgements

This work was supported by CHIMERA BIOTEC GmbH, Dortmund, Germany.

References

Adler, M., Langer, M., Witthohn, K., Eck, J., Blohm, D., and Niemeyer, C.M. (2003a). Detection of rViscumin in plasma samples by immuno-PCR. Biochem. Biophys. Res. Commun. *300*, 757-63.

Adler, M., Wacker, R., and Niemeyer, C.M. (2003b). A real-time immuno-PCR assay for routine ultrasensitive quantification of proteins. Biochem. Biophys. Res. Commun. *308*, 240-50.

Cao, Y. (2002). *In situ* immuno-PCR. A newly developed method for highly sensitive antigen detection *in situ*. Meth. Mol. Biol. *193*, 191-6.

Crowther, J.R. (1995). ELISA; Theory and Practice. Totowa, New Jersey, Humana Press Inc.

Furuya, D., Kaneko, R., Yagihashi, A., Endoh, T., Yajima, T., Kobayashi, D., Yano, K., Tsuda, E., and Watanabe, N. (2001). Immuno-PCR assay for homodimeric osteoprotegerin. Clin. Chem. *47*, 1475-7.

Heid, C.A., Stevens, J., Livak, K.J., and Williams, P.M. (1996). Real time quantitative PCR. Genome Res. *6*, 986-994.

Joerger, R.D., Truby, T.M., Hendrickson, E.R., Young, R.M., and Ebersole, R.C. (1995). Analyte detection with DNA-labeled antibodies and polymerase chain reaction. Clin. Chem. *41*, 1371-7.

Jung, R., Soondrum, K., and Neumaier, M. (2000). Quantitative PCR. Clin. Chem. Lab. Med. *38*, 833-836.

Komatsu, M., Kobayashi, D., Saito, K., Furuya, D., Yagihashi, A., Araake, H., Tsuji, N., Sakamaki, S., Niitsu, Y., and Watanabe, N. (2001). Tumor necrosis factor-alpha in serum of patients with inflammatory bowel disease as measured by a highly sensitive immuno-PCR. Clin. Chem. *47*, 1297-301.

Liang, H., Cordova, S.E., Kieft, T.L., and Rogelj, S. (2003). A highly sensitive immuno-PCR assay for detecting Group A Streptococcus. J. Immunol. Meth. *279*, 101-10.

Maia, M., Takahashi, H., Adler, K., Garlick, R.K., and Wands, J.R. (1995). Development of a two-site immuno-PCR assay for hepatitis B surface antigen. J. Virol. Meth. *52*, 273-86.

Mullis, K.B., and Faloona, F. (1987). Specific synthesis of DNA *in vitro* via a polymerase-catalyzed chain reaction. Methods Enzymol *155*, 335-350.

Niemeyer, C.M., Adler, M., and Blohm, D. (1997). Fluorometric Polymerase Chain Reaction (PCR) Enzyme-Linked Immunosorbent Assay for Quantification of Immuno-PCR Products in Microplates. Anal. Biochem. *246*, 140-5.

Niemeyer, C.M., Adler, M., and Blohm, D. (1999a). High Sensitivity Detection of Antigens using Immuno-PCR. NUNC Tech Note *5*, 35.

Niemeyer, C.M., Adler, M., Pignataro, B., Lenhert, S., Gao, S., Chi, L., Fuchs, H., and Blohm, D. (1999b). Self-assembly of DNA-streptavidin nanostructures and their use as reagents in immuno-PCR. Nucleic Acids Res. *27*, 4553-61.

Niemeyer, C.M., Wacker, R., and Adler, M. (2001). Hapten-functionalised DNA-streptavidin nanocircles as supramolecular reagents in a novel competitive immuno-PCR assay. Angew. Chem. Int. Ed. *40*, 3169-3172.

Niemeyer, C.M., Wacker, R., and Adler, M. (2003). Combination of DNA-directed immobilization and immuno-PCR: very sensitive antigen detection by means of self-assembled DNA-protein conjugates. Nucleic Acids Res. *31*, e90.

Ozaki, H., Sugita, S., and Kida, H. (2001). A rapid and highly sensitive method for diagnosis of equine influenza by antigen detection using immuno-PCR. Jpn. J. Vet. Res. *48*, 187-95.

Ren, J., Ge, L., Li, Y., Bai, J., Liu, W.C. and Si, X.M. (2001). Detection of circulating CEA molecules in human sera and leukopheresis of peripheral blood stem cells with E. coli expressed bispecific CEAScFv-streptavidin fusion protein-based immuno-PCR technique. Ann. N.Y. Acad. Sci. *945*, 116-8.

Sano, T., Smith, C.L., and Cantor, C.R. (1992). Immuno-PCR: very sensitive antigen detection by means of specific antibody-DNA conjugates. Science *258*, 120-2.

Sims, P.W., Vasser, M., Wong, W.L., Williams, P.M., and Meng, Y.G. (2000). Immunopolymerase chain reaction using real-time polymerase chain reaction for detection. Anal. Biochem. *281*, 230-2.

Sugawara, K., Kobayashi, D., Saito, K., Furuya, D., Araake, H., Yagihashi, A., Yajima, T., Hosoda, K., Kamimura, T., and Watanabe, N. (2000). A highly sensitive immuno-polymerase chain reaction assay for human angiotensinogen using the identical first and second polyclonal antibodies. Clin. Chim. Acta. *299*, 45-54.

Walker, J.N. (2001). Real-time and Quantitative PCR: Applications to Mechanism-Based Toxicology. J. Biochem. Mol. Toxicol. *15*, 121-126.

Wu, H.C., Huang, Y.L., Lai, S.C., Huang, Y.Y., and Shaio, M.F. (2001). Detection of Clostridium botulinum neurotoxin type A using immuno-PCR. Lett. Appl. Microbiol. *32*, 321-5.

Zhou, H., Fisher, R.J., and Papas, T.S. (1993). Universal immuno-PCR for ultra-sensitive target protein detection. Nucleic Acids Res. *21*, 6038-9.

4.2

Rolling Circle Amplification in Multiplex Immunoassays

Michael C. Mullenix, Richard S. Dondero, Hirock D. Datta, Michael Egholm, Stephen F. Kingsmore and Lorah T. Perlee

Abstract

Rolling circle amplification (RCA) has been applied to immunoassays (immuno-RCA) in several different formats resulting in significantly enhanced sensitivity. RCA is incorporated in such diagnostics by crosslinking a primer oligonucleotide to the detector antibody of the immunoassay. The primer serves as a docking site for the hybridization of a complementary single-stranded (ss) circular DNA molecule. In the presence of nucleotide triphosphates and a strand-displacing DNA polymerase, the primers are extended to produce long ssDNA transcripts complementary to the circular DNA sequence. These transcripts can be detected by several methods including labeled oligonucleotide probes or direct incorporation of hapten-labeled nucleotides. In immuno-RCA, the ssDNA product remains covalently attached to the detector antibody. This covalent attachment facilitates spatial separation of the amplified signal required in multiplex immunoassay formats, such as flow cytometric beads and microarrays. Here, we present a detailed description of the use of immuno-RCA in flow cytometric bead assays and suggested approaches to applying RCA to immunoassays in other formats.

Background

Antibody-based detection systems for specific antigens are versatile and powerful tools for various molecular and cellular analyses as well as clinical diagnostics (Gosling, 1990). The power of such systems originates from the inherent specificity of antibodies for particular antigenic epitopes. There are numerous examples, however, where important biological markers for cancer, infectious disease or biochemical processes are present at concentrations too low to be detected in body fluids or tissues using conventional immunoassay approaches. With the completion of the human genome sequencing, the number of new biomarkers discovered for disease and drug effects is escalating rapidly. Increasingly, multiplex immunoassays are being called on to provide high-throughput methods for screening potential biomarkers or identifying protein-protein interactions and protein modifications (*i.e.* phosphorylation states; MacBeath, 2002). Multiplex immunoassays are also being used to consolidate identified biomarkers into panels with more comprehensive diagnostic value (Kingsmore and Patel, 2003). Additional advantages of multiplex immunoassays include decreased sample consumption and lower cost compared to conventional monoplex formats.

One challenge facing immunoassay developers is the need to provide highly multiplexed immunoassays with adequate sensitivity and specificity. Most immunoassay amplification systems use enzymes for substrate conversion in an aqueous environment, thus leading to products' diffusion from the site of the antigen-antibody interaction. These amplification systems are not compatible with microarray or cytometric bead platforms which rely on fine spacial separation of diagnostic signal for simultaneous detection of multiple analytes.

With these needs in mind, RCA has been applied, as immunoassay, to the detection of protein antigens (Schweitzer *et al.*, 2000; Wiltshire *et*

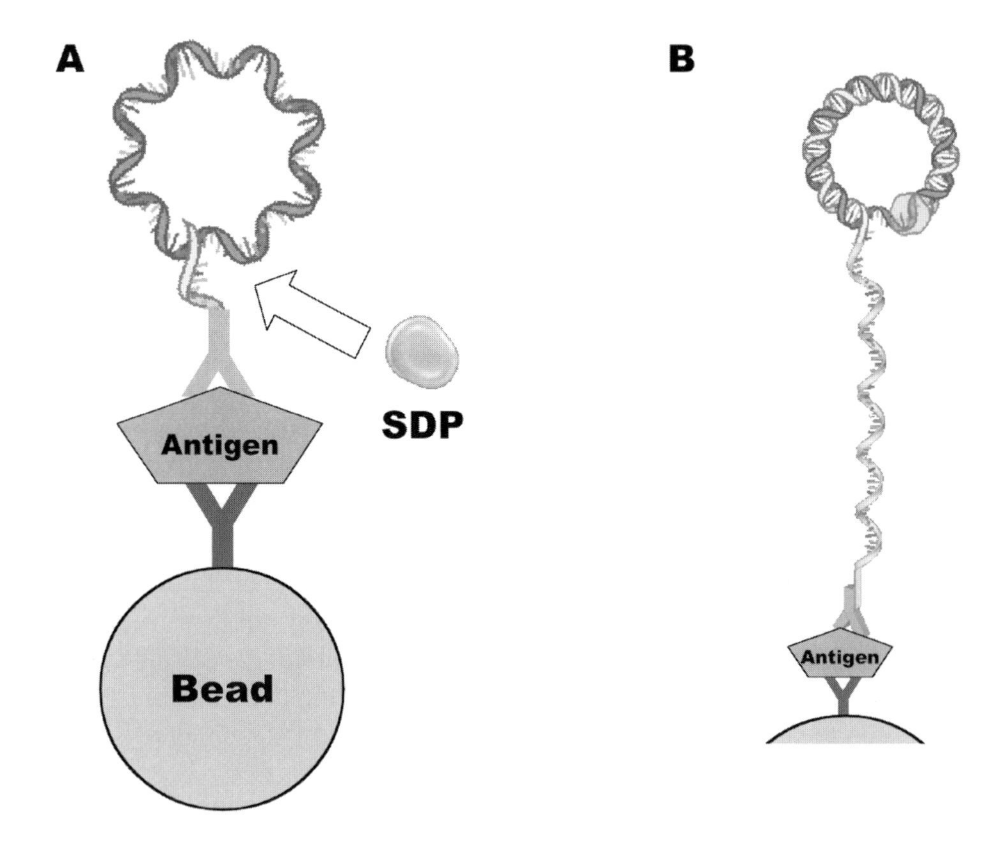

Figure 1. Bead based immunoassay incorporating immuno-RCA. An immune complex is formed on the surface of a capture antibody coated polystyrene bead (A). The immune complex consists of the capture antibody, the antigen and a primer labeled detector antibody prehybridized to a circular piece of ssDNA. After the immune complex is formed a strand displacing polymerase (SDP) is added in the presence of nucleotide triphosphates. The SDP extends the primer oligo around the circular DNA and until it returns to the double primer region of the DNA. When it reaches the double stranded region it begins displacing the double strands and continues to extend the ssDNA product (B).

al., 2000; Mullenix *et al.*, 2001; Demidov, 2004). In immuno-RCA, the 5' end of an RCA primer is attached to an antibody. This way, in the presence of a complementary circular DNA template, DNA polymerase, and nucleotides, the RCA reaction produces a ssDNA transcript complementary to the circular DNA sequence (Figure 1). Thus obtained DNA transcripts can be detected by hybridization to oligonucleotide probes or by binding to antibodies specific for nucleotide analogues incorporated during the RCA reaction. Importantly, the RCA-generated transcript remains covalently attached to the antibody, therefore maintaining spatial localization of signal required by multiplex immunoassays.

The initial reports on immuno-RCA described an improved sensitivity in magnetic bead and ELISA formats, while establishing compatibility with microarray immunoassays (Schweitzer *et al.*, 2000). The value of providing an amplification technique compatible with multiplex immunoassay systems was demonstrated by the development of microarray-based diagnostics for the detection of allergen-specific IgE (Wiltshire *et al.*, 2000). In these analyses, individual allergens were spotted and immobilized on glass slides and the arrays were incubated with patient serum to capture particular IgE. The captured antigen was subsequently detected by a conjugate of the anti-IgE antibody with a RCA primer. Such assays were shown to provide superior sensitivity, specificity and positive predictive values over analogous commercial ELISA assays (Wiltshire *et al.*, 2000). In a separate study, multiplexed allergen-specific IgE-detecting arrays exhibited very good correlation with several commercial monoplex assays (correlation coefficients ≥0.9) and the multiplex assay performance matched the level of correlation obtained when comparing commercial assays for the same analytes (Mullenix *et al.*, 2001). Both reports suggested that microarrayed immuno-RCA assays were capable of providing effective methods for simultaneously screening hundreds of different allergen specificities.

These new assays continued to advance with the introduction of 'sandwich' format assays for highly sensitive detection of multiple analytes in biological matrices. In this format, capture antibodies specific for different analytes are spotted onto microarrays. The capture antibody binds an analyte present in the sample matrix, which is subsequently detected with a second biotinylated detector antibody. RCA is introduced through the use of the conjugate of a secondary anti-biotin antibody with RCA primer that is employed to detect the biotinylated detector antibodies. The success of RCA in this platform lead to its proposed use in protein-profiling applications (Schweitzer and Kingsmore 2001; Schweitzer and Kingsmore 2002). Schweitzer and coworkers later described a combination of two immuno-RCA microarrays capable of detecting a total of 51 individual cytokines, chemokines and soluble receptors in serum and plasma (Schweitzer *et al.*, 2002). The arrays were utilized to gain an understanding of cytokine production during the TNFα or LPS induced maturation of Langerhans cells (dendritic cells found in the epidermis of the skin). The multiplexed assays confirmed the induction of several previously reported cytokines and also provided novel findings demonstrating the involvement of several new cytokines in the maturation process.

Microarrayed immuno-RCA assays also continued to advance with the development of an extended panel of 150 analytes spread across multiple arrays (Shao *et al.*, 2003). The low level of sample consumption and high level of multiplexing capability coupled with the sensitivity provided by immuno-RCA result in a microarray platform suitable for its use in biomarker discovery, disease diagnosis, pharmacodynamic studies and drug efficacy surrogate development. Immuno-RCA microarrays have proven to provide robust and reproducible results enabling their value in both research and diagnostic applications.

Immuno-RCA has also been applied to cellular assays providing improved sensitivity for immunohistochemistry and flow cytometry applications (Gusev *et al.*, 2001; Raghunathan *et al.*, 2002). As with microarrays, RCA is well suited here in part because the amplification product remains covalently linked to the immune complex. The linkage allows amplified signal to travel with the target cell in flow cytometry or to remain localized, thereby enhancing image quality in immunohistochemistry applications. As a result, immuno-RCA was capable of boosting signal intensity 4-fold in immunohistochemistry applications and ~40-fold in flow cytometry applications.

Microspheres are a proven solid-phase format for the development of immunoassays used in both research and clinical diagnostics. These tiny particles can be manufactured to incorporate very precise quantities of fluorescent dyes (McDade *et al.*, 1997; Fulton *et al.*, 1997). Microspheres containing controlled amounts of multiple fluorescent dyes are easily distinguished using a flow cytometer. The individual bead types can be used to prepare separate immunoassays and later be combined in a multiplex assay capable of simultaneously detecting multiple analytes. The cytometric bead array products from Becton Dickinson and Luminex take advantage of this approach for multiplexing. As with flow cytometry applications, immuno-RCA is particularly well suited to cytometric bead assay formats because the amplification product remains tethered to the target molecules captured by the microsphere and does not interfere with the intrinsic fluorescence of the microspheres (Mullenix *et al.*, 2002). Immuno-RCA was applied to bead format immunoassays from several manufacturers and shown to provide an average sensitivity improvement of ~30-fold while maintaining specificity in multiplex assays.

The cytometric bead immunoassay has significant potential to bring immuno-RCA into widespread use since conventional flow cytometers are widely available and the relatively inexpensive cytometric bead immunoassay instruments allow access to the technology at a fraction of the cost of full featured flow cytometers. Reagents and beads are readily available for immunoassays on either conventional or Luminex flow cytometers. In addition, immuno-RCA can be applied with simple modifications to commercial multiplex assay kits. In this chapter, a method for the application of immuno-RCA to multiplex cytometric bead immunoassays is provided and an example of its use in an 11-plex bead immunoassay is given. It should be noted that only minor modification of the protocol would be necessary to adapt immuno-RCA to most other immunoassay formats. Importantly, RCA offers improved sensitivity while maintaining the multiplexing capability of the Becton Dickinson cytometric bead assays and Luminex fluorescent microsphere immunoassays.

Materials and Methods

Oligonucleotides

Various oligonucleotide probes were synthesized by Integrated DNA Technologies. The 5'-thiol-labeled primer oligonucleotide utilized in the preparation of antibody conjugates contained a poly A spacer region with the sequence 5'-thiol-AAAAAAAAAAAAAAATGTCTCAGT

AGCTCGTCAGT. Prior to use in conjugation reactions this primer was resuspended in water with 150 mM dithiolthreitol (DTT). DTT was removed by desalting in a PD-10 (Pharmacia) column equilibrated with ice cold water. Reduced thiolated oligonucleotides were stabilized by lyophilization and stored frozen at -20°C. Circular ssDNA was prepared from linear oligonucleotide with the sequence 5'-ACTGACG AGCTACTGAGACATGTACAATCGGACCTGTGAGGTACTACC CTAATCGGACCTGTGAGGTACTACCCTAACTT. The linear DNA was circularized as previously described by enzymatic ligation after hybridization to a splint oligonucleotide 5'-TCGTCAGTAAGTTAG (Lizardi *et al.*, 1998).

Antibody-Primer Conjugate

The primer oligonucleotide was covalently coupled to a monoclonal anti-biotin antibody (Ab; Jackson Immuno Research Laboratories Inc.) by the method previously described (Schweitzer *et al.*, 2000). Briefly, Ab was treated with a 10-fold molar excess of sulfo-GMBS (Pierce) in the dark for 30 min at 37°C, followed by 30 min at room temperature. Unreacted sulfo-GMBS was removed by gel filtration over a PD-10 column (Amersham Biosciences) equilibrated with sodium phosphate (pH 7.5)/150 mM NaCl. The Ab was then concentrated in a Centricon YM-30 (Millipore) at 4°C. An amount equal to 28 nmol of sulfo-GMBS-activated Ab and 142 nmol of 5'-thiolated oligonucleotide was conjugated in a volume of 825 µl for 2 h at room temperature and overnight at 4°C while mixing on a rotating platform. Conjugate was purified by size exclusion chromatography on Superdex-200 (Pharmacia) at 4°C to remove free oligonucleotide.

Conjugation of Capture Antibody to Beads

All antibodies were purchased from commercial vendors (R&D Systems, Peprotech) and were coupled to carboxylated microspheres (Luminex) using 1-ethyl-3-3(3-dimethylaminopropyl) carbodiimide hydrochloride (EDC; Pierce). In one well of a pre-wetted 96-well multi-screen filter plate (Millipore), ~$6 \cdot 10^5$ beads were suspended in 80 µL of 0.1 M sodium phosphate buffer (pH 6.2) and mixed at 1,100 rpm on a titre plate shaker (Lab-Line Instruments) for 30 sec with 10 μL of 50 mg/mL freshly prepared solution of sulfo NHS in the same buffer. Immediately after that, 10 μl of freshly prepared 50 mg/mL EDC in the same buffer was added and again mixed at 1,100 rpm for 30 sec. The beads were incubated for 20 min at room temperature on the shaker platform set at 300 rpm. They were then washed three times with 200 μL of phosphate buffer saline (PBS; Gibco) using vacuum filtration on a multi-screen filter plate manifold (Millipore). Ab is dissolved in PBS at concentration 100 μg/mL. 100 μL of Ab is immediately added to the beads followed by resuspension of the beads for 30 sec at 1,100 rpm on the titre plate shaker. The mixture is then incubated in the plate for 2 h at room temperature with shaking at 300 rpm. At the end of the incubation, 15 μL of 0.25 M ethanloamine in PBS is added to each well followed by mixing at 1,100 rpm for 30 sec and incubating for 30 min at room temperature while shaking at 300 rpm. The beads are washed 4 times with 200 μL of PBS containing 0.05% Tween-20, 0.05% proclin (PBST) by vacuum filtration and resuspended in 150 μL of PBS with 10 mg/mL bovine serum albumin and 0.1% proclin preservative. Ab-coated beads are stored at 4°C until use.

Multiplex Bead-Based Assay Procedure

Prior to adding any reagents, 0.22-micron 96-well filter plates were pre-wet with 100 μL of PBS containing 0.05% Tween-20, 1% BSA, 0.05% proclin (PBST-BSA). Fifty μL of solution containing 10^3 beads of each analyte specificity were added to all wells and the beads were filtered to remove storage buffer. The beads were washed twice with 200 μL of PBST. All serum and plasma samples are diluted 4-fold in an assay diluent consisting of AD1 diluent (ImmunoChemistry Technologies) diluted 2-fold in PBS containing 0.05% Tween-20 and 1% BSA. Fifty μL of each diluted sample was added to separate bead-containing wells and the plates were shaken at 1,100 rpm for 30 sec to re-suspend the beads. The plates were incubated in the dark for 1 h at room temperature while shaking at 300 rpm. The beads were filtered and washed twice in 200 μL of PBST.

Fifty μL of detector antibody mix (each detector antibody concentration was previously optimized separately) was added to each well and the plates were shaken at 1,100 rpm 30 sec to re-suspend the beads. The plates were incubated in the dark for 1 h at room temperature while shaking at 300 rpm. During detector antibody incubation, the anti-biotin RCA primer conjugate/circle complex was formed by incubating a mixture of 0.25 μg/mL conjugate with 100 mM circle in PBS containing 0.05% Tween-20 and 2 mM EDTA for 30 min at 37°C. After the detector antibody incubation, the beads were filtered and washed twice with 200 μL PBS containing 0.05% Tween-20. Fifty μL of the pre-annealed primer conjugate/circle complex was added to each well and the plate was shaken at 1,100 rpm for 30 sec to resuspend the beads. The plates were incubated for 30 min in the dark at room temperature while shaking at 300 rpm. The beads were filtered and washed twice with 200 μL of PBS containing 0.05% Tween-20 and once with 200 μL of RCA wash buffer containing 200 mM potassium glutamate, 35 mM HEPES-potassium hydroxide (pH 7.5), 20 mM magnesium acetate and 0.01% BSA.

Fifty μL of RCA mix containing 100 μM concentration each of dATP, dCTP and dGTP (Promega), 22.5 μM concentration of dTTP (Promega) and 2.5 μM biotin-16-dUTP (Roche) in the RCA wash buffer were added to each well and the plates were shaken at 1,100 rpm for 30 sec to resuspend the beads. The plates were incubated for 1 h in the dark at 31°C. The beads were filtered and washed twice with 200 μL of PBS containing 0.05% Tween-20. Biotin-labeled RCA product was detected with 50 μL/well of 1.5 μg/mL streptavidin-B-phycoerythrin (Jackson ImmunoResearch Laboratories). After addition of the streptavidin-phycoerythrin the plates were shaken at 1,100 rpm speed for 30 sec to resuspend the beads. The plates were incubated for 30 min at 37°C. The beads were filtered and washed twice with 200 μL of PBS containing 0.05% Tween-20. The beads were resuspended in 150 μL of this solution and analyzed on Luminex-100 cytometric bead analyzer. Four-parameter curve fitting was applied to cytokine standard curves and regression analysis was used to convert data into pg/mL values.

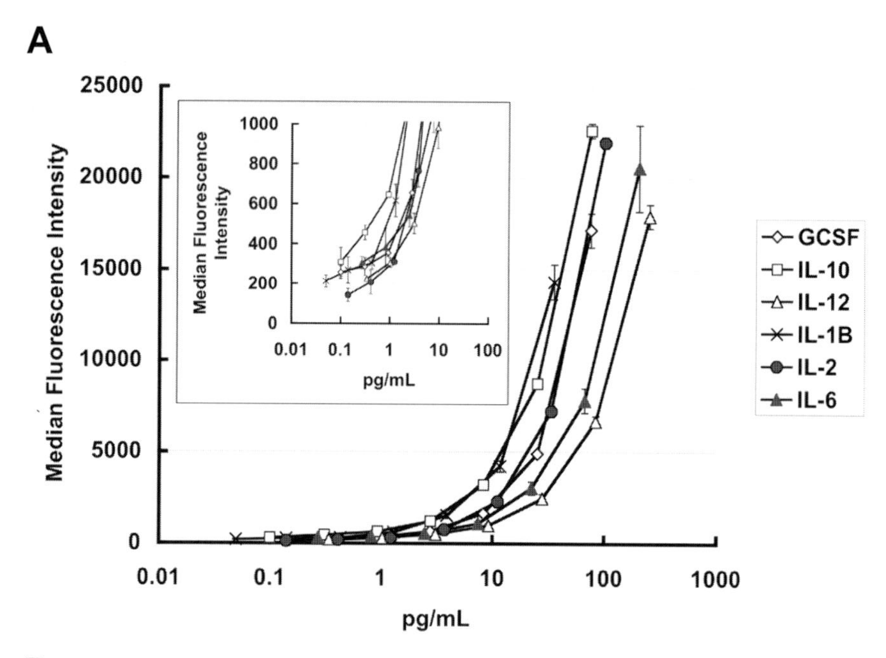

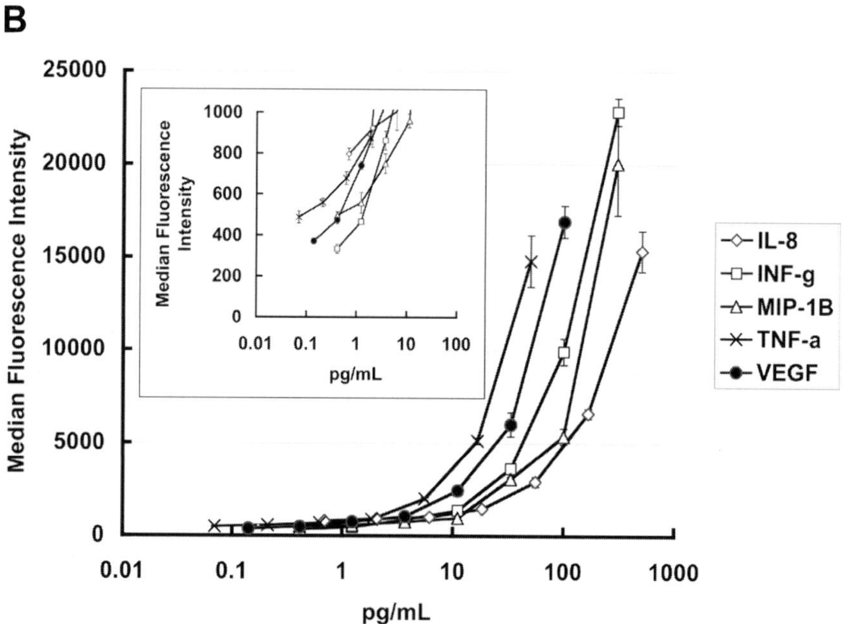

Figure 2. A and B: Dose response curves for 11-plex assay. A mixture of 11 cytokines was serially diluted in assay diluent and run in the immuno-RCA 11-plex cytometric bead assay. Dose response curves are provided on two fluorescence intensity scales to cover the linear range of the assay with the lower concentrations of analyte being represented in the inset figures.

Table 1. Minimum detection limits of the immuno-RCA 11-plex cytometric bead assay.

Cytokine	MDL (assay buffer)	MDL in serum[a]
GCSF	0.10	0.40
IL-10	0.10	0.40
IL-12	0.34	1.36
IL-1B	0.05	0.20
IL-2	0.14	0.56
IL-6	0.27	1.08
IL-8	0.69	2.76
INF-g	0.41	1.64
MIP-1B	1.00	4.00
TNF-a	0.21	0.84
VEGF	0.14	0.56

[a]Serum and plasma require a 4-fold dilution in assay buffer. The minimum detection limits provided occur at the 95% confidence interval above background.

Example

To prove the use of immuno-RCA in multiplex cytometric bead immunoassays, capture and detector antibodies for 11 cytokine analytes were selected. Each of the capture antibodies were coated onto Luminex beads with different fluorescence signatures. The detector antibodies were screened individually for non-specific crossreactivity in the absence of antigen with each of the 11 bead types. The sensitivity at the 95% confidence interval above background was determined for each of the cytokines in titration experiments in assay diluent (Figure 2 and Table 1). Sensitivity ranged from 0.05 to 1 pg/mL and each of the assays provided a greater than 2.5 log dynamic range. Sensitivity in serum and plasma is 4-fold higher due to the required dilution in assay diluent.

To validate specificity, each of the cytokines was spiked individually at 1000-fold above the minimum detection limit of the least sensitive assay (MIP-1β, 1 pg/mL). None of the 11 cytokines produced signals distinguishable from background on any of the 10 non-cognate bead types resulting in crossreactivity of less than 0.1% for cytokines

Table 2. Crossreactivity of 135 analytes with the 11-plex immuno-RCA cytometric bead assay.

Crossreactivity[a]	Analytes tested
<0.1%	ANG, ACE-2, AFP, AgRP, BLC, b-NGF, BTC, CCL28, CD-141, CD27, CD30, CD40, CNTFR, CRP, CTACK, D-Dimer, DR6, EGF, ENA-78, Eot, Eot-2, Eot-3, E-Selectin, Fas, FasLigand, FGFbasic, FGF-4, FGF-7, FGF-9, FGFacid, Follinstatin, Fractalkine, GDNF, GM-CSF, GRO-b, GRO-g, HCC-1, HCG, HGF, HVEM, ICAM-1, ICAM-3, IGFBP-2, IGFBP-1, IGFBP-3, IGFBP-4, IGFBP-6, IGF-I, RIGF-II, IL-10Rb, IL-12p40, IL-13, IL-15, IL-18, IL-1ra, IL-1SRII, IL-2Ra, IL-2Rb, IL-2Rg, IL-2sRa, IL-3, IL-4, IL-5, IL-5Ra, IL-6, IL-7, IL-8, IL-9, I-TAC, Leptin/OB, LIFRa, L-Selectin, LYMPHOTACTIN, MCP-2, MCP-3, MCP-4, M-CSFR, MIF, MIP-1a, MIP-3a, MIP-3b, MMP-7, MMP-8, MMP-9, MPIF-1, NAP-2, Neutrophilelastase, OSM, P-selectin, PAI-II, PDGFRa, PECAM-1, PIGF, Proein S, Prolactin, Protein C, RANK, SCFR, SDF-1b, ST2, TGF-a, TIMP-1, TIMP-2, TRAILR1, 4-1BB, TRAILR4, TSH, VEGF-R2, ALCAM
<1.0%	AR, BDNF, Flt-3L, GCP-2, HCC4, I-309, IL-17, IL-1a, IL-1b, IL-2, MCP-1, M-CSF, MIG, MIP-1b, MIP-1g, NT-3, NT-4, PARC, RANTES, SCF, sgp130, TARC, TNF-a, TNF-b, TNF RI, VEGF

[a]Crossreactivity is calculated by dividing the minimum detection limit of the least sensitive assay (MIP1β, 1 pg/mL) by the concentration of the crossreactant required to produce detectable signal in the assay. The analytes were grouped according to overall crossreactivity with all 11 assays.

detected in the multiplex assay. The crossreactivity to cytokines not included in the multiplex assays was also evaluated through the addition of mixtures of 25-30 cytokines in which each cytokine was present at concentrations of 100 or 1000 pg/mL. This approach demonstrated that the assay had less than 0.1% crossreactivity with 109 different cytokines and less than 1% crossreactivity with 26 cytokines (Table 2). Specificity with complex antigen mixtures was validated in experiments in which all 11 cytokines were added at concentrations corresponding to the mid-range biological levels in serum or plasma. Eleven different detector mixes were prepared, each with a different

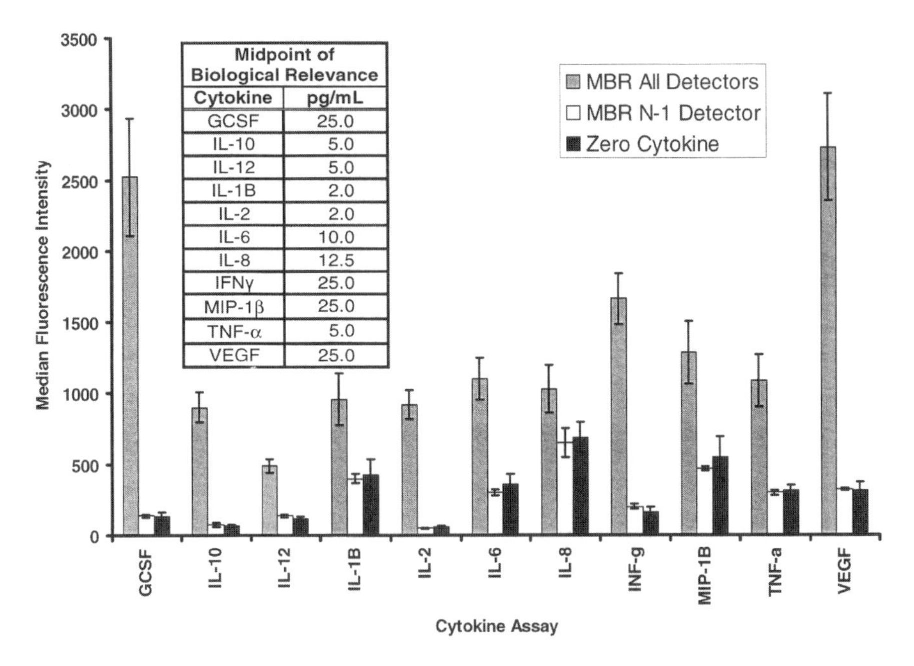

Figure 3. Specificity in complex antigen mixtures of the 11-plex immuno-RCA cytometric bead assay. Specificity with complex antigen mixtures was demonstrated by adding all 11 cytokines to the assay at concentrations corresponding to the mid-range levels in serum or plasma (inset table). The captured antigens were screened with 11 different detector mixes, each with a different single detector antibody removed. Gray bars represent the signal intensity achieved at the concentrations corresponding to the mid-range level of each analyte in serum or plasma. White bars represent the signal intensity when the detector antibody for that feature has been removed from the detector antibody set. The cross-hatched bars represent the signal intensity achieved when no analyte is added to the assay. Error bars are represent the standard deviation around three replicate data points.

single detector antibody removed (Figure 3). When a given feature's detector antibody was removed, the signal intensity of the feature dropped to background indicating specificity in complex mixtures of analytes.

The 11-plex assay is intended for use in serum and plasma. In order to minimize matrix effects, serum and plasma samples are diluted 4-fold in assay diluent. To validate the multiplex assay in biological matrices, all 11 antigens were added simultaneously to assay diluent or samples of serum and plasma prepared by pooling 5 or more clinical

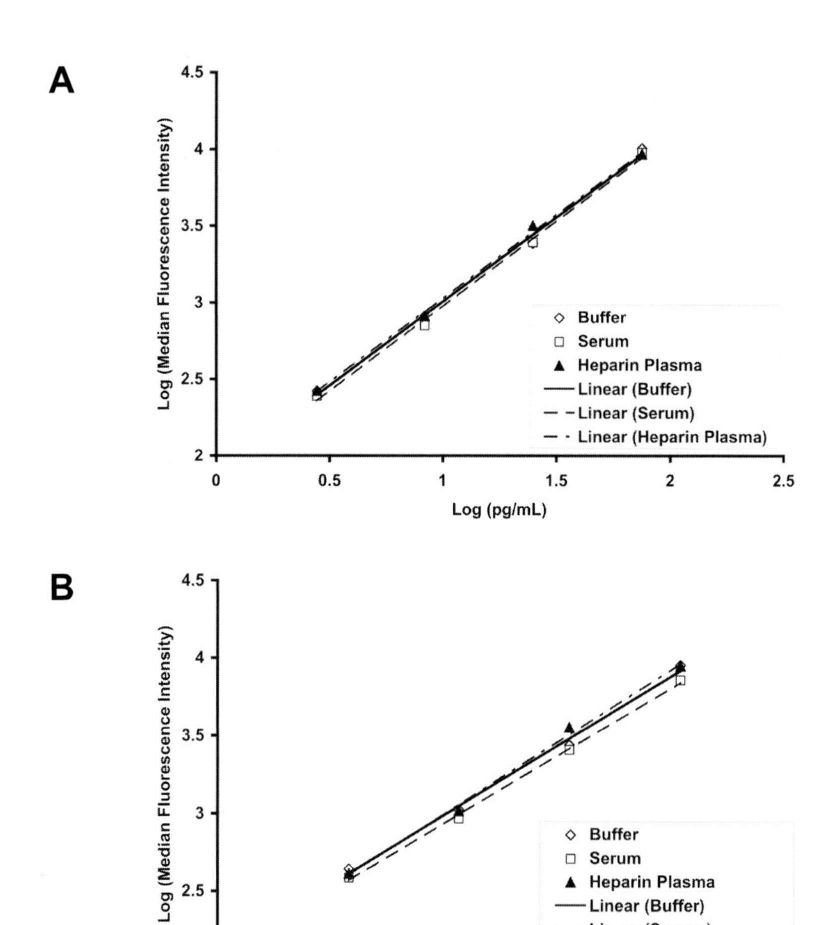
A
4.5
4
3.5
3
2.5
2
0
0.5
1
1.5
2
2.5
Log (Median Fluorescence Intensity)
Log (pg/mL)
Buffer
Serum
Heparin Plasma
Linear (Buffer)
Linear (Serum)
Linear (Heparin Plasma)
B
4.5
4
3.5
3
2.5
2
0
0.5
1
1.5
2
2.5
Log (Median Fluorescence Intensity)
Log (pg/mL)
Buffer
Serum
Heparin Plasma
Linear (Buffer)
Linear (Serum)
Linear (Heparin Plasma)

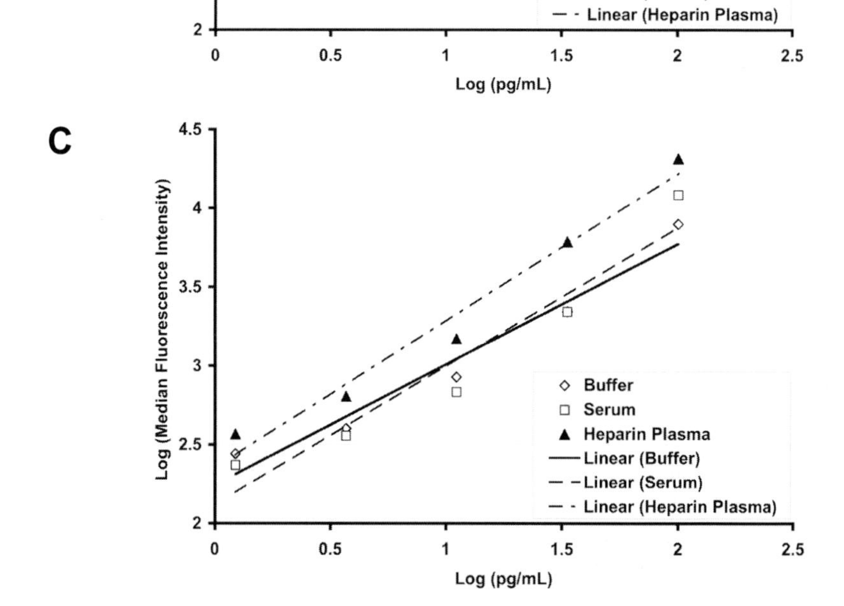
C
4.5
4
3.5
3
2.5
2
0
0.5
1
1.5
2
2.5
Log (Median Fluorescence Intensity)
Log (pg/mL)
Buffer
Serum
Heparin Plasma
Linear (Buffer)
Linear (Serum)
Linear (Heparin Plasma)

samples. All three matrices were initially diluted 4-fold in assay diluent and further serially diluted in assay diluent. The dilutions were run in the 11-plex assay and shown to provide parallel dose response curves indicating the absence of any serum or plasma matrix effect (Figure 4).

Discussion

Multiplex immunoassays are finding an ever increasing number of applications in clinical diagnostics, proteomics and pharmacology. In clinical diagnostics, multiplex assays are meeting the demand for simultaneous determination of complex analyte profiles in immunology or microbiology. Specific examples include multiplex assays for drugs of abuse, cardiac markers, and autoimmunity (Buechler *et al.*, 1992; Garcia *et al.*, 2000; Robinson *et al.*, 2002; Rouquette *et al.*, 2003). Also driving the need for multiplex immunodiagnostic assays is the constant need for cost reduction. Combining commonly used panels of immunoassays into a single multiplex assay increases laboratory throughput and also reduces the sample volume requirement. Finally, novel, multiplexed panels of immunoassays are being developed by many groups that offer the potential to guide diagnostic, prognostic or therapeutic decision making in complex clinical situations where conventional, monoplex immunoassays lack sufficient predictive value.

The completion of sequencing human genome has led to increased interest in the scientific discipline of proteomics. Effort is now focused on developing our understanding of the various functions of the proteins encoded in our gene sequences. Adding to the complexity of proteomics

Figure 4. Validation of the 11-plex immuno-RCA cytometric bead assay in serum and plasma. All 11 antigens were added simultaneously to assay diluent or samples of serum and plasma prepared by pooling 5 clinical samples. All three matrices were initially diluted 4-fold in assay diluent and further serially diluted in assay diluent. The dilutions were run in the 11-plex assay and data were log transformed, plotted and fit with a linear curve function to demonstrate parallelism. Three of the 11 analytes are plotted as representative examples (IL-2, 4a; IL-12, 4b; IFNγ 4c).

is the variety of translational and post-translational modifications and physical translocations that occur during the lifecycle of proteins and which are intimately involved in biological consequences of protein expression. Complexity is further increased in proteins with multiple functional regions, the ability of various proteins to undergo structural changes triggered by binding events, the synergistic and antagonistic effects of concomitantly expressed proteins, and the extreme dynamism of many protein events. Multiplex immunoassays have the potential to play an important role in precisely defining complex protein expression patterns in response to disease, drug treatments and environmental factors. As we further our understanding of the patterns of protein expression, new targets for clinical diagnostics and drug development will be identified.

The compatibility of immuno-RCA with multiplex immunoassays offers the opportunity to transform the two major multiplex platforms. With microarrays, immuno-RCA is enabling in that it allows microarray assays to achieve sensitivity comparable to conventional monoplex ELISA assays. Without a signal amplification procedure many microarray immunoassays are not be sensitive enough to detect ambient concentrations of analytes in, for example, serum or plasma. With cytometric bead assays, immuno-RCA provides multiplex immunoassays with a level of sensitivity typically only achieved in highly optimized monoplex immunoassays using conventional amplification systems. Cytometric bead immunoassays coupled with immuno-RCA can provide the requisite sensitivity to measure previously undescribed protein expression patterns and can impact clinical diagnostics by lowering detection limits for critical targets such as blood borne viruses or antibody responses.

The use of immunoassays for monitoring biomarkers in the drug development process is well established (Frank and Hargreaves, 2003; Colburn, 2003). Immunoassays for biomarkers can accelerate the drug approval process by providing information related to drug mechanism of action and pharmacodynamics, by providing surrogate end-points, predicting drug efficacy and by providing companion diagnostics that can be used to guide individualized drug selection and dosage schedules. Multiplex immunoassays are expected to continue the expansion of this association (Kingsmore and Patel, 2003; Shao *et al.*, 2003).

Immuno-RCA has the potential to measure previously undetected biological responses through the development of highly sensitive multiplex immunoassays. Uncovering these biological processes will undoubtedly provide new information leading to improved diagnostics or drug development opportunities. We anticipate increased application of immuno-RCA to ultrasensitive and multiplexed immunoassays and widespread application in studies in clinical diagnostics, proteomics and drug development.

Acknowledgements

We would like to thank R. Murli Krishna, Ramou Sivakamasundari, John Feaver and Martin Sorette for lending to us their expertise in the immuno-RCA cytometic bead assay platform.

References

Buechler, K.F., Moi, S., Noar, B., McGrath, D., Villela, J., Clancy, M., Shenhav, A., Colleymore, A., Valkirs, G., and Lee, T. (1992). Simultaneous detection of seven drugs of abuse by the Triage panel for drugs of abuse. Clin. Chem. *38*, 1678-84.

Colburn, W.A. (2003). Biomarkers in drug discovery and development: from target identification through drug marketing. J. Clin. Pharmacol. *43*, 329-341.

Demidov, V.V. (2004a). Rolling-circle amplification (RCA). In: Encyclopedia of Diagnostic Genomics and Proteomics. J. Fuchs and M. Podda, eds. Marcel Dekker, New York (in press).

Frank, R. and Hargreaves, R. (2003). Clinical biomarkers in drug discovery and development. Nat. Rev. Drug. Discov. *2*, 566-580.

Fulton, R.J., McDade, R.L., Smith, P.L., Kienker, L.J. and Kettman, J.R. (1997). Advanced multiplexed analysis with the FlowMetrix™ system. Clin. Chem. *43*, 1749-1756.

Garcia, L.S., Shimizu, R.Y. and Bernard, C.N. (2000). Detection of Giardia lamblia, Entamoeba histolytica/Entamoeba dispar, and Cryptosporidium parvum antigens in human fecal specimens using the triage parasite panel enzyme immunoassay. J. Clin. Microbiol. *38*, 3337-3340.

Gosling, J.P. (1990). A decade of development in immunoassay methodology. Clin. Chem. *36*, 1408-1427.

Gusev, Y., Sparkowski, J., Raghunathan, A., Ferguson, H., Montano, J., Bogdan, N., Schweitzer, B., Wiltshire, S., Kingsmore, S.F., Maltzman, W., Wheeler, V. (2001). Rolling circle amplification: a new approach to increase sensitivity for immunohistochemistry and flow cytometry. Am. J. Pathol. *159*, 63-69.

Kingsmore, S.F. and Patel, D.D. (2003). Multiplexed protein profiling on antibody-based microarrays by rolling circle amplification. Curr. Opin. Biotechnol. *14*, 74-81.

Lizardi, P.M., Huang, X., Zhu, Z., Bray-Ward, P., Thomas, D.C., Ward, D.C. (1998). Mutation detection and single-molecule counting using isothermal rolling-circle amplification. Nat. Genet. *19*, 225-232.

MacBeath, G. (2002). Protein microarrays and proteomics. Nat. Genet. *32,* Suppl; 526-32.

McDade, R.L. and Fulton, R.J. (1997). True multiplexed analysis by computer-enhanced flow cytometry. Med. Device. Diagn. Ind. *19*, 75-82.

Mullenix, M., Wiltshire, S., Shao, W., Kitos, G., and Schweitzer, B. (2001). Allergen-specific IgE detection on microarrays using rolling circle amplification: correlation with *in vitro* assays for Serum IgE. Clin. Chem. *47*, 1926-1929.

Mullenix, M.C., Sivakamasundari, R., Feaver, W.J., Krishna, R.M., Sorette, M.P., Datta, H.D., Morosan, D.M., and Piccoli, S.P. (2002). Rolling circle amplification improves sensitivity in multiplex immunoassays on microspheres. Clin. Chem. *48*, 1855-1858.

Raghunathan, A., Sorette, M.P., Ferguson, H.R. and Piccoli, S.P. (2002). Rolling circle amplification technology as a potential tool in detection and monitoring of cancer by flow cytometry. Clin. Chem. *48*, 1853-55.

Robinson, W.H., DiGennaro, C., Hueber, W., Haab, B.B., Kamachi, M., Dean, E.J., Fournel, S., Fong, D., Genovese, M.C., de Vegvar, H.E., Skriner, K., Hirschberg, D.L., Morris, R.I., Muller, S., Pruijn, G.J., van Venrooij, W.J., Smolen, J.S., Brown, P.O., Steinman, L. and Utz, P.J. (2002). Autoantigen microarrays for multiplex characterization of autoantibody responses. Nat. Med. *8*, 295-301.

Rouquette AM, Desgruelles C, Laroche P. (2003). Evaluation of the new multiplexed immunoassay, FIDIS, for simultaneous quantitative determination of antinuclear antibodies and comparison with conventional methods. Am. J. Clin. Pathol. *120*, 676-81.

Schweitzer, B., Wiltshire, S., Lambert, J., O'Malley, S., Kukanskis, K., Zhu, Z., Kingsmore, S.F. (2000). Immunoassays with rolling circle DNA amplification: a versatile platform for ultrasensitive antigen detection. Proc. Natl. Acad. Sci. USA. *97*, 10113-10119.

Schweitzer, B. and Kingsmore, S. (2001). Combining nucleic acid amplification and detection. Curr. Opin. Biotechnol. *12*, 21-27.

Schweitzer, B. and Kingsmore S.F. (2002). Measuring proteins on microarrays. Curr. Opin. Biotechnol. *13*, 14-19.

Schweitzer, B., Roberts, S., Grimwade, B., Shao, W., Wang M., Fu, Q., Shu, Q., Laroche, I., Zhou, Z., Tchernev, V.T., Christiansen, J., Velleca, M., Kingsmore, S.F. (2002). Multiplexed protein profiling on microarrays by rolling-circle amplification. Nat. Biotechnol. *20*, 359-365.

Shao, W., Zhou, Z., Laroche, I., Lu, H., Zong, Q., Patel, D.D., Kingsmore, S., and Piccoli, S.P. (2003). Optimization of rolling-circle amplified protein microarrays for multiplexed protein profiling. J. Biomed. Biotech. *5*, 299-307.

Wiltshire, S., O'Malley, S., Lambert, J., Kukanskis, K., Edgar, D., Kingsmore, S., Schweitzer, B. (2000). Detection of multiple allergen-specific IgE on microarrays by immunoassay with rolling circle amplification. Clin. Chem. *46*, 1990-1993.

Index

S

T

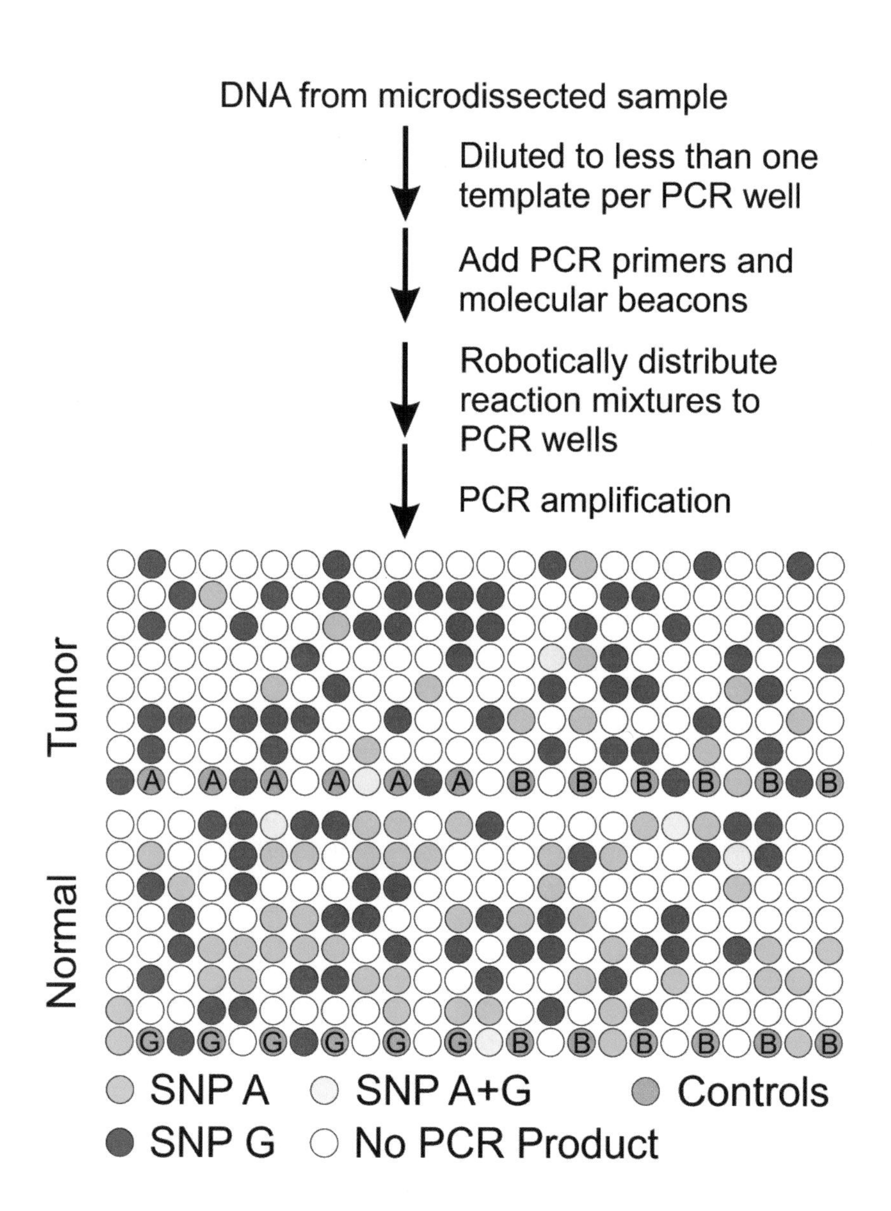

Chapter 2.4

Figure 2. Schematic representation of digital PCR analysis of allelic status. Genomic DNA was isolated from tumor and autologous normal breast tissues and diluted to the extent that single molecules can be individually amplified in separate PCRs. Green and Red wells contain PCR products from only one of the parental alleles. Wells with background fluorescence are colorless and yellow wells contain both alleles (green + red = yellow). Grey wells contain control samples used for confirming the expect value of fluorescence (A for SNP A, G for SNP G, B for blank control).

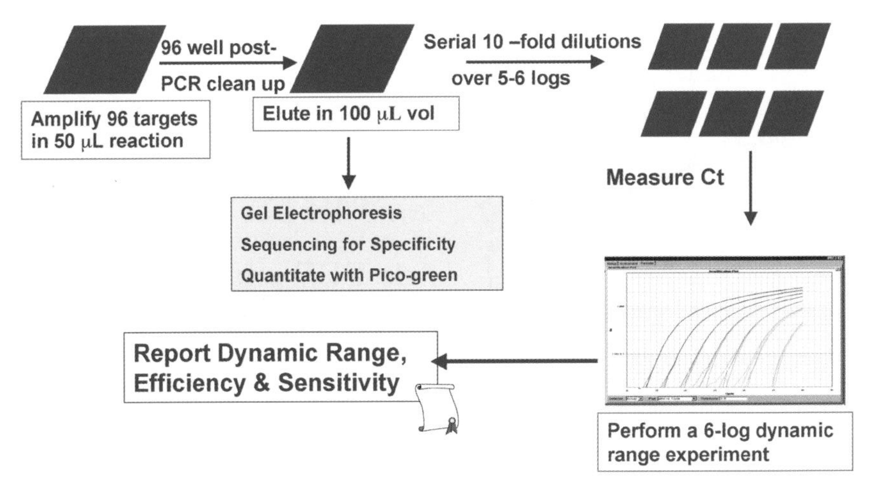

Chapter 2.5
Figure 1. Flow chart depicting experiments on measuring assay efficiency, dynamic range and sensitivity of real-time PCR. Amplification of target amplicon from the URR pool is followed by purification of the PCR product, which is then serially diluted and subjected to real-time PCR in buffers lacking Uracil N-Glycosylase. The Ct values obtained over a 6-log dilution range are used to determine amplification efficiency. Quantitation by Picogreen staining of the PCR amplicon before real-time PCR is used to determine absolute DNA concentration necessary for sensitivity determination and agarose gel electrophoresis is employed to verify amplicon size.

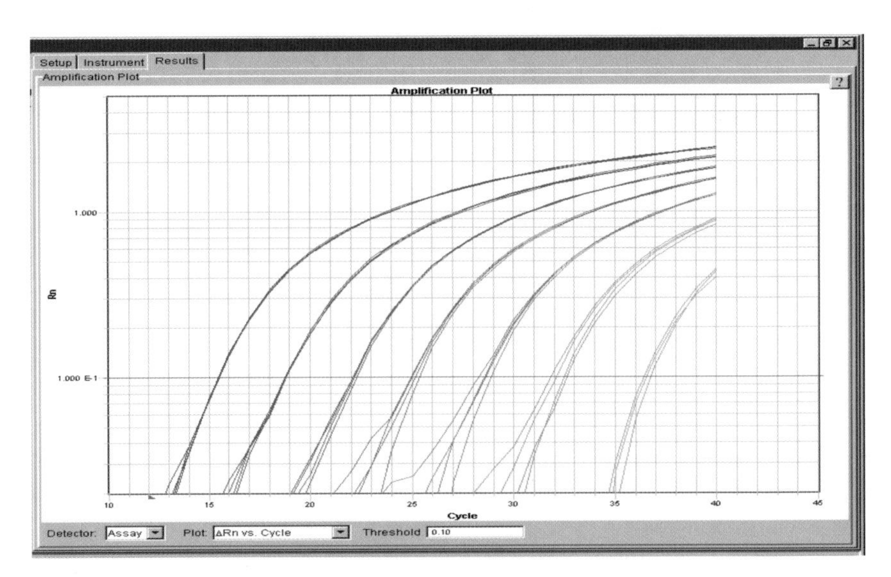

Chapter 2.5
Figure 2. Typical amplification plots obtained in real-time PCR using serially diluted DNA samples. The cycle number (X-axis) corresponding to the point when the amplification plot rises above the threshold set at 0.10 (red line) is the Ct value.

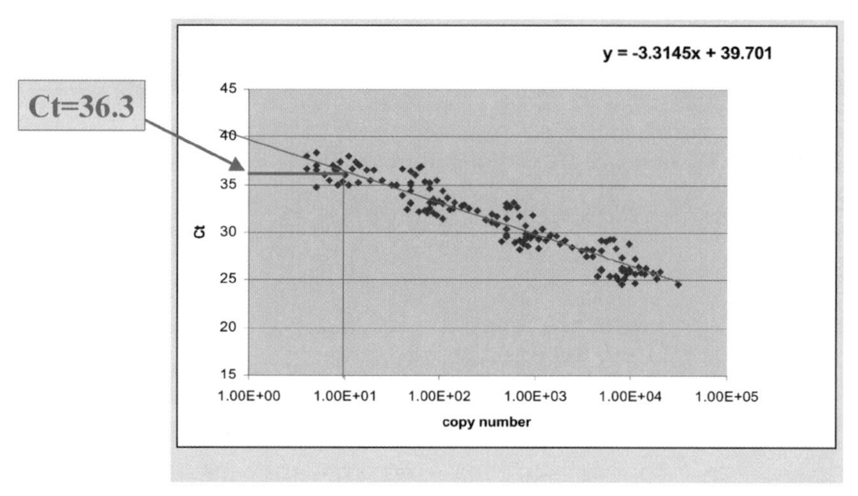

Chapter 2.5

Figure 3. A plot of the amplicon copy number (X-axis) *vs* the Ct value obtained for the corresponding dilution. The results are a compilation of data from 40 different assays. The average Ct obtained for 10 copies is around 36. The slope of the graph is 3.3 indicating an average efficiency value of 100%.

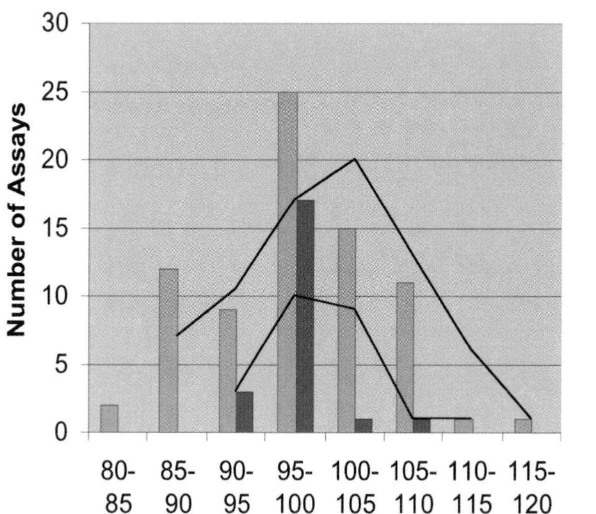

Chapter 2.5

Figure 4. Amplification efficiency distribution for GAPDH target. Measurements correspond to a 2-log (blue) and 5-log (red) dilution series of cDNA. The data show that 2-log range dilution results in amplification efficiencies 82-112%, with average = 97.6% (number of assays 76). At the same time, 5-log range dilution results in amplification efficiencies 92-105%, with average 98.4% (number of assays 22).

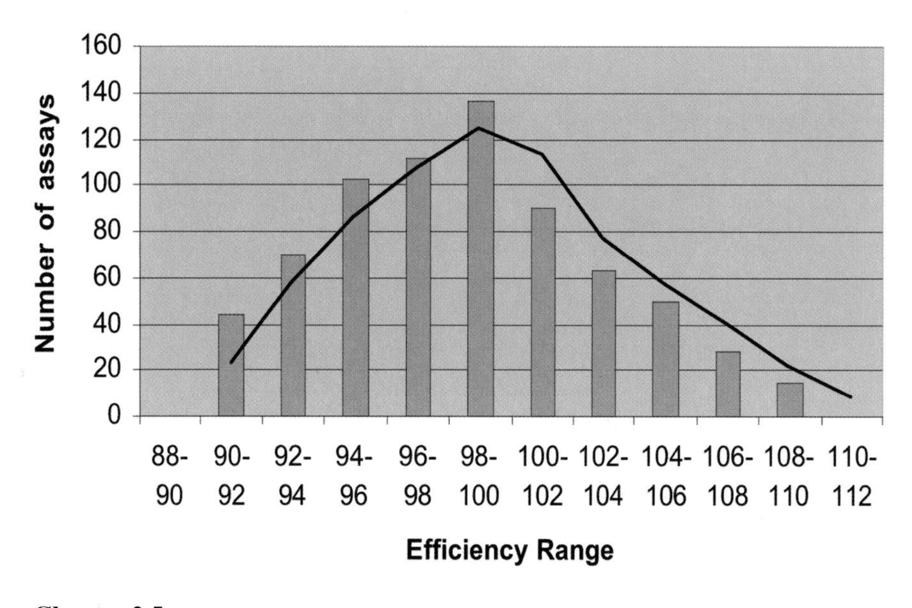

Chapter 2.5
Figure 5. Distribution of amplification efficiencies for about 700 real-time PCR assays. The data display normal distribution with a mean around 100%.

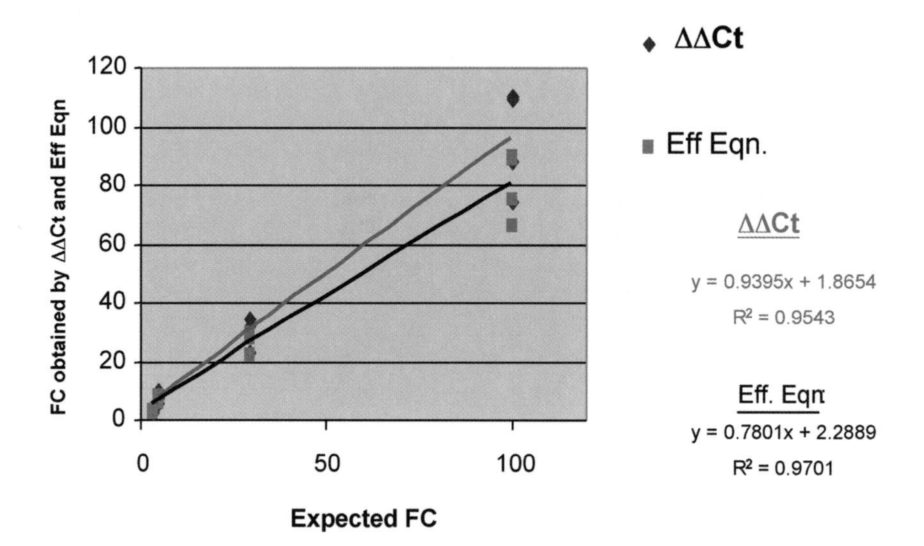

Chapter 2.5
Figure 6. Comparison of the two methods for quantitation of fold change (FC) values with the expected FC. (A) ΔΔCt method, (B) method using pre-determined amplification efficiencies for a target (GAPDH) and a control gene (see text for details).

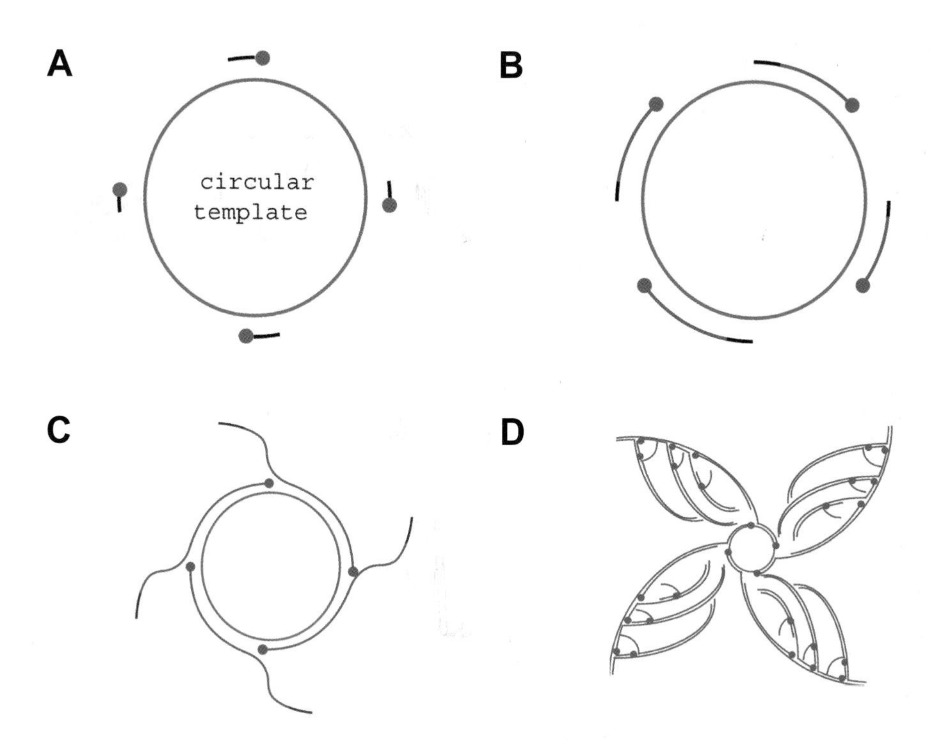

Chapter 3.5

Figure 2. Schematics of the random-primed RCA reaction with circular vector template employing strand displacement. A: Random primers hybridize to multiple locations on DNA circle (single-stranded or denatured plasmid or phage DNA). B: Phi29 DNA polymerase (black sphere) initiates replication from many primers simultaneously synthesizing multiple end-to-end copies. C: Strand displacement allows the polymerase to continue the RCA-type replication. D: Random-priming events subsequently occur on the displaced DNA strands yielding the highly-branched amplification products. This series of displacement and replication events continues until the nucleotide pool is depleted.

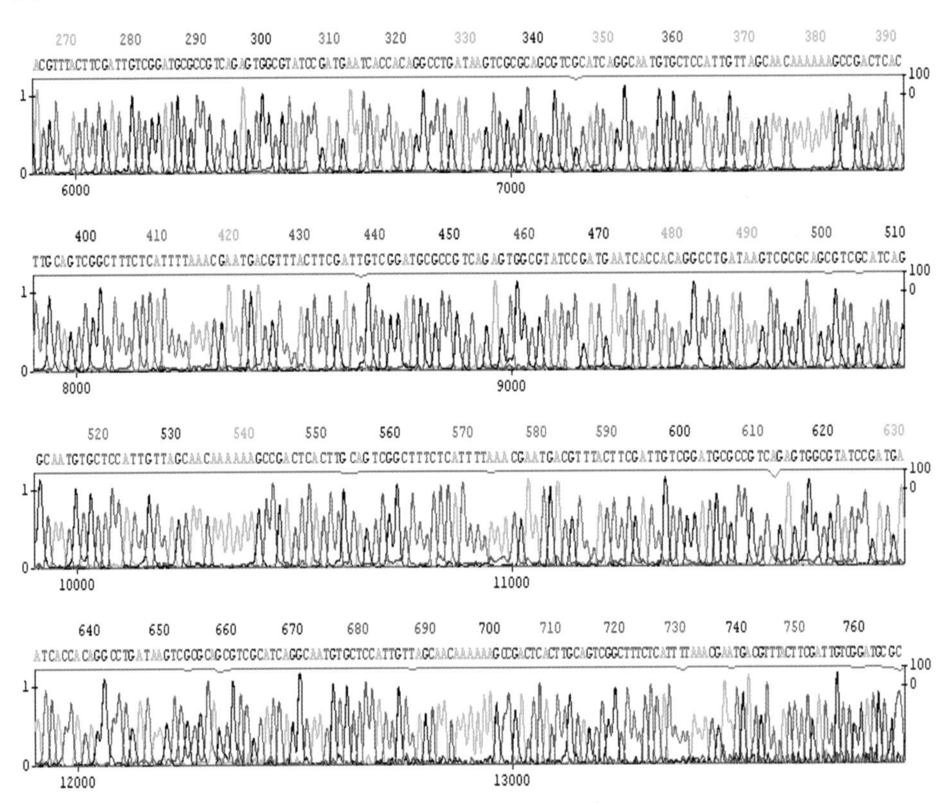

Chapter 3.5

Figure 6. Sequencing traces obtained from the RCA-generated DNA templates (3 kb inserts from the TempliPhi amplified pUC18 plasmids). Both DYEnamic™ ET terminator chemistry (A) and BigDye terminator chemistry (B) were used. Sequencing reactions were cleaned up using the BET SPRI method and run on either the MegaBACE 1000/4000 (A) or ABI3730 (B) capillary sequencer.

B

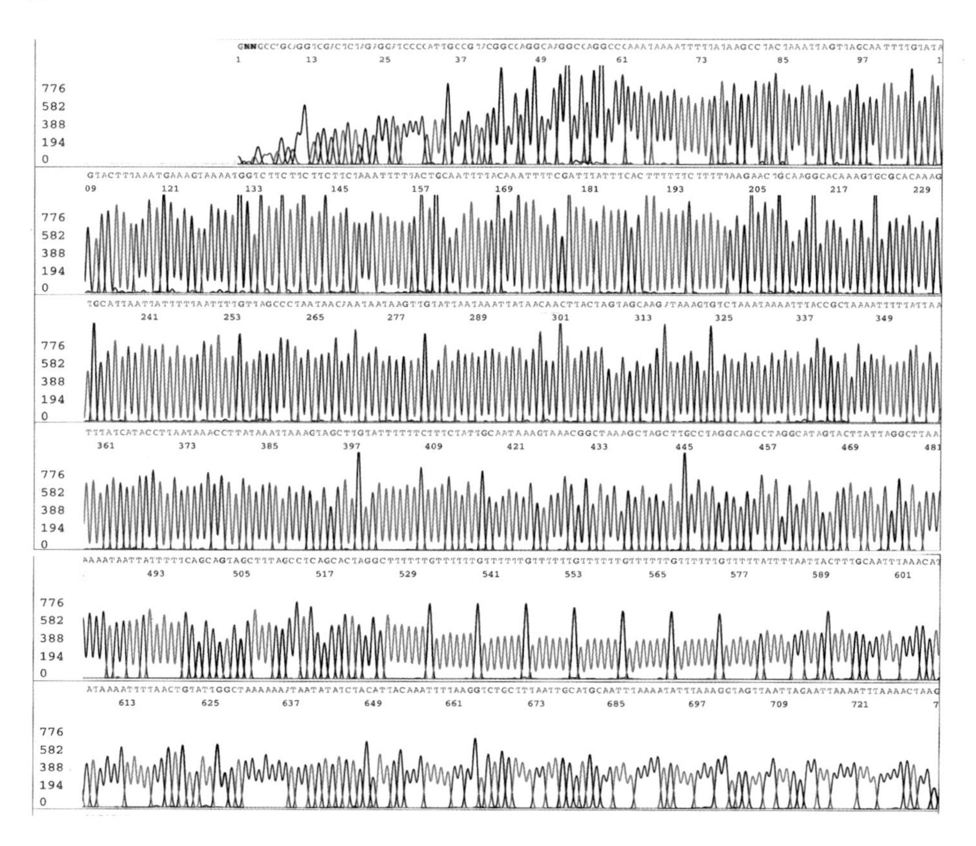
776
582
388
194
0
1 13 25 37 49 61 73 85 97
121 133 145 157 169 181 193 205 217 229
241 253 265 277 289 301 313 325 337 349
361 373 385 397 409 421 433 445 457 469
493 505 517 529 541 553 565 577 589 601
613 625 637 649 661 673 685 697 709 721

Books of Related Interest

Malaria Parasites: Genomes and Molecular Biology	2004
Pathogenic Fungi: Structural Biology and Taxonomy	2004
Pathogenic Fungi: Host Interactions and Emerging Strategies for Control	2004
Bacterial Spore Formers: Probiotics and Emerging Applications	2004
Strict and Facultative Anaerobes: Medical and Environmental Aspects	2004
Foot and Mouth Disease: Current Perspectives	2004
Sumoylation: Molecular Biology and Biochemistry	2004
DNA Amplification: Current Technologies and Applications	2004
Prions and Prion Diseases: Current Perspectives	2004
Real-Time PCR: An Essential Guide	2004
Protein Expression Technologies: Current Status and Future Trends	2004
Computational Genomics: Theory and Application	2004
The Internet for Cell and Molecular Biologists (2nd Edition)	2004
Tuberculosis: The Microbe Host Interface	2004
Metabolic Engineering in the Post Genomic Era	2004
Peptide Nucleic Acids: Protocols and Applications (2nd Edition)	2004
Ebola and Marburg Viruses: Molecular and Cellular Biology	2004
MRSA: Current Perspectives	2003
Genome Mapping and Sequencing	2003
Bioremediation: A Critical Review	2003
Frontiers in Computational Genomics	2003
Transgenic Plants: Current Innovations and Future Trends	2003
Bioinformatics and Genomes: Current Perspectives	2003
Vaccine Delivery Strategies	2003
Multiple Drug Resistant Bacteria: Emerging Strategies	2003
Regulatory Networks in Prokaryotes	2003
Genomics of GC-Rich Gram-Positive Bacteria	2002
Genomic Technologies: Present and Future	2002
Probiotics and Prebiotics: Where are We Going?	2002

Full details of all these books at: www.horizonbioscience.com